EXPLORING THE MOON

Through Binoculars and Small Telescopes

Ernest H. Cherrington, Jr.

DOVER PUBLICATIONS, INC.

NEW YORK

Copyright © 1969, 1984 by Ernest H. Cherrington, Jr.
All rights reserved under Pan American and International Copyright Conventions.

Published in Canada by General Publishing Company, Ltd., 30 Lesmill Road, Don Mills, Toronto, Ontario.
Published in the United Kingdom by Constable and Company, Ltd., 10 Orange Street, London WC2H 7EG.

This Dover edition, first published in 1984, is an enlarged and updated republication of the work originally published in 1969 by McGraw-Hill, Inc. under the title of *Exploring the Moon Through Binoculars*. A new section of photographs, The Lunar Photographs Supplement, has been added at the back of this edition. The Preface to the Dover Edition is another new feature.

Manufactured in the United States of America
Dover Publications, Inc., 31 East 2nd Street, Mineola, N.Y. 11501

Library of Congress Cataloging in Publication Data

Cherrington, Ernest H.
Exploring the moon through binoculars and small telescopes.

Reprint. Originally published: Exploring the moon through binoculars. 1st ed. New York : McGraw-Hill, [1968, c1969] With corrections.
Bibliography: p.
Includes index.
1. Moon—Observers' manuals. I. Title.
[QB581.C44 1984]   523.3   82-18310
ISBN 0-486-24491-1

PICTURE CREDITS

Lick Observatory: pages 14, 16, 18, 22, 24, 26, 28, 30, 46, 82, 90, 96, 102, 116, 122, 128, 132, 146, 150, 156, 160, 164, 166, 170, 182
NASA: frontispiece (ii), pages 74, 75, 76, 187, 188, 214–229
Jet Propulsion Laboratory: frontispiece (x), pages 50, 52, 53, 54, 56, 57, 58, 59, 60, 61, 62, 63, 64, 67, 68, 69, 71
US Air Force: pages 44, 110
US Army: page 48

To Ann Naylor Cherrington

whose inspiration, encouragement,

and assistance contributed much to this book

Preface to the Dover Edition

Many books have been written about the moon, particularly in the 1960s and early 1970s when people looked up into the heavens with new incentive. The exploration of space, formerly the exclusive province of science fiction, had become a startling reality in October 1957 as Sputnik 1 winked its way across the night sky every 96 minutes. A few months later, Congress passed the National Aeronautics and Space Act creating the National Aeronautics and Space Administration (NASA), and plans were begun to send men into space. After the feasibility of such a bold venture had been established, the first goal would be the moon, a natural space station remote indeed in terms of familiar terrestrial distances but a mere first step in terms of planetary distances. The pace of preparation quickened and public interest spread following President Kennedy's challenge to the nation in 1961.

By the mid-twentieth century, professional astronomers generally had abandoned the moon as a dead world of little further interest, but the early 1960s brought a sudden renaissance. Large-scale navigational charting and geologic mapping were begun through large telescopes, and theoretical investigations into the nature of the lunar surface were undertaken. Articles and books appeared in increasing numbers. The original edition of this book was one of them. Like others, it includes fundamentals of the earth-moon-sun relationship, a history of selenography, discussion of lunar features, photographs, and maps. Unlike others, its main function is that of a handbook for use under the night sky with binoculars or small telescope. It not only describes all lunar features visible with the aid of small optics, but it tells observers interested in exploration how to find the features, one after another. Perhaps that is the reason it has been deemed worthy of republication.

The original edition was written and accepted for publication in 1966. It finally reached bookstands in the United States and the United Kingdom in 1969, a few months before the first manned lunar landing. It discussed the early lunar probes and their findings and attempted to forecast some of the Apollo mission objectives as released at the time of writing. Obviously, the vast store of samples, geologic soundings, experiment results, and observational data yielded by the Apollo program was not covered. This edition has been updated to the extent possible for a reprint. Predictions that proved erroneous have been corrected or eliminated. Brief summaries of subsequent unmanned and manned missions have been included, and the Reference Books section has been expanded to include more recent works. The major portion of the text, which is devoted to nightly observations throughout a lunation, remains unchanged except for necessary corrections. However, one important feature has been added—at the end of this edition is a supplement of 16 photographs from Lunar Orbiters 2–5, Surveyor 7, and the Apollo 11 moon landing.

Finally, an apparent inconsistency should be mentioned. We live in an age of changing images and titles. Two of the institutions that appear in the text now have different names. Since it would be incorrect to substitute the present name in the historical context of the reference, it seems best to list those changes in this preface and retain the then-current names in the text. Contributions of the U.S. Army Map Service are recognized throughout the book, but after publication the name was changed to U.S. Army Topographic Command. The other case is that of the well-known NASA Atlantic launch facility east of Orlando, Florida. The cape on which it stands was named Cape Canaveral by sixteenth-century Spanish explorers. Following the assassination of President Kennedy in 1963, the launch facility was officially named the John F. Kennedy Space Center, and it bears that name today. At about the same time, Cape Canaveral was renamed Cape Kennedy by President Lyndon Johnson, and that name was used for a decade. In the mid-1970s, citizens of Florida formally asked that the name revert to Cape Canaveral, and government action was taken to comply with that request.

March 1983   E. H. C.

CONTENTS

Invitation 1

Selecting Binoculars 2

Your Observatory 5

How to Find the Moon 6

The Man in the Moon and Other Features 12

How the Moon Behaves 32

Selenography 38

A Month with the Moon 77

One-Day Moon 80

Two-Day Moon 81

Three-Day Moon 85

Four-Day Moon 86

Five-Day Moon 88

Six-Day Moon 93

Seven-Day Moon 95

Eight-Day Moon 101

Nine-Day Moon 108

Ten-Day Moon 114

Eleven-Day Moon 118

Twelve-Day Moon 126

Thirteen-Day Moon 130

Fourteen-Day Moon 135

Fifteen-Day Moon 137

Sixteen-Seventeen-Day Moon 144

Eighteen-Day Moon 148

Nineteen-Day Moon 152

Twenty-Day Moon 154

Twenty-One-Day Moon 158

Twenty-Two-Day Moon 168

Twenty-Three-Day Moon 173

Twenty-Four-Day Moon 176

Twenty-Five-Day Moon 178

Twenty-Six-Day Moon 179

Twenty-Seven-Day Moon 181

Twenty-Eight-Day Moon 184

The New World of Tomorrow 186

The Charts 189

The Gazetteer 191

Reference Books 205

Index 207

Lunar Photographs Supplement 213

PTOLEMY

GUERICKE   DAVY

ALPHONSUS

LASSELL

ALPETRAGIUS

ARZACHEL

N
W ↑ E
S

Ranger 7 gave man his first close look at the moon on July 31, 1964. It returned this early view of lunar terrain at 8:10 A.M. EST, televised by camera B when the space craft was 1240 miles above the plain on which it crashed 16 minutes and 2 seconds later. The area shown measures 275 miles from left to right and 215 miles from top to bottom. The craters on the right are foreshortened by low-camera angle. The black corner marks that divide the picture into small squares reveal any artificial distortion that may have been introduced.

INVITATION

An invitation to explore the moon through binoculars tonight may seem as absurd as an invitation to cross the Pacific Ocean in an outboard motorboat. But wait! It does make sense. True, we live in a world of huge, powerful, and expensive instruments —giant optical telescopes up to 200 inches in aperture and vast radio telescopes up to 1000 feet in diameter. Yet, of the 4 billion people who inhabit the small planet Earth, only a negligible percentage even have seen such a telescope, and of them only a negligible percentage have had an opportunity to observe with it. Few, indeed, have gazed outward into the universe through a telescope $\frac{1}{10}$ the diameter of the 200-inch Hale telescope. Even if we go down to $\frac{1}{25}$ its diameter, the number of people who have access to such a telescope remains very small in spite of the fact that amateur telescope making has been a popular hobby for the past 60 years.

With binoculars the situation is entirely different. Fifty years ago they constituted a luxury item sold by opticians and pawnbrokers. Today they are sold in large volume by thousands of department stores, discount houses, mail-order firms, drugstores, and other retail establishments. Since the informed and critical buyer can obtain a good "glass" for as little as twenty to forty dollars, no home need be without one in this prosperous country of ours.

But are ordinary binoculars good for anything more than watching birds, horses, and the neighbors across the street? The answer is definitely "Yes!" Carefully selected 7×50 binoculars (magnification, 7 diameters; aperture, 50 millimeters or 2 inches) constitute a far better astronomical instrument than any professional astronomer possessed during the first half-century that followed the invention of the telescope. In fact, even a casual glance at the first quarter moon through 7×50 or 6×30 binoculars may show you more detail of the lunar world than was seen by the great Galileo who invented an astronomical telescope three and a half centuries ago and with it inaugurated the modern age of science through a series of unparalleled discoveries. The largest telescope Galileo ever used had a lens only two inches in diameter. It was mounted in a tube about 4 feet long, and its magnifying power was 33 diameters. However, the lens was of poor quality by modern optical standards, and it evidently resolved only the larger lunar surface features, judging from the drawings of the moon which Galileo made at the telescope and published in his book *Sidereus Nuncius* (1610).

If, therefore, you do not happen to own a private observatory, do not conclude that the thrill of personal observation of the moon, planets, and stars is an experience which you never can enjoy. You need not be restricted to vicarious exploration of outer space through the writings and the photographs of those who have observed. You can make direct personal contact with the moon and other worlds beyond. You can launch your own Space Program right from your roof or back yard. You can become not only well informed about the moon but intimately acquainted with its surface features. You will come to know it as a world which you have explored personally. Then the landings of unmanned space probes followed by the Apollo modules carrying astronauts will be as meaningful to you as the daily news of events around our terrestrial world. The moon is today's frontier—the New World of Tomorrow.

Perhaps you are still skeptical about the effectiveness of binoculars as an astronomical instrument. If so, let me give you an observational fact. In the monumental work *Photographic Lunar Atlas,* edited by Gerard Kuiper (1960), there are listed a total of 670 named lunar features. By actual observation check I have found that 605 of those

1

features can be seen with ordinary 7 × 50 binoculars. Don't expect to resolve all the fine details shown in the illustrations in this book, however, because many of our pictures are enlargements from negatives taken with one of the world's largest telescopes. The smaller named craters can be made out as tiny dark or bright specks, but the larger lunar craters and mountain ranges show well through binoculars and with considerable detail.

Either you already possess the principal piece of equipment needed for your personal exploration of the universe, or you can obtain it easily and at little expense. You are, therefore, virtually ready to begin.

The moon is a good starter for several reasons. Not only is it easier to find and identify than any other inhabitant of the night skies, but it is the most satisfying object that the beginner can observe. It looks the biggest and exhibits lots of surface features. It shows well even in the brilliantly lighted city. It has intrigued the observer and confounded the theory maker for thousands of years. Recently it has experienced a renaissance of professional interest, and the tremendous drive to put a man on its remote surface during the 1960s attracted the attention and stirred the imagination of every thoughtful person.

Just as binoculars serve well in the exploration of the moon, they also reveal to advantage much of the stellar universe beyond our solar system. There are hundreds of double stars, variable stars, star clusters, gaseous nebulae, and stellar galaxies which can be located and examined by the armchair astronomer who holds his instrument in his hands and who knows where to direct it.

SELECTING BINOCULARS

An ocular is an eyepiece—that system of small lenses at the bottom of a telescope (or top of a microscope) with which the observer examines the image formed by the main lens (or mirror) of the instrument. Consequently, a binocular is simply a double eyepiece arranged so the observer can use both eyes at the instrument instead of one.

Such a device is used frequently in connection with microscopic studies but rarely is found attached to an astronomical telescope. The plural form "binoculars" is used to designate any optical device in which two identical telescopes are fastened together side by side for simultaneous visual use by the observer. Generally when we employ the term today we have in mind the modern "prism binoculars" which are characterized by stubby, offset tubes with main lenses about twice as far apart as are the eyepieces.

In principle, a telescope is a simple optical device, as you probably will agree once you have demonstrated its operation to your satisfaction. The basic kit for the laboratory exercise consists of a magnifying glass and a plain white card. The de-luxe kit contains two magnifying glasses of different sizes. On a sunny day pick a spot indoors near the wall opposite a window where the illumination is as poor as possible and from which you can see objects at various distances out-of-doors. Hold the card vertically near the wall and hold the magnifying glass vertically a few inches from the card in the direction of the window. Move the glass slowly toward or away from the window until a small image of the window frame appears on the card. Slight adjustment will bring a sharp picture of the window and the plants or other knicknacks on the sill —all in full color and all inverted. The tree in the yard appears slightly fuzzy on the card, but it can be brought into good focus by reducing the separation between glass and card by a tiny amount. Additional small reductions bring into sharp focus the images of objects located successively farther beyond the window.

The magnifying glass is doing exactly the same thing that the main lens or mirror of a telescope does. It is gathering light from distant objects and using that light to form images of those objects in the focal plane of the lens, which is the plane occupied by the card in your other hand. If you replace the card with a sheet of ground glass or translucent plastic, you can view the image from the opposite side by looking through the translucent sheet as you adjust it between your eyes and the glass while gazing in the direction of the window. Actually, you are looking through a telescope, but it lacks an eyepiece. The smaller magnifying glass in the de-luxe kit will serve that purpose, but you need a third hand to place the second glass between your eye and the sheet that holds the image. Even if you are merely ambidextrous you can complete the exercise, because as soon as you place the eyepiece properly you no longer need the translucent sheet to hold the image. Place the small glass directly in front of your eye, hold the larger glass just beyond it, and move the larger glass slowly toward the window until suddenly you glimpse a distant tree or building inverted. You now have a telescope in your hands. It is not a good one, but it is a telescope, and its fundamental secrets have been revealed to you.

How do we get that inverted image turned right side up again? Each binoculars tube has a bulge in it. Inside that bulge are two double prisms that force the entering light to make four right-angle turns. That maneuver erects the image, and it also makes the instrument more compact since the tube can be cut several inches shorter than would be required if no prisms were used.

In selecting binoculars or telescopes do not yield to the common urge to get the highest possible magnifying power for the investment you plan to make. That is like buying an automobile solely on the basis of the maximum speed which the dealer claims it can attain. Maximum speed is a powerful bragging point for an automobile, but there are various other criteria of much more importance to most drivers. So it is with magnifying power. The novice may announce proudly that he is the owner of a "50-power telescope." Probably it does magnify about 50 times, but if the main lens is only 1½ inches in diameter and the tube is two feet long, it produces only a very dim, fuzzy image. Moreover, the field of view is probably so narrow that it is very difficult to set the telescope on a given star and even harder to hold it there. Usually a question or two addressed to the owner about his observing experience brings out the admission that his telescope "doesn't work very well, but it cost only $9.79." Beware of such bargain instruments! Binoculars offer you the best optical quality and the most utility and convenience per dollar invested of anything available in the broad category of telescopic devices.

Magnification is important, but of even greater importance to the astronomer are light-gathering power and resolving power, both of which increase with increasing size of the *objective* (main lens or mirror). Light-gathering power, in connection with the focal length of the objective (camera F value), determines the brightness of the image. It also sets the limit of star faintness beyond which the telescope will not go. Resolving power indicates the ability of the telescope to separate close double stars and to reveal surface markings on the moon and planets. Magnification of 100 power, for example, makes the diameter of the moon appear 100 times as large as it does to the naked eye, and it also increases the apparent sizes of lunar markings in the same ratio. However, it likewise magnifies 100 times all atmospheric disturbances in the line of sight, and it amplifies a hundred-fold the effect of slight movements of the telescope. If the atmosphere is unsteady and if the telescope is not rigidly mounted, the highly magnified image of the moon will "boil" and shimmer and jump about like a cork on a choppy lake. Under such conditions far better results can be attained with much lower magnification.

Binoculars are more or less available in a wide variety of sizes with magnifications ranging up to 40 power and lens diameters up to 125 millimeters (five inches). The larger ones are heavy instruments that must be supported on a tripod or pier like a telescope. The most popular sizes are 6 × 30, 7 × 35, and 7 × 50. During World War II the U. S. Navy subjected binoculars of all feasible magnification and aperture combinations to an exhaustive series of performance tests. The conclusion was that the best all-purpose hand glass for night use is the 7 × 50 size. For similar reasons the 7 × 50 is recommended first for general astronomical viewing, but the 6 × 30 is a very close second in performance. In fact, the smaller glass actually is preferred by some observers since it is lighter and easier to hold steady.

One may not be able to measure precisely the degree of excellence of binoculars at the store counter, but with a few simple tests he can spot the poor ones to be rejected. First, note that the two telescopes or oculars are fastened together by a pair of hinges on an axle running between the two and parallel to the line of sight. By grasping the instrument with both hands you can turn it about these hinges, thus changing the distance between the two eyepieces until the separation matches that of your eyes. Looking at a distant wall, make this adjustment until you get a single circular field of view of maximum brightness and of uniform brightness with no flitting shadows near the center or periphery. You will find a graduated scale and index mark at the end of the axle near the eyepieces. Through experimentation at home on the night sky you will discover the best setting for your particular eye separation, and thereafter you can set the binoculars at that reading each time you remove them from the carrying case. If you wear glasses, there is a good chance that you will not need them when using binoculars. Deviations in focus of your eyes can be compensated by focusing the instrument unless you are very nearsighted or have severe astigmatism. Glasses and binoculars can be used together if necessary, but from experience I recommend that you use them separately if at all possible.

If you are looking at center-focus binoculars you will observe that near the base of one of the eyepieces (usually the right one) there is a graduated scale and an index mark. Place the large lens cap over the objective lens of this particular ocular, and direct the instrument toward a printed card or sign as far away as possible down the store aisle. Look through with both eyes open and relaxed as if you were gazing at an object miles away. Now turn the central focusing knob until the card, or other detailed object selected, comes into sharp focus. Don't focus your eyes. Keep them relaxed, and let the ocular do the adjusting until you can see the object sharply. Next remove the lens cap from the right ocular and place it over the left ocular lens. This time look at the same object, but do not touch the central focusing knob. Do all the focusing by turning the focusing eyepiece of the right ocular until you get a good, sharp image. Now remove the lens cap and look with both oculars at the same object. You should get an excellent image. Moreover, the image should be bright—almost as bright as it appears to the naked eye. Check this carefully. If the picture through the binoculars is conspicuously weaker than it is without them, the glass is a poor one. Don't buy a dim glass. It might do for a bright object such as the moon, but for star study it would be inadequate.

Individual focus binoculars have no central focusing knob, but a graduated scale and index mark is found at the base of each eyepiece barrel. To focus an instrument of this type place the lens cap over the objective of the right ocular, look at a distant object, and rotate the eyepiece of the left ocular until a good image is obtained. Then move the lens cap to the left ocular, look at the same object, and rotate the eyepiece of the right ocular until you get a good image. The binoculars should then be in focus. After some practice at home, focusing on stars to get the smallest possible image, you can note down the setting for each eyepiece and thereafter check the proper focus for your eyes before going out to observe.

When a telescope or any other ocular is focused properly, the rays of light emerging from the eyepiece toward the eye are parallel rays as far as any given point in the picture is concerned. This is the same condition that holds for the light rays that come to you directly from an object a few miles or more away when you look at it without optical aid. That is why you should relax your eyes completely when focusing binoculars or telescope.

With the two oculars properly separated for your eyes, and with each correctly focused, you are now ready for a most important test. Take the binoculars to a location in the store or outside where you can look at an object at least 300 feet away. Check the focus again. Since the light leaving the eyepieces is in the form of parallel rays, you should be able to move the binoculars slowly forward toward the object at which you are looking without losing the picture. Of course the field of view will narrow rapidly as you move the instrument away from your eyes, but the central part of the picture should remain in view undiminished in size. When you have advanced the binoculars five or six inches from your eyes you still should have a clear view of what appeared at the center of the field when the instrument was in the normal position at your eyes. Hold steady and close one eye, leaving the other open. Then open the first eye and close the second. Alternate rapidly from eye to eye and note what happens to the picture. If nothing changes, the binoculars have been aligned properly by the manufacturer, and the instrument is acceptable. If the picture jumps up and down or from side to side as you switch from one eye to the other, or if you see a double overlapping image with both eyes open, the optic axes of the two oculars are not parallel, and use of the instrument will produce eyestrain. It must be rejected even though the individual oculars may be of excellent quality.

A few words about the care of your astronomical binoculars may be in order here. Of course I do not have to caution you never to let your fingers touch any of the lenses, but watch your friends. Some people, when in the presence of an optical instrument, seem to be seized with an uncontrollable urge to touch a finger to the lens if it is within reach. Others are just downright clumsy and can't pick up a magnifying glass without grabbing it in the middle of the lens instead of by the handle. Still others apparently wonder whether or not their finger will leave a print. It always does, and it always will, and every fingerprint reduces somewhat the efficiency and quality of the instrument. Like testing the law of gravity by jumping from the roof, this is an experiment that hardly needs to be performed again.

When not in use, keep the lens caps on both ends of your binoculars, and store them in the carrying case. That procedure will keep the lenses clean for a long time. When dust eventually does accumulate on the eyepiece and objective lenses, get a one- or two-ounce syringe at the drugstore. Use it to expel quick air blasts onto the lens, taking care that the nozzle does not touch the glass. If that does not suffice, the lens must be washed. Never touch a dusty lens with a dry cloth no matter how soft and clean the cloth may be. Dip a piece of clean cotton in water and touch it lightly to the lens turning the cotton over in your fingers until the entire lens is wet. Then dip a fresh piece of cotton and swab lightly, holding the binoculars in such a way that water cannot run inside. Finally blot dry by light application of a dry piece of cotton or a freshly laundered handkerchief. Any lint that remains may be blown off with the syringe.

On cold nights with high humidity (when you can see your breath) it may be difficult to keep condensed moisture off the lenses. Hold and move your binoculars so that you do not breathe toward them. Even with that precaution, vapor from the atmosphere may condense on the outer lenses and vapor from your moist eyeballs may condense on the eye lenses. Do not wipe it off. Use the syringe to blow it dry. If such treatment becomes necessary every few minutes, the best remedy is to close down operations and wait for a better night.

When observing or carrying binoculars out of the case always use the neck strap. This practice saves observing time and insures against damage should the glass slip from your grasp.

YOUR OBSERVATORY

The term observatory may sound a bit pretentious in view of the scope of astronomical operations here contemplated, but you are going to see a great deal more with your simple equipment than did the astronomers of old who operated the enormous ancient observatories of the Far East. The location of your observatory will be governed by the position and size of trees, buildings, lights, and other distracting conveniences of civilization in the vicinity of your home. If necessary, you may be able to select various observing sites for different parts of the sky. Pick the spot that gives you the best sweep of the sky—front yard, back yard, or apartment-house roof. If a street lamp or a neighbor's light shines in your eyes, its effect often can be nullified by a portable screen or a small tarpaulin hung over a clothesline. In fact a 6-by-8 or 5-by-6-foot canvas tarpaulin attached to a frame of aluminum pipe or 2-by-2's and provided with legs can screen out a host of offending luminaries and bring good observing conditions to a spot that otherwise would be bathed in impossible glare.

A garden variety of chaise lounge will be found convenient for lunar (and stellar) observations particularly if it has an adjustable back and if the arms are correctly located to support your elbows as you hold the binoculars to your eyes. If the arms do not quite fit, perhaps they can be supplemented by cushions or pads of some sort. Always sit down or lie down when observing because the closer your eyes and arms are to your point of contact with the solid earth, the easier it will be to hold the binoculars steady. For the same reason don't tilt your chair back on its legs or rock it. The better you can imitate the relaxed rigidity of a concrete pier as you hold the glass, the better will be your view. Support yourself well, but relax your muscles and forget your problems as you explore the moon.

You also will need a light for reference to the charts. At this point the beginner often makes a mistake that wastes much of his observing time and may soon tire his eyes as well. He goes out equipped with a powerful flashlight or a desk lamp with a 100-watt bulb in it. These show the charts well, but they also throw back into the eyes so much light that one is essentially blinded for several minutes as far as observing the night sky is concerned. Since flashlights are designed to illuminate objects 10 or more feet distant, they are invariably too bright for chart-reading purposes. If you must use one, wrap several layers of dark-colored cellophane over the head to cut down the glare. Much better is an extension cord with a "Nite Lite" plugged into the end. The Nite Lite consists of a small-base miniature white light bulb of the Christmas-tree type mounted in a plastic plug-in base and provided with a switch and a small plastic shade. The complete unit is only four inches long, burns seven watts, and may be obtained at a variety store. It should be placed from one to two feet above the chart with the shade turned so that no direct light strikes the eye. For lunar observations it may be left burning all evening, but for stellar viewing it should be turned off when not in use. If you happen to be a Do-It-Yourselfer you can make a convenient observing desk for lap or table from a clipboard, two nine-inch lengths of perforated steel pipe strap, some bolts with wing nuts, an extension cord, a single electrical outlet, and a Nite Lite. A gooseneck desk lamp is also convenient for this purpose, but the regular bulb must be replaced with the weakest one you can buy—a seven-watt standard base bulb (white glass $1\frac{1}{4}$ inches in diameter). Extensive reading with such feeble illumination certainly is not recommended, but when one is observing the night sky, a seven-watt bulb provides adequate lighting for checking charts, noting location directions, and reading short descriptions of objects.

The equipment of the observatory may be completed with the addition of a card table or convenient substitute therefor.

HOW TO FIND THE MOON

Over sixty years ago the world-renowned experimental physicist Robert Wood composed, illustrated (with "woodcuts"), and even lettered a little volume of humorous verses entitled *How to Tell the Birds from the Flowers*. After one of its numerous reprintings, an astronomer friend of mine asked for a copy at a bookstore. Upon hearing the title, the inexperienced but confident young clerk drew himself up and pontificated: "Sir! That is ridiculous!" To which my friend replied: "It is intended to be."

The title of this chapter is not intended to be ridiculous. Nor is it implied that you may experience difficulty in finding and identifying the moon on nights when it is in the sky. Even though the Space Age promoters may bring about astronomical disaster by filling the skies with artificial satellites that go blinking and beeping among the stars in endless procession—even then no observer should hesitate in identifying *the* moon. My purpose here is to discuss some of the motions of the moon so you will be able to determine on any given date approximately when the moon can be expected to rise, to be in position for observation, and to set.

When the ardent Romeo at Juliet's balcony seeks to convince the lovely girl that no longer can he exist apart from her he swears his love "by yonder blessed moon . . . , That tips with silver all these fruit-tree tops." At that point Juliet interrupts with: "O, swear not by the moon, th' inconstant moon, That monthly changes in her circled orb, Lest that thy love prove likewise variable." Shakespeare here portrays the moon as the traditional symbol of fickle fluctuation although he does refer to the periodicity of its changing aspect. This has been man's general impression down through the ages. Probably most people today, if they were to think seriously about the moon for a few moments, would

recall from a general science course that the moon travels in a closed path around the earth, that it completes its course in about one month (the origin of the term), that it changes from a thin crescent to full and back again, that on some nights it shines brightly while on others it is absent, and that the United States was committed to spend many billions of dollars to put a man on it before 1970. In most cases that would be about all in spite of the fact that during the 1960s and 1970s far more research on the moon was carried out and far more information about it was published than in all the previous history of mankind.

The moon is believed to be about 4,500,000,000 years old, give or take a few hundred million years. During most of its existence it has been associated with the earth as our satellite. Many are the theories of how it got there and what it has been doing since it arrived, but it seems likely that the moon has behaved just as it does today ever since the first primitive human being looked up at it several million years ago. Like the earth and other planets, it is regarded as a cold body which emits no light of its own making but shines simply by reflecting some of the sunlight that happens to fall upon it. The moon, like the earth, is approximately spherical in shape, and its diameter is about $\frac{1}{4}$ that of the earth. Its surface area is, therefore, about $\frac{1}{4}$ by $\frac{1}{4} = \frac{1}{16}$ that of the earth, and its volume is about $\frac{1}{4}$ by $\frac{1}{4}$ by $\frac{1}{4} = \frac{1}{64}$ that of the earth. If it were made out of the same ingredients as the earth we would expect it to have about $\frac{1}{64}$ the mass of the earth (the more precise value is $\frac{1}{49}$ since the moon is a little bit larger than $\frac{1}{4}$ the diameter of the earth). However, astronomers long ago found that the mass of the moon is only $\frac{1}{81}$ that of the earth, which tells us that our satellite is composed of lighter (less dense) materials than is the earth.

Perhaps you best can visualize the relationship between the moon and the earth in space if you set up a scale model. Take a golf ball to represent the moon. The dimples on the ball remind one of the lunar craters except for their uniform size and regular geometrical pattern. As a matter of fact, those dimples are in the same relation to the size of the golf ball as the largest craters are to the size of the moon. One of the dimples, then, represents the great crater Clavius, a conspicuous feature when the moon is about nine days old. Clavius is the number-two crater in size. Thus all but two of the craters will be smaller than the dimples on our model, and most of them will be much smaller. For the earth you need a ball six inches in diameter. If you haven't a ball that size, inflate a small balloon or cut out a circular disk of cardboard. Next measure off across the room a distance of 15½ feet. Then place your six-inch earth at one end of this line and the golf ball (1⅔ inch) moon at the other. This is a true model of the earth-moon system in space (Fig. 1). If you place your eyes beside the earth ball, the golf ball appears exactly the same size as does the moon when we see it in the sky, and if you strain to see those dimples you can appreciate the need of some optical assistance in viewing the lunar craters.

You should not forget the sun in your little project of model making. Not only does the sun illuminate the moon and make it visible to us, but the sun lights and warms the earth and makes it possible for us to be here to observe the moon. At this point you run into a difficulty in model making. Things get big and far away in a hurry. You will not be able to find a ball around the house big enough to represent the sun. Probably the house itself isn't big enough. For the sun you would need a huge sphere 56 feet in diameter! After you have found a 56-foot sphere or have made one out of an old surplus Echo Satellite, continue with the homework. You will want to place it in proper relationship to the earth-moon system now installed in the family room. To do this you will need to carry the 56-foot sphere down the street 6000 feet or a little more than 1.1 miles if you use your car to make the delivery.

The golf-ball moon has a mass of about 1½ ounces. Therefore, from what we already have noted our model earth should have about 81 times as much mass, or 7½ pounds. If you had made the six-inch sphere out of good clay mud it would have about the right relative mass.

For the next exercise in our lunar laboratory as-

signment you will need a light, rigid bar 15½ feet long and, say, ¼ inch in diameter, which will not bend and which has no mass because we wish to keep the problem as "simple" as possible. Astronomical distances are always measured from the center of one body to the center of another, so place one end of the bar at the center of the golf ball and the other end at the center of the earth globe. You now have a sort of dumbbell consisting of two unequal balls at opposite ends of a long rod. By experiment you should be able to find a point along the bar at which a support would maintain the system in perfect balance. That point marks the center of mass of the system. Since the earth is 81 times as massive as the moon, the center of mass should be found at a point $\frac{1}{82}$ the distance from the center of the earth ball to the center of the moon ball. In our model, the center of mass lies 2¼ inches from the center of the six-inch ball toward the golf ball. Finally, if you give the rod a push and start the dumbbell revolving around the balance point or center of mass, you will bring to life our static model and convert it into a dynamic model of the earth-moon system turning in space.

Let me here call attention to the astronomical distinction between the terms *rotation* and *revolution*. Each has a precise meaning, and they should not be used interchangeably as they frequently are in general conversation. *Rotation* describes the turning of a body on its axis; *revolution* describes the movement of a body in its orbit. Thus, the earth rotates once each day and so produces the sequence of day and night. Also, the earth revolves around the sun once each year and so produces the sequence of the seasons.

If you didn't quite make it with the 56-foot sun globe and the 15½-foot inflexible but massless rod,

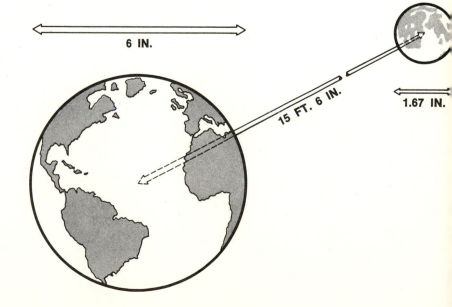

FIGURE 1. Model of the earth and moon in space. A 6-inch ball and a golf ball separated 15½ feet provide a scale model of the two bodies at average distance.

6 IN.

15 FT. 6 IN.

1.67 IN.

don't flounder. If they exist in your mind's eye, they accomplish the purpose fully as well and far more conveniently than real ones would. You have a golf ball and a 6-inch ball on the floor separated by a distance of 15½ feet center to center, and can picture that model in slow revolution about a point between them located only 2¼ inches from the center of the earth globe. For this dynamical model you need a rather large room, or back yard, because the moon is going to revolve in a circle nearly 31 feet in diameter while the earth revolves in a circle only 4½ inches in diameter. So it is not strictly true to say that the moon revolves around the earth. Both earth and moon revolve around the balance point between their centers, but the small-mass moon has to do most of the traveling in obedience to the mandates of the law of gravitation.

Another statement that is not quite true is the one just made about the moon and the earth revolving in circular orbits. In science truth is always our goal but rarely our attainment. We move from error toward truth by successive approximations. Just as parallel lines are said to meet at infinity, we, as the human race, might expect to arrive at absolute truth only if given infinite time in which to flounder intelligently. All the great minds of antiquity believed the moon's orbit to be a circle. So did Nicolaus Copernicus whose unorthodox views shocked the intellectual world of the mid-sixteenth century and forecast the modern concept of the solar system. But the Copernican system didn't work any better than the ancient Ptolemaic system until John Kepler eradicated from it the vestigial errors of ancient thinking. Kepler's first law of planetary motion, published in his book *Commentaries on the Motions of Mars* (1609), states that the orbit of each planet is an ellipse with the sun at one of the foci. This law applies to satellites just as it applies to planets. Consequently, the orbit of the moon is a large ellipse, and the orbit of the earth is a small ellipse of the same shape. The center of mass of the earth-moon system is located not at the center of the orbits but at the common focus of those two orbits (Fig. 2).

We now have our earth-moon model slowly revolving, but, if we wish to be precise, we must picture the distance between the two balls changing from a minimum of 14 feet 3 inches to a maximum of 16 feet 4 inches and back to the minimum again during a complete revolution through 360 degrees. However, if this Keplerian refinement seems too complicated, set it aside for the present and stay with Copernicus. If you hold to a circular orbit with radius of 15 feet 5 inches for your model you won't be off more than about 7 per cent at worst, and you will see the moon just as clearly through your binoculars.

But the earth rotates on its axis, as Galileo is said

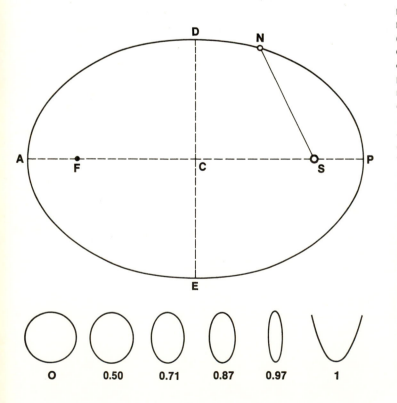

FIGURE 2.   The elliptical orbit of a planet. The planet N travels in the closed curve PDAEP. PA is the major (maximum) axis, and DE is the minor (minimum) axis. One focus is at S close to the center of the sun; the other focus F is empty. SN, the radius vector of the planet, varies from a minimum SP (perihelion) to a maximum SA (aphelion). CP, the semimajor axis, is called the mean distance because it is the average of the perihelion and aphelion distances SP and SA. It is also equal to SD and SE, D and E being the points on the orbit midway between perihelion and aphelion. The length of the semimajor axis determines the size of the orbit. The shape is determined by the eccentricity which is the ratio of CS to CP. For an ellipse the eccentricity can have any value between zero and unity. If the eccentricity is zero, the orbit is a circle (lower left figure), the foci S and F having moved inward to coincide with C. As the eccentricity increases, the foci S and F move farther apart, and the ellipse becomes flatter. If the eccentricity is unity, the orbit is a parabola (lower right figure), F and A having receded to infinity. A body moving in a parabola never returns to perihelion. All parabolas are the same shape just as all circles are the same shape, but ellipses can have any shape between the limiting cases of the circle on the one hand and the parabola on the other. Four ellipses of different shape are shown together with their eccentricities.

to have muttered following an official recantation during his trial for the heretical crime of teaching the Copernican theory. So it does, and you must introduce that motion into our model. Picture the six-inch earth ball rotating on an axis through its center that makes an angle of about 67 degrees (¾ of a right angle) with the plane of the floor. Let the direction of rotation be counterclockwise as you stand looking down on the ball, and let the period be one minute which will represent our 24-hour day. Put a mark on the ball to represent your observatory. Look about out-of-doors and pick out the most distant large building or similar object you can see to represent the sun (56-foot sphere 6000 feet away). When your mark on the earth ball is turned toward the distant building it is day, when turned away it is night. Since we have chosen one minute to represent 24 hours, the moon will revolve in its orbit around the earth in the same counterclockwise direction with a period of almost half an hour (the month).

You now have everything in motion, and confusion is rampant unless you are a most unusual person. It could be worse, because the earth, as we all know, revolves also in an orbit around the sun. However, we are going to ignore that further refinement for the present. Another complicating factor which exactitude would require is a plane for the earth and the sun inclined about five degrees to the plane of the floor which we have used as the plane of the earth-moon system. If these two planes coincided, we would have an eclipse of the sun every time the moon is new and an eclipse of the moon every time the moon is full. Because of the existence of two fundamental planes rather than one, the moon usually passes above or below the sun when new, and above or below the earth's shadow when full. As for the confusion, don't let it deter you. That is the normal experience for the first time around. The picture will clear up a little bit later.

As the earth turns silently on its axis the sun appears to rise above the eastern horizon, sweep across the sky, set at the western horizon, and swing around the other side of the earth bringing the regular sequence of day and night. Simultaneously the moon travels along continuously in its big orbit around a point not far from the center of the earth. Let us begin to trace the cycle when the moon is *new*. This is an ancient term that doubtless had its origin in prehistoric times when it actually was believed that a new moon was provided for each cycle. In like manner we still speak of the *age* of the moon

as the number of days, hours, and minutes since the event of new moon.

New moon occurs at the instant in which the moon is a few degrees exactly north or south of the sun in the sky (or in front of it on those occasions when we have a solar eclipse). Generally speaking, north in the sky is approximately the direction from the body in question to Polaris, the North Star. South is the opposite direction. When speaking of new moon and other lunar phases, north is the direction from the body to a point in the sky about halfway from Polaris to Vega. That is the north pole of the ecliptic. Do not be concerned with this further technical refinement which is mentioned primarily for the purpose of reassuring those experts who may fear that I am misleading you through oversimplification. Henceforth, "north" will be delineated well for us by the direction from the body to Polaris, a bright star only one degree removed from the north celestial pole. The new moon is invisible, except during an eclipse, for two reasons. First, it appears very close to the sun and is lost completely in the glare of scattered light that gives us a bright daylight sky. Second, its sun-illuminated side is turned entirely away from us. The new moon sets at about the same time the sun does. Consequently, it is already below the horizon when the evening sky begins to darken. (Fig. 3).

As the earth turns through the night and into the next day the moon moves forward steadily in its orbit. Its direction of revolution, as we have seen, is the same as that of the earth's rotation—the direction we call eastward. Thus the moon on the following day is found east of the sun in the sky, and every hour carries it farther eastward by an amount equal to its apparent diameter, or ½ degree. At sunset the one-day-old moon is about 12 degrees east of the sun and will set shortly after it. By this time a tiny fraction of the illuminated half is turned toward us, and under good conditions we should see a very thin sliver of a crescent moon close to the horizon shining feebly in the twilight glow after sunset. In a few minutes it will follow the sun below the horizon. Again the earth turns us through the night, and the moon rises unnoted about an hour after sunrise. All day long it follows the sun westward across the sky but steadily lags behind because of its eastward orbital motion around the earth.

This evening we definitely should see the moon if the weather is fair and if we have an open view to the west with no extensive obstructions rising more than a few degrees above the horizontal

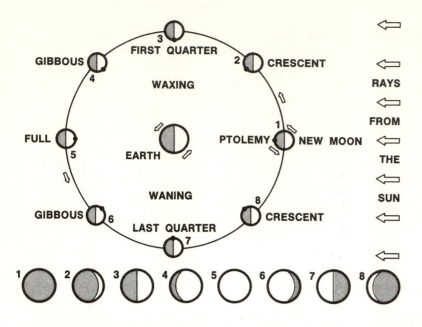

FIGURE 3. Rotation, revolution, and phases of the moon are explained in this schematic diagram of the moon's orbit around the earth. The illuminating sun is located far to the right, and the moon's positions at intervals of three and two-thirds days are shown. The illumination of the moon as we see it in the Northern Hemisphere at each numbered position is shown at the bottom of the diagram. For the Southern Hemisphere invert the drawing. The directions of orbital revolution and axial rotation for both bodies are indicated by small arrows. The protrusion on the earth-ward side of the moon represents the large crater Ptolemy which is always near the center of the lunar disk we see. If you begin at 1 and glance quickly around the orbit, centering your attention on Ptolemy at each position, you will see the moon rotate on its axis in a period equal to that of its revolution.

plane. Half an hour after sunset the thin crescent should be found low in the west, and observations should be possible for about ¾ of an hour before distant trees or horizon haze force a halt. The two-day-old moon is about the "youngest" that the casual observer ever notices. Consequently, the thin crescent in the evening sky is erroneously but generally referred to as "the new moon," to be viewed by good-luck seekers.

On successive evenings the moon is seen in the western sky shortly after sunset, progressively higher above the horizon and farther removed from the point where the sun has set. This is the *waxing* crescent stage since each night brings a slightly larger fraction of the illuminated half into our view, and the crescent appears to be growing. When four days old the moon remains above the horizon long enough after sunset to be seen against a fairly dark sky. One then can observe the interesting phenomenon called "the old moon in the new moon's arms." In addition to the brilliant crescent, the rest of the lunar disk is seen to be shining with a faint glow which our European friends call "the ashen light." The horns or tips of the crescent seem to reach out like arms to encircle the dimly-lit old-moon portion of the disk. After you have become familiar with the more prominent features of the lunar surface, you probably can trace some of them at this stage on what is supposed to be the dark part of the disk.

Whence comes this strange, feeble light? The cold moon cannot provide it. Nor can the sun, since it is now lighting the opposite side. As you already have guessed, this weak illumination is earthshine—

second-hand sunshine reflected from the earth back onto the dark side of the moon. Since both the earth and the moon get all their light from the sun, each one presents phases to the other, and the phases are complementary. Thus when the moon is between the sun and the earth with its bright side turned entirely away from the earth, the bright side of the earth is turned entirely toward the moon. An astronaut standing on the dark new moon would see in his black sky the full earth, and a spectacular sight it would be! We have seen that the surface area of the moon is only about $1/16$ that of the earth, so the earth would serve as a mirror nearly 16 times as large as the moon. Moreover, the clouds, which at any time cover a considerable portion of the earth, are much more efficient reflectors than the dull rocks that make up the surface of the moon. Consequently, the full earth is around 80 times as bright as the full moon. Imagine how well the countryside would be illuminated by 80 full moons in the sky or one full moon with nine times the diameter of the one we know. If it were possible for us to see the one- or two-day moon against a dark sky background, the earthshine effect then would be quite imposing.

We now require a definition which is not new but may need to be reviewed. The *celestial meridian* is the great circle which passes through the north point on the horizon, the *zenith* or point overhead, and the south point on the horizon. It thus divides the sky into east and west halves. Ordinarily we say that a celestial body is rising when it is moving up into visibility from below the horizon. Actually, a celestial body continues to rise as long as it increases its altitude above the horizon. In this more general sense, then, all bodies east of the celestial meridian are rising, and all bodies west of the celestial meridian are setting, each having attained its greatest altitude when it crossed the celestial meridian. Since atmospheric opacity and turbulence increase with *decreasing* altitude, the best astronomical observations are made when the object is near the celestial meridian.

Recalling that the earth makes one complete rotation about its axis (through 360 degrees) each 24-hour day, we see that its rate of turning is 360 degrees divided by 24 hours or 15 degrees per hour. As a result, all celestial bodies appear to move westward across the sky at that rate of 15 degrees per hour. We have seen also that the moon in its orbital motion around the earth moves eastward from the sun about 12 degrees per day. The sun crosses the celestial meridian at 24-hour intervals, but twenty-four hours after the moon has crossed we find it 12 degrees east of the meridian. It will need $^{12}/_{15}$ of an hour more to return to the celestial meridian or roughly ¾ of an hour. So the moon, on the average, crosses the celestial meridian about ¾ of an hour later each day. The same ¾ of an hour delay applies, but even more roughly, to the rising and setting of the moon from day to day. Other factors are involved, and the actual delay of moonrise from day to day, as observed at different stations in the temperate zones, may vary from 10 minutes to 1½ hours.

Between seven and eight days after the new phase the moon has fallen so far behind the sun in the daily westward parade of heavenly bodies that it crosses the celestial meridian at approximately 6 P.M. and sets around midnight. It is now 90 degrees, or one right angle, east of the sun. The crescent has grown steadily until ½ the illuminated surface is turned toward us, and we say that the moon is in first quarter. On successive nights it is found progressively farther east in the sky at 6 P.M., and more and more of the disk is illuminated as it passes through the waxing *gibbous* stage. Fifteen days after new moon the moon reaches a point 180 degrees east of the sun. Since it is opposite the sun in the sky, it rises at sunset, crosses the celestial meridian about midnight, and sets at sunrise. This, of course, is the full moon which bathes the countryside in a soft, flattering light, hiding its blemishes and highlighting its beauty. During the short nights of midsummer it rides low across the southern sky, as seen in Northern latitudes. In midwinter, when the sun follows a low, short daily path across the

sky, the full moon is said to compensate by riding high in the sky and shining throughout the long, cold night.

Following the full phase, the moon does not rise until after sunset, and the delay of roughly ¾ of an hour is continued from night to night. Having waxed for two weeks, it now *wanes,* going through the waning gibbous stage. After another week has passed the moon is 270 degrees east of the sun or 90 degrees west of that body. Again ½ the illuminated surface is turned toward us, and the moon is said to be in last quarter. This time when we view it near the celestial meridian in the northern hemisphere it is the left half that is bright whereas at first quarter it was the right half of the disk that was illuminated. The moon now rises approximately at midnight and reaches the celestial meridian at about 6 A.M.

If you are going to follow our satellite through the waning crescent stage back to new moon again you either will have to stay up late or get up very early. In the first week of the moon we looked for it in the western to southern part of the sky after sunset. Now, in its last week, we must look for it in the southern to eastern part of the sky before sunrise. Two days before new moon the thin crescent may be seen low near the eastern horizon at dawn. Perhaps the next morning it can be detected briefly before the rising sun overwhelms its feeble light.

The orientation of the crescent and gibbous moons in the sky seems to be a point of frequent confusion among nonscientific people. Except for those relatively rare moments when a total lunar eclipse is in progress, one *limb* or edge of the disk always will be filled out and bright while the opposite limb will appear cut off to a more or less extent and lost in darkness. Naturally the bright limb will be the one on the side nearer the sun. If we face south when looking for the moon from any station in the Northern Hemisphere, the illuminated limb, as we have noted, will be the right-hand one between new and full moon, but the left-hand one from full on to new moon. This condition requires that the horns (or *cusps*) of the crescent *always* point away from the direction of the sun. In fact if you extend the direction of the horns backward when viewing the crescent moon, the line you draw in space will pass through the sun wherever it may be below the horizon. Thus the visible crescent is always so oriented that the horns point upward away from the horizon and never down toward it as cartoonists usually prefer to draw them. You perhaps have heard the old folks' weather rule to the effect that we shall have a dry or a wet month depending on whether the "new moon" (two-day crescent) is tipped near the horizontal to "hold water" or tipped near the vertical to "spill water." I don't know whether or not this rule has any validity. I do know that one can get the same predictions by looking at the calendar and noting what month it is. The angle which the horn line of the two-day moon makes with the horizon changes regularly from month to month in an annual cycle.

If the relationship between the moon's phases and its *elongation* or angular distance from the sun in the sky is not perfectly clear, try this little demonstration. Pick up that golf-ball-moon model and darken exactly one hemisphere with crayon, glass-marking pencil, or even lipstick. This time your head will be the earth, and the distance to the moon will be restricted to your arm's length. Choose a window or a lamp to represent the sun and stand a few feet from the opposite wall of the room. Hold your "moon" between thumb and forefinger of your left hand toward the "sun" and turn the ball so the entire shaded half is toward you. This is new moon. Now swing your arm slowly to the left, turning the ball as it goes so that the light half is always toward the sun. You are now viewing the waxing crescent stage that characterizes the first week of the moon. You should see the thin crescent appear and grow steadily until it fills exactly ½ the disk when you have moved the ball to the first quarter position 90 degrees away from the direction of the sun. Continue in the same manner through the gibbous phases to full moon, on to last quarter, and back to new.

I started out in this chapter to tell you how to find the moon, a technique which may have appeared obvious at the outset but doubtless seems much more complicated now. I am sure we have had enough theory and should get on to the practical application. Here is the secret: Look around the house until you find a calendar that has those little full-moon and half-moon faces filling the empty date spaces at the beginning or end of the month. The faces don't matter, but the code messages in fine print under them contain the necessary data. The one I am looking at reads as follows: "FM5 LQ12 NM19 FQ27." From what we have learned we now should be able to expand this cryptogram into a complete monthly forecast of where and when to look for the moon. During the first four nights of the month it will be in the waxing gibbous stage and will be found in the southeast part of the sky at sunset, conveniently located

for early-evening observation. On the fifth day the moon will be full and will rise at sunset from a point on the eastern horizon approximately 180 degrees from the setting sun. It will be in position for observation all night long but will be placed best around midnight when near the celestial meridian. On the sixth it will rise near the same point on the horizon about ¾ of an hour after sunset. On successive nights it will rise later and later. On the twelfth it will be in the last quarter phase, will rise about midnight, and will climb up to the celestial meridian by sunrise the next morning. On the next night it will rise about ¾ of an hour after midnight and won't make it to the celestial meridian until after daylight. It then will be in the waning crescent stage and will continue to rise later each morning. By the eighteenth it will be a thin crescent observable at best only for a few minutes close to the horizon after dawn. On the nineteenth the moon will be new and not observable. The following night the thin crescent may be glimpsed low in the west soon after sunset. On the twenty-first the 2-day moon will be visible for perhaps an hour near the western horizon after sunset. On successive nights the crescent will wax thicker, and the evening observation period will grow longer. On the twenty-seventh the moon will be in the first quarter phase, ½ illuminated, and on the celestial meridian at sunset. During the remaining evenings of the month it will be found in the southeast at sunset progressing through the gibbous phase toward full moon again. Those four numbers and eight letters contain quite a lot of information, and you are now in position to read it all out each month of the year.

THE MAN IN THE MOON AND OTHER FEATURES

The most obvious features of the moon's face are, of course, the shadowy dark markings that avoid the southern third of the disk, nearly fill the western hemisphere, and cut a broad swath southeasterly across the eastern hemisphere. (For the points of the compass on the moon's surface refer to the chart section.) With vague, undulating borders they meander one into another, all being connected except for one that stands out conspicuously near the eastern limb a little north of the east point. For many thousands of years nomads, hunters, farmers, and outdoor people in general have observed those large markings and speculated about them. Such people of earlier generations, with little education and simple interests, were much more familiar with the appearance and the ways of the moon than are today's city dwellers into whose floodlighted canyons its rays seldom penetrate. The Man in the Moon, the Old Woman, the Rabbit, and doubtless many other figures were "discovered" by the more lively imaginations and passed on from generation to generation. One rarely hears mention of such childish creations in the sophisticated Space Age. More familiar are the tales of the Apollo Program which carried 24 astronauts to the moon and landed 12 of them on its surface.

Long ago it was guessed that the dark areas are depressions, and their shadowy appearance was somehow associated with that concept. The guess has proved to be correct, but various inferences that followed from it have been found wanting. The large depressions resemble the ocean basins of the earth. Since all but one are interconnected just as are the oceans and most of the seas of the earth, it is not surprising that someone in the early years of lunar studies recognized the analogy and called the dark areas seas or *maria* in Latin, which was the language used in learned writing and discourse for many centuries. Thus we have such peaceful names as Mare Serenitatis and Mare Tranquillitatis and the sweet Mare Nectaris. Near the North Pole what could be more appropriate than Mare Frigoris? There are eleven easily-identified maria plus a huge area in the western hemisphere known as Oceanus Procellarum. In addition there are twelve relatively inconspicuous maria located along the limb. Even though some of these actually are extensive, they usually appear foreshortened into thin streaks when seen at all.

In keeping with the watery nomenclature, some of the smaller dark areas have been designated *lacus* (lakes), *sinus* (bays), and *paludes* (marshes). So we find such startling or fanciful names as Lacus Mortis, Palus Epidemiarum, and Sinus Iridum. The maria and their smaller subsidiaries constitute the coarse features of lunar topography. Most of them can be picked out and identified with the naked eye, but they are much more interesting when examined through binoculars because of the variety of detail found on many of them.

The brighter areas of the moon are covered with a great profusion of craters, the more conspicuous ones being named after astronomers, mathematicians, and other scientists. They evidently first were seen in the first decade of the seventeenth century by Galileo and his less-famous astronomical contemporaries Thomas Harriot in England and Simon Marius in Germany. Their crude, little, prototypal telescopes revealed those surprising features, and their observations and descriptions inaugurated the science of Selenography. Today it would be hard to find in the Western world a person who has not looked at various lunar photographs showing far more detail than any seventeenth-century scientist ever saw. If that statement appears exaggerated, I need only remind you that during the first few years of the Space Age the eruptive, wrinkled face of the ancient moon largely displaced the pretty girl as an advertising eye catcher. The fat commercial sections of magazines approximated photographic lunar albums for a time until acute banality set in. Nevertheless, I have yet to meet or hear of a person who looked for the first time at the moon through a small telescope or binoculars without expressions of wonder and delight. That first view always carries with it the thrill of exploring another world.

No final count has been made of the number of craters on the moon. In 1878 Julius Schmidt, well-known lunar cartographer, published a map showing 32,856 craters, according to his count. Since most of Schmidt's work was done with a telescope of only six inches aperture, it is likely that a comparable visual survey with a large modern telescope would reveal millions. They range in size from the enormous ring plain Bailly, 200 miles across, down to depressions a few hundred feet in diameter which are revealed only by the largest telescopes under the best observing conditions.

So much for earth-based observation. From mid-1964 through 1972, camera-equipped robot space probes followed by manned landings mapped the unknown far side of the moon and explored in increasing detail numerous selected small lunar areas containing craters from a few feet in diameter all the way down to hemispherical pits no larger than 1/30 inch across found in surface rocks. Many craters have central mountains, the larger examples being seen easily through binoculars. Many have smaller craters on their floors or walls, showing that some were formed earlier than others. Some of the great craters such as Copernicus, Tycho, and Theophilus stand out sharply like a newly-minted coin while others such as Catharina and Walter appear softened and indistinct like the markings of an old coin worn almost smooth (Figs. 4, 5, 6). This is striking evidence that the period of crater formation was not a short one.

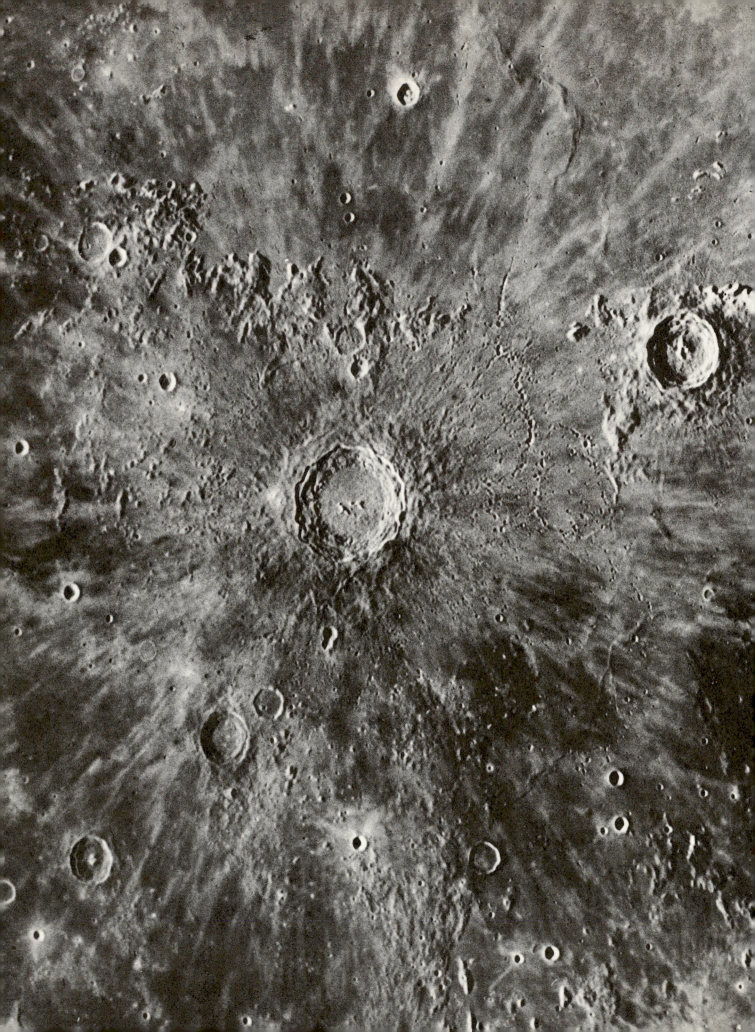

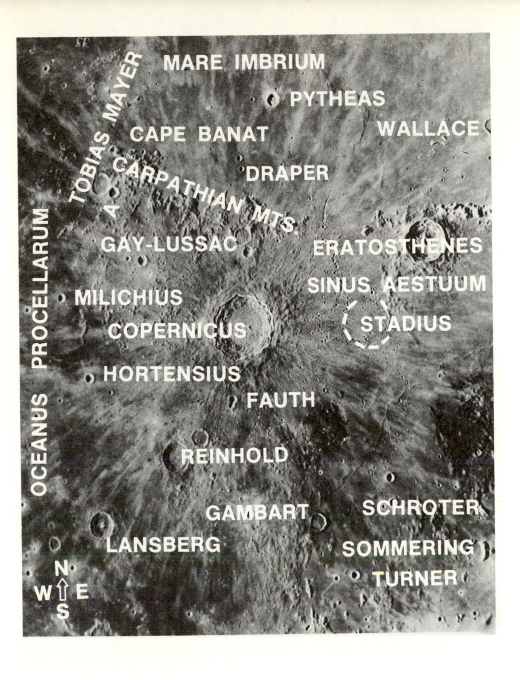

FIGURE 4. The crater Copernicus and its surroundings in afternoon illumination. Examples of almost every
type of lunar surface feature are shown on this plate
taken with the 36-inch Refracting Telescope. On November 19, 1969, Charles Conrad and Alan Bean
touched down the Apollo 12 lunar module 3 crater
diameters southeast of Lansberg within 65 yards of
Surveyor 3 which had been soft-landed on April 17,
1967.

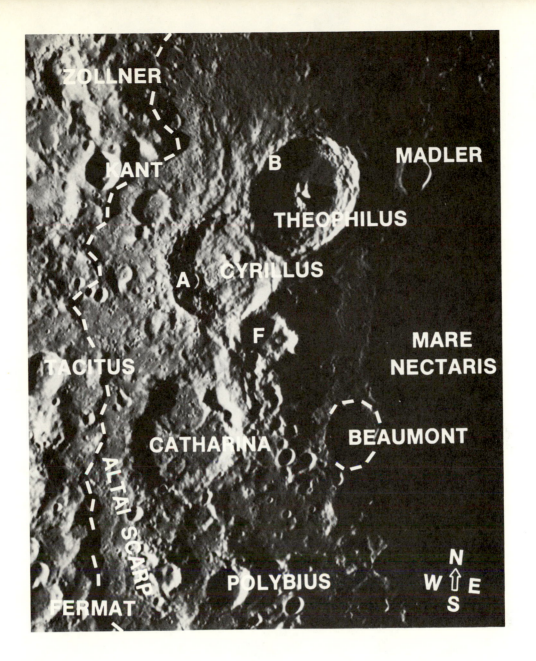

FIGURE 5. Three great craters at sunset—young Theophilus, old Cyrillus, and ancient Catharina. From a plate taken with the 120-inch Reflecting Telescope.

FIGURE 6. The large young crater Tycho and the huge old crater Maginus under afternoon sun. (120-inch Reflector.)

In general, craters are best seen when near the *terminator*—the sunrise or sunset line that separates the bright and dark hemispheres of the moon. When the terminator lies within 200 miles of a crater, the sun as viewed at the crater is no more than 10 degrees above the horizon. Under such conditions even slight elevations cast shadows which are intensely black owing to the absence of a light scattering atmosphere. There is strong contrast between the black shadows and the bright areas in direct sunlight. Craters thus stand out sharply and give the impression of great height and steepness of wall and central mountain. That is an illusion which was not appreciated fully by all the selenographers of the past. Generally speaking, the craters are rather flat, with gentle to moderate slopes. For example, the prominent crater Theophilus was measured with high precision at the McMath-Hulbert Observatory some years ago, and a cross-sectional diagram was prepared from the measurements. It shows that if we let a dime represent the outline of the crater, the coin is about ⅓ *too thick* to show proportionately the elevation difference between the wall crest and the crater floor. Perhaps even more striking is the fact that if an observer should stand on the floor at the center of the great crater Clavius he would not be able to see any portion of the walls which rise 16,100 feet above that floor. All the walls would lie beyond and below the horizon owing to the curvature of the moon's surface, and the observer would have the impression of standing on a limitless, flat plain.

Quite a few of the more prominent craters and some of the small ones are known as *ray craters* because of their curious appendages. Rays are bright streaks that emanate more or less radially from certain craters. They show to the greatest advantage when the sun's altitude above the horizon is high at the crater, and they usually disappear entirely when the crater is near the terminator. Most of them extend from about 10 to 200 miles outward, and their widths range up to around 10 miles. They show best on the maria where their white material contrasts sharply with the dark background. They run across old craters and sometimes even up and down mountains with no apparent change in brightness or width. They must be very thin because none has ever been observed to cast any shadow whatsoever. By far the most outstanding ray system is associated with the crater Tycho (Fig. 6). Tycho, easily seen in detail with binoculars, is located 200 miles west of a point about ⅓ the way from the south limb to the center

of the disk. Although it is 56 miles in diameter, it might be overlooked easily when viewed near the terminator because of the abundance of craters in the region, some of which are considerably larger. Within a few days of full moon, however, no one could miss Tycho, which then lives up to Thomas Webb's appellation "the metropolitan crater of the moon" (*Celestial Objects for Common Telescopes*). Its broad, brilliant rays extend outward in all directions for many hundreds of miles, one apparently crossing the faraway Mare Serenitatis and running a total distance of 1500 miles. In fact, Tycho and its amazing rays give the full moon the general appearance of a peeled orange, the crater marking the point where the sections meet. The rays are important clues to the process of crater formation.

In addition to the craters, sometimes called "ring mountains," the moon has a considerable number of the more conventional mountains which resemble those we know here on earth. Most of them are found in ranges named after those of Europe and Asia Minor which, evidently, were the only ones known to the early lunar feature namers. The outstanding range is the APENNINES (Fig. 7), which begins at a point midway between the center of the disk and the northern limb and stretches some 450 miles in a shallow arc southwestward toward Copernicus. These mountains, some of which have been measured to heights of nearly 20,000 feet, present a magnificent sight when near the terminator (moon's age about 7 or 21 days). Stripped of its contrasting shadows when the moon is near full, the range still is seen easily as the peaks of its crest shine out over the bright background like a string of sparkling jewels. The Apennines rise gradually through a series of foothills from the dark plain of Mare Vaporum on the southeast. The ascent continues northwesterly for 100 to 150 miles until the crest is reached near the boundary of Mare Imbrium. There the peaks appear to rise almost abruptly 2½ to 3½ miles above the plain, the straight edge of their base broken only near the center of the range by what looks like an enormous rectangular rubble pile 100 miles long, 30 miles wide, and about one mile high.

Two other fine ranges are associated with the Apennines in delineating Mare Imbrium. If the arc of the crest is extended northeastward across the 25-mile "inlet" that joins Mare Imbrium and Mare Serenitatis, the Caucasus Mountains are reached. They run northward 200 miles toward Aristotle (Fig. 8). Their foothills mingle with those of the

Alps, which continue the elliptical curve 200 miles northwestward toward Plato (Fig. 9). West of Plato the rough boundary is characterized by two 150-mile scallops, the second and more prominent of which is formed by the Jura Mountains marking the periphery of Sinus Iridum. Most of the other ranges are less conspicuous, some consisting of only a few hills. Individual mountains that do not appear to be part of a range are rare. The best situated examples are found near the northeast edge of Mare Imbrium, south of Plato and south of the Alps. They are Pico and Piton, respectively, both of which are bright objects easily seen through binoculars (Fig. 9).

Where mountains are found valleys naturally are expected, and they occur in abundance. However, only those that are prominent because of unusual extent attract our attention as we view our satellite from a distance of nearly ¼ of a million miles. The most striking is the one which cuts centrally through the Alps perpendicular to the line of the crest. The ALPINE VALLEY (Fig. 9) is 1 to 13 miles wide and 110 miles long. It runs the entire width of the range, and it is perfectly straight with a smooth floor except where it cuts through the highest mountains and where landslides may have provided the rocky debris seen there. It may be observed through binoculars under good conditions when the moon is about seven to nine days old. The observer is strongly impressed by the artificial appearance of this feature. It is inconceivable that the Alps could have been formed in a way that would leave such a broad, straight highway through their center. One is tempted to speculate that the valley was carved out in a later epoch by an enormous projectile that plowed through the range, its impact energy grinding to dust and heating to vapor the mountain masses before it and scattering the pulverized remains as an undetected blanket over a vast area. Several eminent authorities tell us that the Alpine Valley probably was so formed. Others maintain that such an explanation is improbable and that this and other valleys are but cracks in the lunar surface.

The RHEITA VALLEY, near the southeast limb and about 150 miles southwest of the bright ray crater Stevinus, is a more readily noted example. A conspicuous feature of the three- to four-day moon, it appears as a black streak running south from the terminator almost to the limb, an apparent distance of 180 miles which stretches into an actual 300 miles when the effect of foreshortening is taken into account. The valley proper is straight, 230 miles long, about 15 miles wide, and some 2000 feet deep. From its south end another rectilinear groove 100 miles long and 9 miles wide continues in a more southerly direction. Before first quarter it is almost completely washed out because of lack of shadow on its shallow floor. Two weeks later, when the sunset terminator falls across it, the valley again is seen easily through binoculars. It is usually black, but when the sun's rays strike it at the proper angle it appears definitely brighter than its surroundings. While the Alpine Valley is evidently a relatively young formation, the Rheita Valley must be very old since several craters cover it and the walls of several more encroach upon it.

Somewhat akin to valleys but generally much less conspicuous are the gaping cracks, known as clefts or *rills,* that split the lunar surface in many places. Near the center of the disk between Mare Tranquillitatis and Mare Vaporum run the ARIADAEUS RILL (Fig. 10) and the HYGINUS RILL resembling parts of a great canal dug to connect the two "seas." The former stretches 150 miles in a straight line composed of short linear segments having only slight differences of heading and an occasional offset. It is uniformly three miles wide but of unknown depth throughout most of its course, which takes it through several hills. The latter, somewhat shorter and narrower, undergoes a 30-degree change in heading as it passes through the small crater Hyginus near the midpoint. It thus consists of two approximately linear segments. Rills show best when the terminator is near one end.

What might be termed the complement of a rill is a *ridge* or wrinkle. They occur in abundance around the edges of or across the maria. Although they attain widths of several miles, they have very little central height and usually go unnoticed on the plains. However, when near the terminator and approximately parallel to it they reveal themselves as one slope appears light and the other black. On such occasions they may be seen through binoculars meandering along the "twilight" zone of poor earthward reflectivity. Most of the ridges are unnamed, a prominent exception being the Serpentine Ridge which crosses eastern Mare Serenitatis (Fig. 11).

Promontories and capes constitute another category of named lunar formations. As might be expected, they are the tips of the bright *continental* areas that jut out onto the dark maria. Easy examples are Laplace Promontory and Heraclides Promontory that separate Sinus Iridum from Mare Imbrium on the east and southwest, respectively.

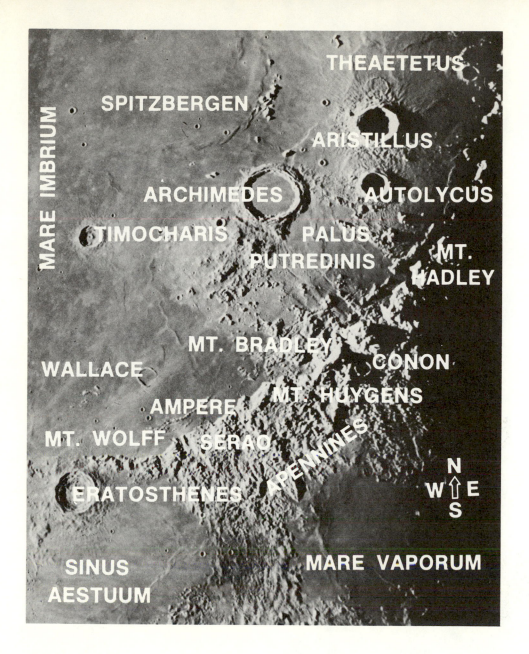

FIGURE 7.   The Apennines, the moon's finest mountain range, in early evening. The Apollo 15 astronauts landed on Palus Putredinis in the valley to the left of the H in "Hadley." (36-inch Refractor.)

PROTAGORAS

MARE FRIGORIS

ALPINE VALLEY

GALLE

TROUVELOT

EGEDE

ARISTOTLE

MITCHELL

ALPS

EUDOXUS

LAMECH

CAUCASUS

ALEXANDER

CALIPPUS

N
W   E
S

MARE SERENITATIS

FIGURE 8. Between mountains and plains—the craters Aristotle and Eudoxus with evening illumination. (120-inch Reflector.)

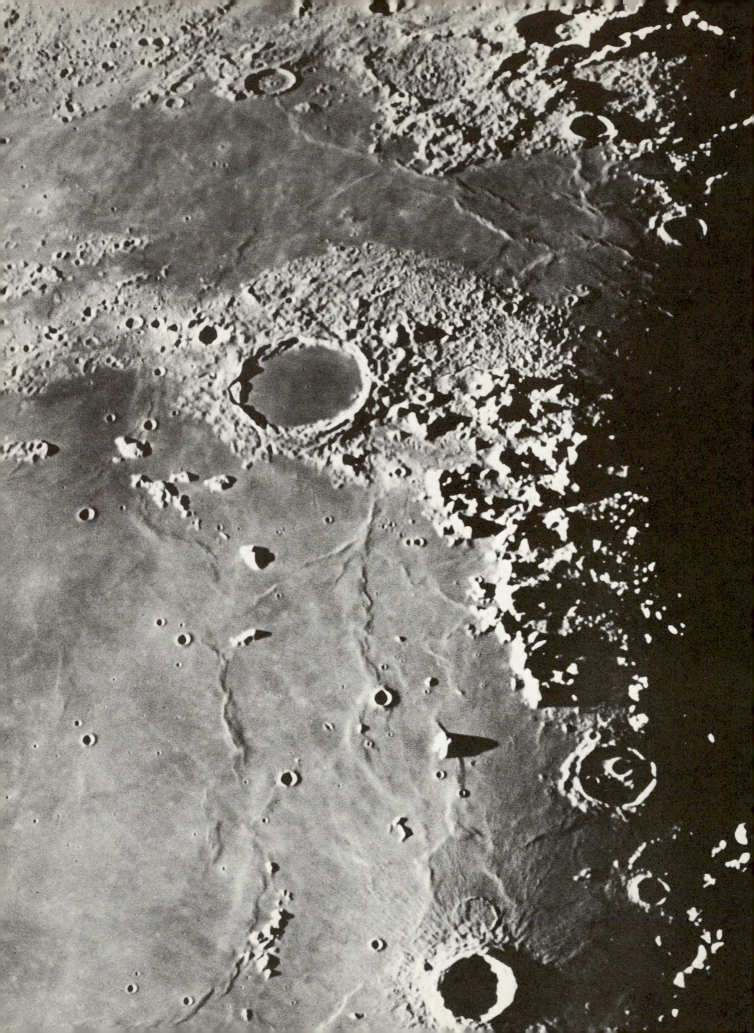

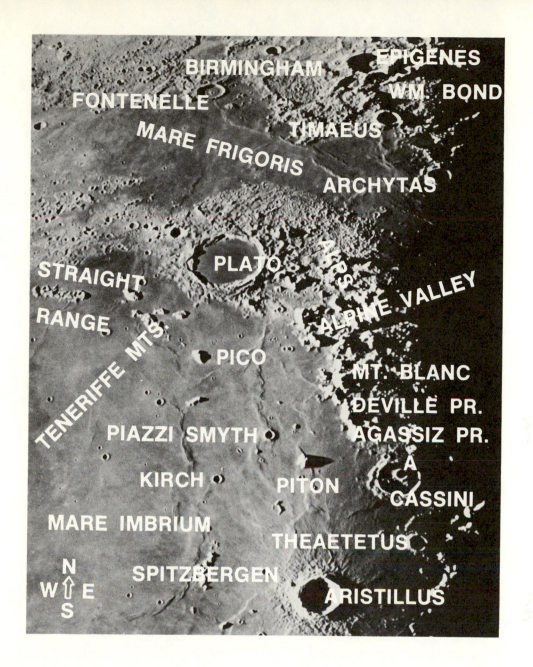

FIGURE 9.   The black crater Plato and the Alps at
the sunset terminator. (36-inch Refractor.)

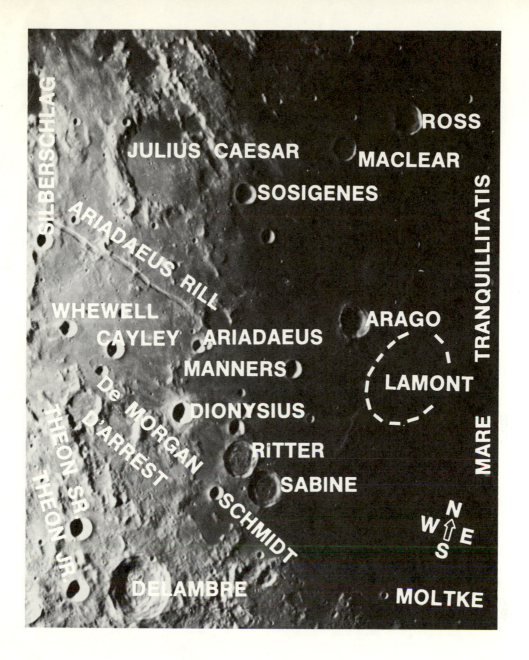

FIGURE 10. The dilapidated crater Julius Caesar on the edge of Ranger territory. The moon probe Ranger 6 landed east of the crater, under the last S in the label "Sosigenes," in February 1964. Ranger 8 landed under the M in the label "Mare" a year later. The first manned lunar landing by the Apollo 11 astronauts in July 1969 occurred 3½ crater diameters east of Sabine. (120-inch Reflector.)

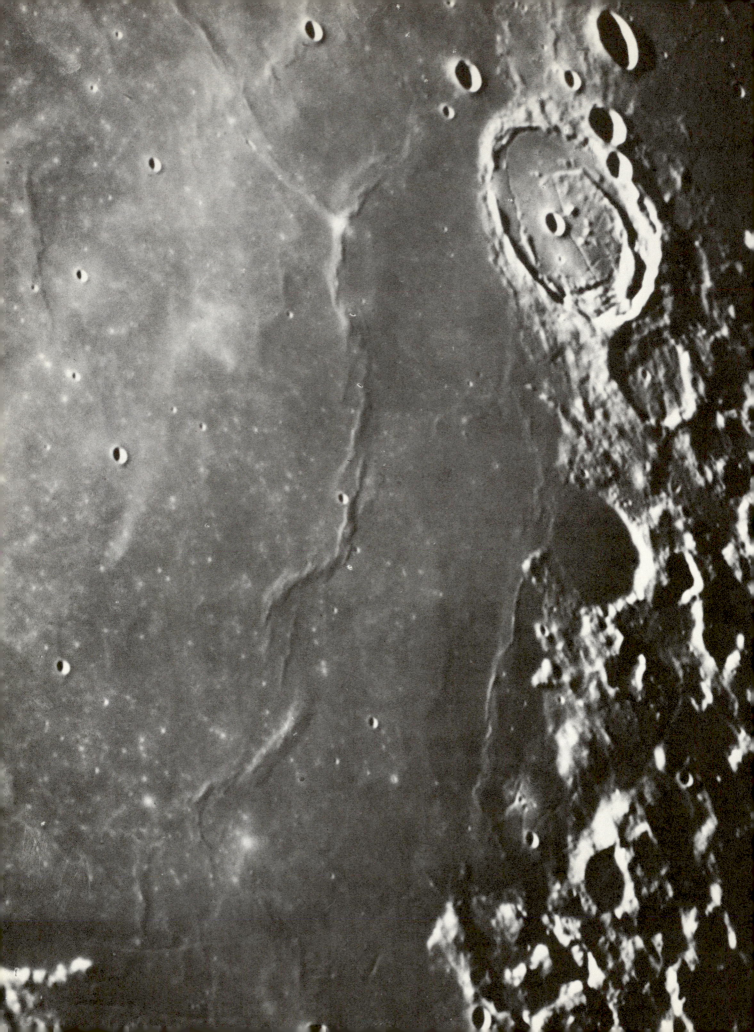

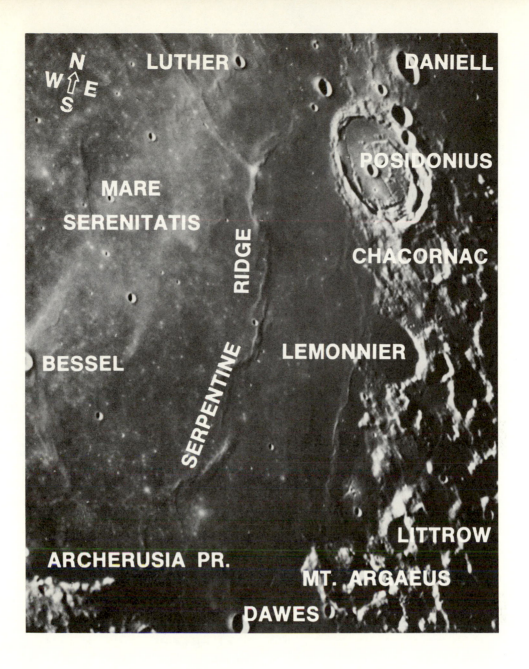

FIGURE 11. The Serpentine Ridge and Posidonius. Smooth Mare Serenitatis contrasts with the cratered upland on the right. On December 11, 1972, the Apollo 17 lunar module carrying Eugene Cernan and Harrison Schmitt landed near the crater Littrow on the edge of Mare Serenitatis. (120-inch Reflector.)

They may be spotted through binoculars whenever they are in sunlight, but under certain angles of illumination they show to best advantage.

One additional type of formation should be mentioned although I shall be hard put to it to find more than one example visible through binoculars. These seldom observed features are known as *domes*. A dome might be described from an observational point of view as an extroverted crater. Such "bumps" are found abundantly in some areas. They are rarely circular but globular in outline, and known examples range from 2 to 40 miles across. Many have one or more *summit craters* near their highest points. While the larger ones are big enough to be resolved easily through binoculars, they are difficult to see. Their shape and the fact that they seldom are more than a few hundred feet high prevent them from casting appreciable shadows except when the sun is very near the horizon. Their resemblance to terrestrial shield volcanoes, pointed out some 80 years ago by astronomer William Pickering and confirmed 60 years later by geologist Eugene Shoemaker, indicates that they have something important to say to us about the history of the moon. One of the larger domes lies 140 miles northeast of Kepler, which it resembles in size only, and a number of smaller examples dot the Procellarum plain about midway between Kepler and Copernicus. Rumker, the most prominent one, is a lumpy group of numerous individual domes located near the northwest limb. Usually lost on the shadowless plain, it is seen easily through binoculars when the terminator is nearby.

HOW THE MOON BEHAVES

As you doubtless realize, the moon keeps the same side always toward the earth. As it swings through its monthly orbital cycle, the portion of the sunlit half turned toward us changes continually to produce the phases, but it is always the same physical hemisphere of the moon that faces us. Regardless of whether it is illuminated or dark, Mare Crisium is always near the eastern limb, Grimaldi is near the western limb, Ptolemy is just south of the center of the disk, etc. Had this not been so, the ancient concept of the Man in the Moon probably would not have developed, and the modern Soviet scientific feat of sending Lunik 3 to photograph the far side in October, 1959, would not have aroused so much interest.

Does the moon rotate on an axis as does the earth? Rotation is standard behavior for the bodies of the solar system and those of the outer universe as well. It is an effective way of storing energy, it provides stability, and we might even say it is an expected property of matter that accumulates in large quantities. All planets rotate, and their periods are precisely known, with the possible exception of our nearest planetary neighbor Venus. Galileo discovered that the sun rotates, and 300 years later Frank Schlesinger found that stars do likewise. So the moon should conform to the general pattern and rotate on an axis. "But it can't," is a frequent objection, "or we would see the other side." This seems to be a valid argument. When, as a bystander, we watch a merry-go-round carry its load of beaming children in musical circles, we certainly do see every part of its circumference. When it comes to observing the motions of the moon, however, we are not in the position of a bystander. We are involved in those motions as part of the earth-moon system, and our observation station is never more than a few thousand miles from the active focus of the moon's orbit. In effect, we are riding on the merry-go-round ourselves, and our point of view regarding what is taking place is quite biased. In order to view the moon's motions as a bystander, we would have to board a space ship and blast off into the far yonder. After traveling several million miles we might settle down to observations of the earth and the moon to keep ourselves occupied during the many months a journey to another planet would take. Then, indeed, we would see all sides of the moon during a month.

The moon does rotate on an axis, an axis that is inclined 83½ degrees to the plane in which its orbit lies. Its direction of rotation is the same as its direction of revolution around the earth (eastward), and the periods of rotation and revolution are exactly the same. That is why we never see the far side. For those who still may be unconvinced, the following lunar laboratory exercise is recommended: Get a can of peas or artichokes or anything else out of the kitchen. That will be the moon, and you will be the earth. Grasp the can by its base and hold it upright at arm's length so that the brand label is toward you. Now cause it to revolve in an orbit around you by turning yourself around in place as you stand. The can will make a complete revolution, but you will see only one side of it. Enlist the assistance of an unbiased observer who will serve as the bystander across the room. After you have revolved the can in the manner described through 360 degrees, ask the bystander if he saw all sides of it. If he is really unbiased, I believe he will have to answer in the affirmative. As in the case of the moon, the can of peas will have revolved in an orbit and rotated on an axis, both in the same direction and both in the same period.

Is it a coincidence that the moon's periods of rotation and revolution are identical? In the vast outer universe, which the scientist must explain in terms of natural law, is there room for what appear to be fortunate accidents? The answer to the second question is yes, and the moon provides an example. We have seen how enormous the sun is compared with the moon (56-foot sphere and a golf ball). In fact, the diameter of the sun is 400 times that of the moon. Yet it happens that the average distance of the sun from the earth is also about 400 times that of the moon. Consequently, both sun and moon appear the same size in the sky (diameter ½ degree), and we are permitted to enjoy perfect eclipses of the sun on those occasions when the new moon actually passes in front of the sun. If the moon's disk appeared considerably smaller, total solar eclipses would be impossible, and we would know much less than we do about the atmosphere of the sun. If the moon's disk appeared considerably larger, it would cut off not only the light of the sun's disk during eclipse but also much of the solar atmospheric light as well. The fact that it just fits is a coincidence unpredictable through natural law. The other planets of the solar system have a total of 42 satellites known to us, but none of them has such a relationship to the sun.

It is not a coincidence, however, that the moon rotates and revolves in identical periods. Long ago the moon undoubtedly turned much faster on its axis and showed all of its surface to the earth. At that time there were no people here to observe it and probably no living creatures of any kind. We know that the moon raises tides in the solid surface of the earth as well as on the oceans, and, of course, the earth does the same thing to the moon. It takes a staggering amount of energy to do all that work every day, and that energy has to come from some source. One source available was the rotational energy of the moon which was thus continually dissipated through tidal friction. Consequently, the rotational speed slowed down, and

the lunar day grew longer. After many millions of years the rotational period lengthened until it was exactly equal to the period of revolution. Thereafter, no further change was required. This problem was investigated mathematically in the latter part of the nineteenth century by George Darwin, who predicted the same fate for the earth. The length of our day is increasing slowly because of loss of rotational energy through tidal friction, and the time will come when the earth will keep the same face toward the moon. Then some of the people will have the moon in their sky all the time, and other people will never see it unless they travel to the side of the earth that is toward it. But be not dismayed. You have plenty of time left to get acquainted with the moon. It will take a great many millions of years to bring about such a radical change.

Having gone to some length to convince you that the moon keeps the same side always toward the earth, I now am going to exercise again the professor's prerogative and point out that this is not the whole truth. If it were, we would have no possibility of seeing more than 50 per cent of the moon's surface. Actually we can do better, and through a series of observations we can see 59 per cent of the surface. This comes about through certain earth-moon relationships known by the ancient term *librations*. The most obvious is the *libration in latitude*. We have seen that the moon's axis of rotation is not exactly perpendicular to the plane of its orbit around the earth since it is set at an angle of 83½ degrees rather than 90 degrees to that plane. Even though the moon rotates rather slowly about that axis it still enjoys great rotational stability and keeps its axis at all times pointed in the same direction with respect to the stars. Consequently, on a certain date during each *lunation* (the moon's monthly cycle of phases), the lunar north pole is tipped about 6½ degrees toward us, and we can observe the surface 6½ degrees beyond the north pole. Two weeks later, when the moon has traveled in its orbit around to the other side of the earth, the lunar south pole will be tipped toward us, and we will observe the surface 6½ degrees beyond the south pole (Fig. 12).

The libration in latitude is a phenomenon analogous to the seasons on the earth. Since the earth's axis is inclined 66½ degrees to the plane of its orbit around the sun, the sun "sees" 23½ degrees beyond the north pole on June 22 each year and 23½ degrees beyond the south pole on December 21 each year. To illustrate this libration, stick a

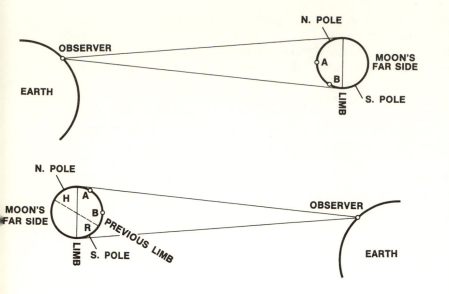

FIGURE 12. The libration in latitude. The upper diagram shows the moon at time of greatest positive libration in latitude— North Pole tipped toward observer (angle exaggerated to bring out the effect). The limb line separates the visible hemisphere (left) from the far side. Two craters, A and B, are equally spaced north and south of the lunar equator. The lower diagram shows the moon about two weeks later (greatest negative libration) after it has revolved 180 degrees in its orbit and rotated 180 degrees on its axis (North Pole—South Pole line). The limb line again separates the visible hemisphere (right) from the far side, and the dashed line shows where the limb line was located two weeks earlier. Section H around the North Pole, visible two weeks earlier, now is hidden while section R around the South Pole, not seen two weeks earlier, now is revealed.

knitting needle through the center of an orange, or a tennis ball if you don't want a juicy demonstration. Mount the satellite by sticking the needle through a candy box or squeeze bottle so it will stand at an angle of about 30 degrees to the vertical. It is often instructive to exaggerate our demonstrations or diagrams to make sure that the effect to be illustrated is large enough to be seen easily, and six degrees is a rather small angle. Set the mounted ball up at about eye level near the center of the room so you can look at it from various directions. The needle, of course, represents the axis of rotation that always points in the same direction in space. Look at the ball from the side toward which the upper (north) end of the needle points, and you will see some of the hemisphere beyond the north pole. With crayon or lipstick outline the half you see. Next, without changing the direction of the needle, rotate the ball around it 180 degrees to simulate ½ day on the moon (two of our weeks). Now go around on the opposite side of the room (moon situation two weeks later). Note that a portion of what you saw in the first view is now hidden beyond the north limb while an equal area of new territory is now visible at the opposite (south) limb.

The *diurnal* (daily) *libration* follows from the rising and setting of the moon and the fact that the earth's radius is a significant fraction ($\frac{1}{60}$) of the moon's distance from us. If an observer views the moon in the zenith, his line of sight coincides with the fundamental line joining the center of the earth with the center of the moon. He thus sees what we might term the "official" half of the moon turned toward us. On the other hand, if he observes the rising moon on the eastern horizon, he looks at it from a point 4000 miles above the center-to-center line, and he thus is able to see a little farther around the eastern limb compared with what he will see when the moon is in the zenith. Similarly, when the moon is seen on the western horizon, the observer is looking at it from a point 4000 miles below the center-to-center line, and he then sees a little farther around at the western limb (Fig. 13). The effect amounts to only about one degree either way, and the advantage is more theoretical than practical. There are very few places in the world where the seeing and atmospheric transparency near the horizon are good enough to make observations there of any value.

Just as the libration in latitude enables us to see between six and seven degrees beyond the moon's north and south poles at intervals of two weeks, the *libration in longitude* makes it possible to see up to 8 degrees beyond the standard east and west limbs at intervals of about two weeks. For an explanation of this effect we must return to the great difference in fundamental concept of planetary motion between Copernicus on the one hand and Kepler on the other.

We have seen that the ancient philosophers believed the orbit of the moon to be a circle. They believed that all bodies in the universe traveled in circular orbits. They regarded the circle as the perfect figure—perfect because all points on it are equidistant from a given point called the center.

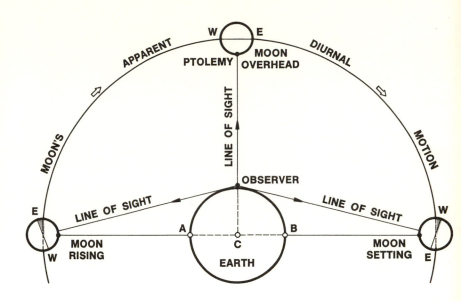

FIGURE 13. The diurnal libration. When the moon is overhead, the observer's line of sight coincides with the line joining the centers of the earth and the moon. If other librations have zero value at that time, the large crater Ptolemy is seen near the center of the lunar disk. When the moon rises, the observer views it from a point 4000 miles above the center-to-center line, and Ptolemy appears displaced toward the west limb. The observer sees the shaded section between his E——W limb line and the dashed line which would be the limb line as viewed by another person at A. Similarly, when the moon sets, the observer sees the shaded section at the west limb which another person at B would not see at that instant.

Naturally the heavenly bodies would choose the perfect figure for their orbits. No arguments could be presented against such logic, and so the idea of circular orbits was accepted without question for more than 2000 years. The ancients also believed that orbital motion must be accomplished at uniform speed. Their reason was based on similar logic. Mortal man, as he pursues his destiny, walks forward, falters, stops, retraces his steps, runs ahead, and throughout his life proceeds at variable speed. Such erratic behavior would be unthinkable for a heavenly body. Only uniform speed would befit the celestial dignity of the moon and the planets which were regarded either as gods or their close relatives. Even such a radical thinker as Copernicus, who tore the solar system apart, taking the earth out of the central position and putting the sun there—even he did not question the firmly-established principles of uniform, circular motion. Nearly a century later Kepler questioned those principles for the simple reason that the original Copernican theory failed to predict accurately the positions of the planets in the sky. As we have noted, Kepler in his first law replaced circular orbits with elliptical ones and placed the controlling body at one focus instead of at the center of the orbit. He then went on to examine the question of orbital speed, and he rejected uniform speed in his second law.

Uniform speed is a satisfactory idea as long as the motion takes place in a circular orbit. Indeed, the circle is a special case of the more general ellipse—an ellipse in which the two foci have moved together at the center (Fig. 2). Any body in the universe that does move in a circular orbit travels at uniform speed, and some examples are known. However, for the general ellipse a little reflection shows that things just aren't going to come out right on the basis of uniform speed.

Let's drop out of the heavens for a moment and sit at the dinner table for a practical illustration of Kepler's problem. We have eight people at dinner, and we wish to climax the occasion with a blueberry pie. Each guest is a very important and sensitive person, and so it is essential that all pieces of pie be identical in size. If we have provided a conventional circular (Copernican) pie, the problem is one of simple measurement. We can determine the circumference with a tape measure and mark off ⅛ths of it around the edge for cuts from the center. That illustrates uniform circular motion. The body traverses exactly ⅛ of its orbit each year, completing it in a period of eight years.

But now suppose we have grown more advanced in our thinking and have ordered an elliptical (Keplerian) pie for dinner. This we must cut not from the center but from one of the foci, and we must cut eight equal pieces. If now we measure the periphery with the tape and lay off ⅛ths of it for cuts from the focus, the pieces will be equal in length along the curved edge, but the guest who receives the piece cut along the longest distance is going to get a lot more pie than he who receives the piece cut along the shortest distance from the focus. If the pieces are to be equal, we must measure much farther along the periphery for the short

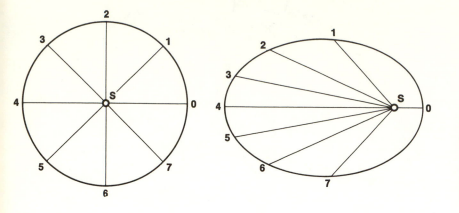

FIGURE 14. Copernican and Keplerian pies (and orbits). Gastronomical observation: Copernicus' pie is the left one and Kepler's pie is the right one. Any slice is equal to any other slice, and the slices of one pie are equal to those of the other pie. Astronomical observation: A planet traveling in either the circular orbit or the elliptical orbit about the sun (S), with a period of eight years, would traverse each section from one numbered point to the next in exactly one year.

cuts than we do for the long cuts. In other words, the body moving in an elliptical orbit, under a central force directed toward the focus, must travel farther per day (faster) when passing close to the focus and slower when passing far from the focus. This was Kepler's problem, and although he worked it out ¾ of a century before Isaac Newton announced the underlying law of gravitation, Kepler saw that there must be some form of equipartition here. He therefore stated in his second law of planetary motion that each planet revolves in its orbit so that the line joining the planet to the sun sweeps over equal areas in equal intervals of time. Returning to the elliptical pie to be cut from one focus, we see that while Kepler's second law requires different angles of cut and different lengths of crust edge for the several slices, it guarantees that all slices will come out the same size (Fig. 14).

Now back to the moon, which, as we have seen, moves about the center of the earth-moon system in obedience to Kepler's laws of planetary motion. When the moon is closest to the earth (at *perigee*), it revolves fastest in its orbit; when it is farthest from the earth (at *apogee*), it revolves slowest; and at intermediate distances it revolves at intermediate speeds determined according to Kepler's second law. Yet the moon must rotate on its axis at perfectly uniform speed unless it is in a state of tremendous upheaval, which does not appear to be the case. Rotation at uniform speed combined with revolution at variable speed is the cause of the libration in longitude which makes the moon appear slowly to rock back and forth during a lunation. When at perigee the moon re-

volves faster than it rotates (both in the same eastward direction), and we begin to see farther around at the east limb. When it is at apogee, it revolves slower than it rotates, and we begin to see farther around at the west limb. Since the effect is cumulative from day to day during each period of gain or loss, the greatest opposite librations in longitude occur, on the average, about one week after perigee and one week after apogee (Fig. 15).

The effect of the libration in longitude may be observed by simply noting the position of Mare Crisium with respect to the eastern limb from night to night between new and full moon during one or more lunations. That oval dark spot, easily seen with the naked eye, varies from the position of nearly touching the limb (large negative value) to approximately its width from the limb (large positive value). The libration in latitude may be checked by observing the changing position of the long, narrow Mare Frigoris with respect to the northern limb. Be sure to make the observations yourself, but you can get an idea of what to look for in the way of librational tilting from the photographs in the chart section. For movement of Mare Crisium with respect to the east limb compare Charts I and III. For libration in latitude compare Charts III and VI. In the second comparison note particularly the relative positions of the crater Philolaus near the north limb and the crater Moretus near the south limb.

Another effect of the moon's elliptical orbit is a continuous periodic change in its apparent angular diameter. While the average distance between the centers of the earth and moon is 239,000 miles, the

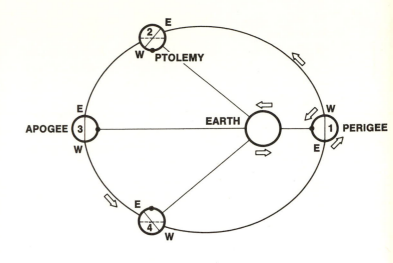

FIGURE 15. The libration in longitude. Positions of the moon at intervals of one-quarter revolution (about one week) are shown beginning at 1 (perigee). The E——W line is the average limb position separating the visible hemisphere (left) from the far side, and the protrusion on the moon designates the large central crater Ptolemy. One week later the moon has advanced to 2, but it has rotated on its axis only 90 degrees. The new E——W line thus differs from the average limb position (dashed line) and permits observation beyond the average east limb (positive libration). At 3 (apogee) the moon has revolved 180 degrees and rotated 180 degrees, and the average limb position is observed again. One week later the moon has rotated another 90 degrees but has advanced only to position 4. Again the E——W limb line differs from the average limb position (dashed line), and this time it permits observation beyond the average west limb (negative libration). The ellipticity of the moon's orbit is exaggerated greatly here to bring out the effect. The above ellipse has an eccentricity of 0.50 whereas the moon's true orbit has an eccentricity of only 0.05, making it a rather close approximation to a circle.

actual distance may vary all the way from 221,000 to 253,000 miles. The moon, of course, exhibits the largest angular diameter when at perigee, and it may appear as much as 14 per cent or 1/7 smaller in angular diameter when at apogee. Such changes are not apparent to the eye from night to night, but any optical device for the measurement of small angles will show what is taking place, and photographs of the moon taken at extreme orbital positions with the same telescope form a striking exhibit when fitted together along a diameter.

The brilliance of the full or gibbous moon is deceiving. One receives the impression that it must be highly efficient as a reflector of sunlight. How else could it send us so much light? Actually, if one set out to design the poorest possible mirror that could be constructed out of the materials available in the universe, he would be hard put to it to come up with anything that would beat the moon as an underachiever. Its average *albedo* (fraction of incident light that is reflected) is 0.072. This means that only 7 per cent of the light that hits the moon is reflected away, and that is low performance indeed. In contrast, the cloud-covered planet Venus has an albedo of 0.76 which makes it more than 10 times as efficient as our satellite. Among the darkest features are Sinus Medii and Mare Nubium with 5 and 6 per cent reflectivity, respectively, and the brightest spot of notable size is the crater Aristarchus with 18 per cent reflectivity.

Not only does the albedo vary from spot to spot on the moon, but the reflectivity of certain spots changes greatly during the lunation. So striking are some of the fluctuations that during the nineteenth century many announcements were made of what the observer considered to be newly-formed features. A bright marking always looks larger than a dim one, and so the size of certain craters, mountains, etc., often varied from one hand-made chart to another. Endless arguments ensued, but until good photographs of the moon under many different angles of illumination became available each observer was his own authority. It is the general opinion of leading selenographers today that all reported changes in size or shape of surface features that have come to their attention are probably results of the changing appearance of an unchanging object under different angles of solar illumination. This does not mean that the moon is absolutely dead. Certain visible changes definitely are going on there, as we shall see, but the building of craters and the raising of mountains big enough to be discovered with telescopes of moderate aperture probably ceased long before the moon was seen by man for the first time.

I have stated that craters generally are seen best when near the terminator. The statement is correct, but it does not apply to every crater. MESSIER and WILLIAM PICKERING are an exceptional pair. Those two small craters, about six and seven miles across, respectively, are found close together in an east-west line in the western part of Mare Fecunditatis. They usually are accompanied by a double ray resembling a comet's tail that can be traced about

75 miles west to the border of the mare. I have looked for those pits many times with either the sunrise or sunset terminator nearby, but always in vain as far as revelation through binoculars is concerned. However, when the moon is near full, the two little craters shine so brilliantly that binoculars disclose them as a close pair of tiny bright dots. Incidentally, the west crater of the pair, William Pickering, is named for the astronomer who studied the two extensively around 1900 and reported striking changes during the lunation which he attributed to the formation and melting of ice. Such changes of appearance certainly occur, but likely they result from variations in reflectivity only.

An even more striking example of this effect is the tiny crater LINNE, found near the western edge of Mare Serenitatis some 40 miles from the "strait" that runs between the Apennine and Caucasus Mountains into Mare Imbrium. Even the best photographs usually show Linne as a small hazy spot, like a star out of focus. The crater itself cannot be much more than one mile in diameter although the light spot on which it appears is five times that size. I have looked at it with a telescope of 3½ inches aperture under favorable conditions with the terminator nearby and have seen nothing more than a small, faint, amorphous light spot. Yet when the moon is near full, it is seen easily through binoculars as a bright dot. Much has been written about Linne. It is the most famous of the "changed" areas. Selenographers in the first half of the nineteenth century reported it as a well-formed, deep crater five miles in diameter. About 100 years ago Schmidt could find nothing there but a "white cloud" or bright patch. Later observers found a small crater such as we sometimes see there today. The mystery of those conflicting observations persists.

If you have difficulty locating the tiny objects Messier, William Pickering, and Linne, you can observe very easily a striking example of the opposite effect—a large, conspicuous crater which disappears completely when the moon is near full. A truly conspicuous feature of the moon when it is between seven and eight days old is the enormous crater MAGINUS (Fig. 6) at the terminator far down in the southern hemisphere. It is 100 by 115 miles across and 16,400 feet deep. It is an old crater, and it has many smaller craters in its floor and walls, some of which can be seen through binoculars. With its bright inner west wall extending out over the dark side, and its black inner east wall casting a narrow black shadow on the edge of its gray floor, Maginus not only is striking as viewed

through binoculars, but it can be seen with the naked eye. Yet, a few days before full moon, the vast crater fades away completely. For about a week no trace of it can be found no matter how large a telescope one may use to search the area. As the sunset terminator approaches, Maginus reappears and again becomes prominent. Many of the major craters lose contrast and seem to fade into the bright background at full moon, but no others quite can match the disappearing act of Maginus.

SELENOGRAPHY

Selenography, the study of lunar features, properly begins with Galileo and his observations of 1610 described and illustrated in his *Sidereus Nuncius* (*The Starry Messenger*), which he wrote and published in that same year. However, it long has been the fashion to belittle his contributions in the field and to refer with amusement to his drawings, none of the features of which, it is claimed, can be identified with anything on the moon. It is my belief that Galileo's drawings are reasonably accurate and that the confusion has arisen out of some unfortunate mistakes made by the publisher of his book. Also I take exception to the dubious honor generally accorded Galileo of having named the large dark areas maria because he thought them seas filled with water. If this legend be true, it is curious that he does not use the term in his book, particularly when he goes to some length to distinguish clearly between the well-known large dark areas (maria) and the smaller ones (craters) which he had just discovered. In translation he writes: "Now these spots, as they are rather dark and of considerable extent, are plain to everyone, and every age has observed them, wherefore I shall call them *great* or *ancient* spots to differentiate them from other spots which are smaller but so abundant that they are scattered over the entire surface of the moon." Certainly the term maria would have simplified his definition had he thought of it or heard of it. It is more likely that Kepler originated the watery sea concept. In the preface to his book *Dioptrice* (1611) he writes of the northern lowlands covered with water drained from the mountainous southern hemisphere, introducing his statement with the sweeping claim:

"Nothing is more certain than that. . . ." While much of the preface is devoted to extensive quotations from Galileo and to discussion of his observations, Kepler is speaking for himself in this passage.

Galileo's book contains five sketches of the moon, three of which are reproduced in Figure 16. This is a photograph not of the original *Sidereus Nuncius* but of a reprint included in Pierre Gassendi's *Institutio Astronomica* published in 1653, eleven years after Galileo's death. The drawings in the reprint are reproduced faithfully from the original and in the same size, being slightly more than three inches in diameter. They are circular in outline, and the distortion shown here is my fault for recording two pages on a single photograph. When I took the picture some years ago, the book, which is in the collection of the Lick Observatory, was three centuries old, its leather binding was very stiff, and the opened pages showed considerable curvature toward the inner margin. The fourth sketch resembles closely the lower one on the left-hand page, and the fifth sketch is of the crescent moon, probably about 26 days old. It is unfortunate that Galileo neglected to record the dates of his drawings. He did so on the drawings he made during the same period of his newly discovered four satellites of Jupiter. Perhaps the well-known repetition of the cycle of lunar phases made it seem unimportant to him to identify the particular lunation in which he observed. His sketches are sufficiently detailed to illustrate the discoveries and other observations discussed in the text, and, no doubt, it never occurred to him that future generations might question his drawings or refuse to take them seriously. Actually, the apparent motions of the moon are so numerous and so complicated that only for a matter of seconds at intervals which may be large will the circumstances of a given observation be repeated exactly. There are perhaps hundreds of thousands of photographs of the moon preserved around the world today, but I doubt that any two could be found which are identical unless they were taken simultaneously from stations close together.

The left upper drawing is a good general representation of the moon as it appears in an inverting telescope when about 23 days old. Look at it upside down, and it will compare well with what you will see through binoculars if you look at the right moment, except that a great amount of detail is missing. Instead of working for hours slavishly recording every visible mark, Galileo evidently was content to sketch quickly those features which seemed to him important at the moment and to let the rest go or indicate them roughly. But Galileo did not have an inverting telescope! The one which

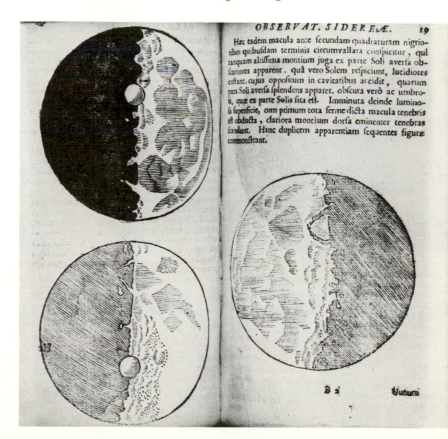

FIGURE 16. Galileo's drawings of the moon made from observations with his first telescope in 1610 and described in his book *Sidereus Nuncius* published that year.

he invented, used, and described in his book differed radically from the few others known to have been in use at the time. It produced an erect image, and it showed the moon in the same orientation in which you see it through your binoculars. This obvious discrepancy between the reproduction of Galileo's drawing and what he must have seen through his telescope provides a clue, I believe, to the reason why identification of the represented features has been difficult. I suggest that this illustration was engraved correctly from the original drawing and that the Venetian printer of 1610 accidentally inverted it as he made up the page. Thus the modern directions for the left upper drawing are north at the bottom and east to the left.

The left lower drawing also represents the 23-day moon, but it has been reflected rather than inverted. The right-hand drawing represents the seven- to eight-day moon, and it appears in the same orientation as the left lower illustration. The last two show north at the top and south at the bottom, as Galileo saw them, but left and right have been interchanged. In other words, these two drawings are mirror images of the moon as Galileo must have seen it. Since they differ in technique from that of the left upper illustration, I suggest that they were made by a different engraver—an apprentice who probably copied the details of the drawing correctly but forgot the basic rule of the engraver to reverse the picture as he cuts the plate. For them the modern directions are north at the top and east to the left. They represent the moon at the phases noted as seen through an inverting telescope with a prism diagonal ahead of the eyepiece—an optical system unknown to their creator.

Is it incredible that such a glaring error could be committed in the publication of a book? I can only suggest that it may have happened in 1610, but I can assure you that it did happen on at least two occasions in recent years. In 1961 a photograph of the well-known Arizona Meteorite Crater was printed inverted in the first run of a widely-used textbook of astronomy. In 1963 there appeared a book about the moon devoted largely to the problems, progress, and plans for lunar exploration. The preface has three full-page photographs of the moon, and all three are printed not as the moon is seen but with right and left interchanged—in mirror image!

Galileo is discussing the right-hand drawing (Fig. 16) in the first sentence of the paragraph printed above it in which he writes: "This same spot is seen before the second quarter surrounded by darker boundaries which resemble the highest mountain range on its side turned away from the sun. . . ." By "second quarter" he refers to the second quarter of the lunar cycle or the period between the events which we term the first quarter moon and the full moon. Thus he refers to the seven- to eight-day moon and not the last quarter moon, a common misinterpretation which has added to the confusion. The "spot" here referred to is not a crater but the dark-bordered, lightly shaded area along the terminator shaped like a slice of pie and located about ¼ the way from the top of the picture to the bottom. The spot is Mare Imbrium, and the dark mountainous boundaries are the shadows of the Apennine, Caucasus, and Alps Ranges which enclose Mare Imbrium on the east. Other less-conspicuous irregularities along the terminator could represent the dark eastern floors of the major craters Clavius, Maginus, and Plato since they occur at about the expected locations.

Consider now the left upper drawing referred to by Galileo later in the paragraph where he writes: "After a time, the illuminated portion of the surface having decreased until almost all of the above-mentioned spot is shadowed in darkness, the brighter ridges of the mountains transcend the darkness to a high degree." By "after a time" Galileo evidently means 15 to 16 days later when the sunset terminator has advanced over the eastern part of Mare Imbrium. We see the shaded mare about ¼ the way from the bottom to the top, the left end in black shadow, and the bright crests of its bordering Apennines and Alps extending over the dark side as described. But what about the great round crater at the terminator just above the center of the drawing?

The first erroneous assumption, made long ago about the drawings on the left, is that since Galileo showed only one crater he must have represented a generally conspicuous crater. Thus Tycho was proposed. However, Tycho does not appear at the terminator until about one day later than the phase represented. Then it is seen nearer the top of the drawing; it is much smaller than the enclosure shown; and the terminator is not straight but definitely convex to the right. Copernicus has been proposed, but Copernicus scarcely is larger than Tycho, and the moon is definitely a crescent when the terminator reaches that crater. This is the most illogical identification I have found because it implies that Galileo was such a poor draftsman that he misplaced Copernicus by nearly half the moon's diameter. Such a picture is not in harmony with the image of the man who probably did more than any other to inaugurate the modern age of experimental science, and the man whose brilliant discoveries destroyed the foundations of the officially established theory that the earth is the fixed center of the universe. Ptolemy and Albategnius have been proposed. They are closer to the right location and larger, but they are much too small also. The proposal of Clavius more nearly meets the size requirement, but its location puts it much farther from the proper position than Tycho.

Galileo's crater evidently is not one of the more prominent excavations, but it must be one of the largest. I suggest that it is a very old crater which generally is quite inconspicuous, a crater which had no official name until 1950, and one to which so little attention has been paid that the author of a 1964 book on the moon referred to it as "unnamed." It is an enclosure that stands out prominently when on the sunset terminator as Galileo observed it more than 350 years ago and as few, apparently, have observed it since. I suggest that the great crater which reminded Galileo of the mountain-ringed country of Bohemia is Deslandres. Its positions with respect to Mare Imbrium and the limb differ slightly in the two drawings on the left, but with large negative librations in latitude and longitude, its location will match either spot. Moreover, its huge diameter on the drawings is only 50 per cent greater than its actual size.

Galileo's pioneer drawings merit much more respect than has been given to them. They show very few features, but they show properly those which attracted his interest at the telescope. The early seventeenth century produced more detailed lunar maps than his, but several decades passed before more accurate ones appeared. The drawings are not perfectly scaled, but they present a far more accurate likeness of the moon than that which appeared 250 years later on the frontispiece of several editions of an Eastern college astronomy textbook used throughout the country during the second half of the nineteenth century.

But why did not Galileo correct or call attention to the "errors" which I have suggested were made in reproducing his drawings? Perhaps he considered the matter unimportant since he had described in detail the locations of his "spots" and mountain ranges. At forty-six he was busy blazing trails, opening up vast realms of observational and experimental science. Perhaps the stubborn refusal of the clerical and political leaders of the day to accept his discoveries or even to look through the telescope burdened him with far greater problems than a printer's error.

One of the better early maps was constructed in 1628 by Michel Langrenus of Belgium, and it was finally published in 1645. It evidently was planned to accompany his book *Selenographia Langreniana,* a work which never got beyond the manuscript stage. He was the first to employ the very practical scheme of naming craters after important people. However, all the names which he used have been dropped except one—his own, which identifies a major crater south of Mare Crisium (Fig. 47). In 1647 John Hevelius of Gdansk published a book of lunar studies consisting of more than 500 pages entitled *Selenographia sive Lunae Descriptio.* The work contained several maps (Fig. 17) showing a great many surface features, 250 of them named by the author according to his own scheme. It was Hevelius' idea that the formations should be named after terrestrial surface features. For example, the prominent crater Copernicus appears as Mt. Etna on his charts. His nomenclature also was short-lived, but he did fare somewhat better than Langrenus in this respect. Several of the major lunar mountain ranges still bear the names given them by Hevelius. The double-exposure effect caused by the overlapping limb rings of his map represents his effort to show the librations of the moon discovered by Gassendi and Galileo. He combined the individual librations in longitude and latitude into a single misconception which he might have omitted to advantage.

By 1651 most of the maria and major craters had received the names they bear today, and they were so designated on Francesco Grimaldi's chart published that year in Giovanni Riccioli's *Almagestum Novum.* The best map of the seventeenth century was published in 1680 by Giovanni Cassini of Paris. Beautifully engraved, it represents the surface of the moon in exaggerated relief, and it compares surprisingly well with modern photographs. Many, indeed, were the astronomers who devoted large parts of their lives to a study of the moon. Professionals and amateurs, colorful characters, extraverts and recluses—all made their particular contributions for which they are remembered in the history of science (Fig. 18). If you would like to pursue this phase of the subject further, I refer you to an excellent 34-page account, *A History of Lunar Studies* by Ernst E. Both, published and sold by the Buffalo Museum of Science, Buffalo 11, New York.

A necessity for every modern selenographer is

FIGURE 17. Lunar map by John Hevelius published in his *Selenographia* (1647). North is at the top and east is to the right. The maria are represented as light areas, and the uplands are shaded. The pentagonal light patch northeast of center labeled PONTVS is Mare Serenitatis. The larger light area west of the latter labeled MARE MEDITER is Mare Imbrium. Numerous other features may be identified although the mapping is as crude as the embellishment is elegant.

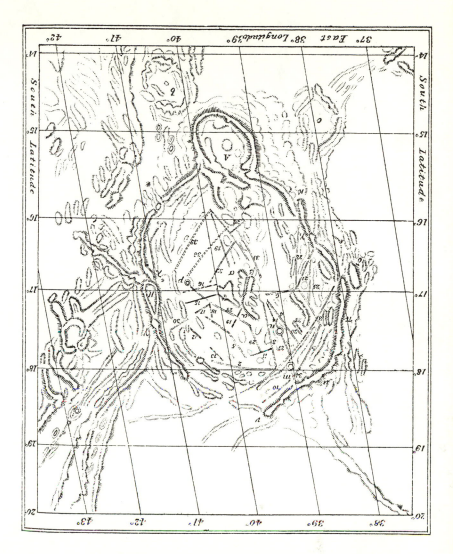

FIGURE 18. The crater Gassendi drawn by Edmund Neison and published in his book *The Moon* (1876)—an example of one of the better pre-photographic lunar atlases. The plate is inverted here to facilitate comparison with Figs. 19 and 20.

the *Photographic Lunar Atlas* previously mentioned. Edited by Gerard Kuiper, with the collaboration of D. W. G. Arthur, E. Moore, J. W. Tapscott, and E. A Whitaker, and published in 1960, it contains the best photographs taken with the world's leading telescopes. They are reproduced in sections, on 220 sheets, each 17 by 21 inches. All have been enlarged to the same scale (about 8½ feet for the diameter of the moon), and at least four different photographs of each area are included showing it under widely varying angles of illumination. It is believed to show every lunar feature that can be seen visually with a telescope of 11 inches aperture under *perfect* atmospheric conditions.

Three lunar mapping projects characteristic of the Space Age have been completed. One is a series of navigational charts of the moon prepared and published by the Aeronautical Chart and Informa-

tion Center of the U. S. Air Force. The scale, 1:1,000,000 or one inch to 16 miles, is the same as that of the terrestrial Regional Aeronautical Charts well known to every pilot. All surface features are shown down to those about ½ mile across, contour lines are drawn at intervals of 300 meters (985 feet), and the topography is represented vividly in shaded relief. Moreover, areas near the limb, where craters appear elliptical owing to foreshortening, have been rectified so the craters are shown in their true circular form as an astronaut would see them from directly above (Figs. 19 and 43). Charts in the vicinity of the original Apollo touchdown zone on Oceanus Procellarum were prepared first, and some 40 sections now are available. These cover most of the moon's disk and represent about ⅗ the rectified mapping required for the visible side. All but the polar areas and the

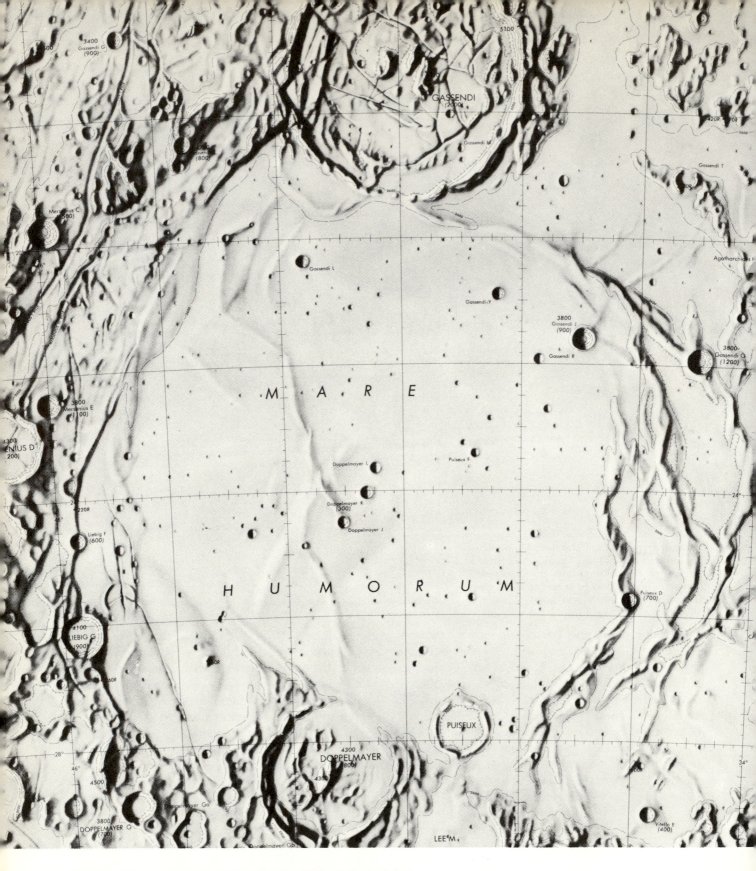

FIGURE 19. A portion of Lunar Aeronautical Chart (LAC) 93 showing the crater Gassendi and Mare Humorum reduced to two-thirds actual chart size. Compare the shape of Gassendi on this rectified chart with its apparent (foreshortened) shape as seen from the earth in Figs. 18 and 20.

east and west limbs has been charted in detail. Ultimately a set of 144 charts covering the entire moon, back as well as front is contemplated. All charts published since March, 1964, have an additional feature which increases their usefulness. The shaded relief drawing is printed not only on the face of the chart but also separately on the back of the sheet. The innovation provides the user with the equivalent of a detailed, large-scale photograph of the region in addition to the navigational chart. Each of the LAC's, as the charts are designated, covers an area of about 115,000 square miles and is printed in several colors on a sheet 29 by 22 inches.

Following the successful flight of the lunar probe Ranger 7 on July 28 to 31, 1964, the Aeronautical Chart and Information Center interrupted its regular production schedule to prepare a series of five charts based on the detailed pictures which Ranger 7 sent back by television. The first edition was issued in October, 1964, under the title *Ranger VII Lunar Charts RLC* 1 through 5. Chart 1 is on the standard scale of 1:1,000,000, and it covers exactly the same area as the regular chart LAC 76 issued in July, 1961, but it reveals a great deal more in the way of fine detail. It also shows the trajectory trace of the Ranger across the region, the point of impact, and data concerning the photographic coverage. Chart 2 shows, with additional detail, a little more than $\frac{1}{4}$ the area on double the scale (1:500,000). The subsequent charts are on scales of 1:100,000, 1:10,000, and 1:1000, respectively. The last is equivalent to 83 feet to the inch, and the chart covers an area 1100 by 1350 feet showing craters down to about five feet in diameter. It also depicts the areas of the last two television frames on the scale of 1:350, or about 29 feet to the inch. Similar sets of charts were prepared for the regions photographed by Rangers 8 and 9, and were issued in 1966. The series *RLC* 6 through 12 depicts the territory passed over by Ranger 8 on scales of 1:1,000,000 down to 1:2000. The Alphonsus neighborhood is covered by the set *RLC* 13 through 17 on scales 1:1,000,000 down to 1:400.

The Air Force also has produced a Lunar Reference Mosaic chart showing the surface features of the entire visible hemisphere for zero libration under afternoon illumination at each station. It is printed with a coordinate grid and is available in three sizes—scale 1:10,000,000 (moon's diameter about 14 inches), double the first size, and four times the first size. On the smallest edition only

major features are labeled, but on the largest one every named feature is identified. The largest size makes an excellent mural. Resembling a photograph of the full moon enlarged to more than 4½ feet in diameter, it has the necessary shading to bring out all relief plus identification labels and coordinate lines.

The second modern project is the first geologic map of the moon. It is being carried forward by the Astrogeology Branch of the U. S. Geological Survey under the direction of Eugene Shoemaker. The scale and the regional divisions are the same as those of the Air Force charts, and all geologic layers and other data which can be gleaned from high-resolution photographs and visual observations with large telescopes are shown and described. The first map of this ambitious series, that of the Kepler region, was published in 1962. It is a beautiful production in 13 colors, and it is packed with information. All or most of the sections now are completed.

The Army Map Service (Corps of Engineers, U. S. Army) has produced several series of splendid topographic lunar maps. On the scale of 1:5,000,000 the right and left halves of the moon's face have been charted in detail on separate sheets 42 by 57 inches. They are available in three forms: Relief Style (contour lines), Gradient Tint Style (a different color for each elevation interval), and Pictorial Relief Style (three-dimensional effect through shading and colors). Thousands of features are labeled by name, letter, or number, and a great amount of information is printed on each map (Fig. 21). The first style tells the whole story to an experienced map reader, the second style is both beautiful and more useful, but the third style is recommended to those who are interested in detailed telescopic study. The Mare Nectaris section of the shaded relief map, covering 650,000 square miles, has been produced as an actual relief map on a plastic sheet 10 by 14 inches. It probably is an excellent representation of the lunar surface with vertical exaggeration of 5:1. Army Map Service also has published, in the three styles, large charts of the central section on 1:2,500,000 scale (Cauchy to Kepler; Cassini to Stofler). Completed in 1967 is a series of six charts covering the entire visible surface on 1:2,000,000 scale. Several sections on the very large scale 1:250,000 also have been issued.

The Air Force charts are available through the Superintendent of Documents, and the others may be ordered from U. S. Geological Survey and Army

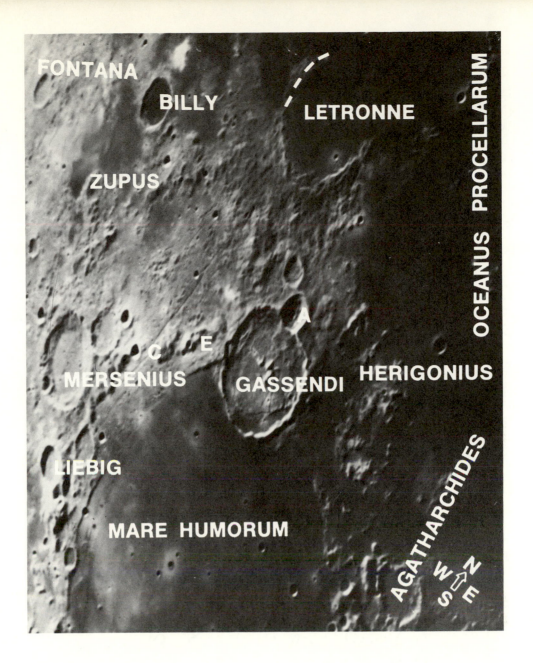

FONTANA

BILLY

LETRONNE

ZUPUS

OCEANUS PROCELLARUM

A

C E

MERSENIUS

GASSENDI

HERIGONIUS

LIEBIG

AGATHARCHIDES

MARE HUMORUM

N
W E
S

FIGURE 20.   The crater Gassendi and the half crater
Letronne in late-afternoon light. Compare with Figs. 18
and 19. (120-inch Reflector.)

FIGURE 21. Sample of the contour treatment and data provided on the lunar maps produced by the Army Map Service. Compare this rectified map with the direct photograph of the region (Fig. 6) which shows the features on about the same scale but foreshortened.

Map Service respectively (all Washington, D. C.). Users must remember that all these charts are rectified. Limb objects have been expanded along the radius of the moon's disk to remove the effect of foreshortening that always accompanies terrestrial observation.

Confusion over the names assigned to lunar surface features has arisen and abated from time to time in cycle with the degree of activity in lunar studies. Thus we have seen in the first half of the seventeenth century at least two complete sets of names introduced and abandoned. During the next 200 years little change was made. The next 100 years saw more, and the last decade has brought a great many proposals for change and addition, some of which have received wider recognition than others.

The International Astronomical Union, which seeks to bring about agreement in such matters, revised an official list of named features in 1935. In 1950 the Union added Deslandres to the list. In 1960, when Kuiper *et al.* brought out the *Photographic Lunar Atlas,* they proposed numerous changes and clarifications most of which were adopted by the Union in 1961. Percy Wilkins and Patrick Moore in their book *The Moon* (second edition, 1961) list 99 new names, which they and others have authored, but very few of them have received wider recognition. In the *Rectified Lunar Atlas* by E. A. Whitaker, G. P. Kuiper, W. K. Hartman, and L. H. Spradley (1963), the authors again proposed many changes and a long list of additions. Almost all of the additions are names for major craters in the limb regions, long neglected but being charted then for astronautical exploration. Those needed additions have been incorporated in this book since they have been approved by the Union in 1964, and many have been published on the U. S. Air Force's Lunar Aeronautical Charts.

However, I question the importance of certain name changes proposed by Kuiper *et al.* in 1963, some of which had been changed through their proposals of 1960 and three years later were to be changed back to the original or to a third spelling. I see no useful purpose served by erasing the name of the late William Pickering from the little crater officially assigned to him for more than three decades. While Pickering had rather fanciful ideas concerning changes on the moon, he did devote a great deal of attention to lunar studies and particularly to his crater which, according to the proposal, would revert to its earlier subdesignation Messier A. Most of the proposed new names serve the opposite purpose of removing such subdesignations. It might be argued that it is confusing to have two lunar craters named Pickering, but Kuiper *et al.* have not proposed the removal of the name Herschel from any of the three craters so named after William, Caroline, and John. As a matter of fact, in 1960 they proposed that the official designation of the second Herschel crater be changed from "Herschel (Car.)" to "Herschel, C." That change introduced real confusion with the slightly smaller crater long officially subdesignated Herschel C because of its proximity to the William Herschel enclosure. The same objection applies to the deletion of J. Cassini and Schneckenberg.

Purists may rejoice at the proposal of Kuiper *et al.* to Latinize a long list of names for features that have been known for generations by their English equivalents. However, my purpose here is to facilitate acquaintance with the moon on the part of nonprofessionals, a purpose that scarcely would be advanced by changing the Apennine Mountains to "Montes Apenninus," the Alpine Valley to "Vallis Alpes," the Straight Wall to "Rupes Recta," etc. I shall, therefore, retain in this book the old bourgeois forms which appear easier to grasp. Actually, for that reason, I have made a number of changes in the opposite direction. It seems rather pointless to learn that a large crater near the center of the lunar disk is designated Ptolemaeus when it has been named for the great philosopher known for many generations in this country as Ptolemy. The name Baco doubtless is familiar to specialists in thirteenth-century life, but Bacon means more to most of us. In short, I have tried always to use the more common or familiar form of the name where there appeared to be a choice. Also I have omitted umlauts and accent marks from foreign names in the interest of simplification. Really, now, if we are going to be consistent about this business of Latinizing feature labels we shouldn't introduce 66 new unlatinized names. For example, the newly-labeled feature created out of the northern half of Otto Struve's double enclosure and named in honor of the late Henry Norris Russell should be designated Russellius, a classicization probably transcending even that of Russell's 1897 Princeton diploma.

On July 31, 1964, radio, television, and the press headlined the news to virtually every inhabited spot on the face of the earth that the U. S. space probe Ranger 7 had "televised the first close-up pictures of the moon today and they look extremely good." During the last 17 minutes of its 68½-hour

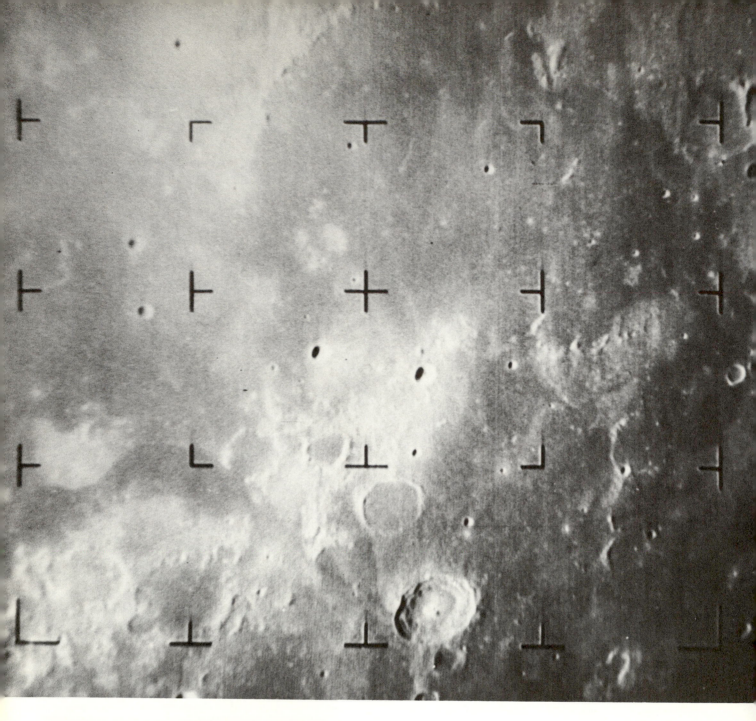

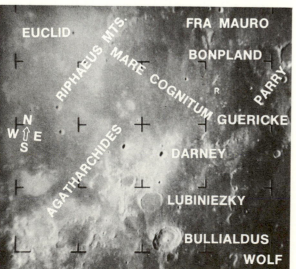

EUCLID
RIPHAEUS MTS.
FRA MAURO
BONPLAND
MARE COGNITUM
R
PARRY
GUERICKE
N
W E
S
AGATHARCHIDES
DARNEY
LUBINIEZKY
BULLIALDUS
WOLF

FIGURE 22. Wide-angle A camera view of the lunar surface televised by Ranger 7 on July 31, 1964, at 8:17 A.M. EST, 8 minutes and 50 seconds before impact on Mare Cognitum at the point marked R. Taken from an altitude of 715 miles, the photograph covers an area that measures 342 miles from left to right.

flight from Cape Kennedy to self-destruction on Mare Nubium Ranger 7 sent back not just one or a dozen or three dozen fuzzy pictures but 4308 sharply focused photographs revealing excellent detail! An astronomer hardly would agree with one news-writer's catchy claim that this accomplishment represented the greatest advance in lunar studies since Galileo, but I shall have to admit that in this case the exaggeration is relatively mild. Those photographs, taken at heights above the lunar surface ranging from 1300 miles down to 1400 *feet,* through the absolute calm of the airless interplanetary medium inaugurated a new era in selenography.

Fifteen hours after the exposures were made nearly a quarter of a million miles away, sample pictures were brought by commercial television into the homes of millions of Americans (see frontispiece). They were described by Principal Experimenter Gerard Kuiper who opened his remarks with the statement: "This is a great day for science, and this is a great day for the United States. What has been achieved today is truly remarkable. We have made progress in resolution of lunar detail not by a factor of 10, which we had hoped would be possible on this flight, nor by a factor of 100, which would have been very remarkable, but by a factor of 1000. This means that the moon, which to the unaided eye is seen at a distance of about 240,000 miles, and which in a good telescope can be brought to an equivalent distance of 500 miles, has been brought in this Ranger 7 experiment to a distance of half a mile."

For every excellent photograph of the moon or planets which an astronomer takes, he generally has made dozens if not hundreds of individual exposures, all the rest of which vary from mediocre to poor in quality of resolution. He selects a night when atmospheric conditions are the best, and he makes a long series of exposures in the hope that he will be fortunate enough to expose one of his plates during one of those rare and unpredictable instants of exceptionally good seeing when fine detail can be recorded. Consequently, Kuiper spoke for all astronomers when he voiced his amazement upon examining several hundred Ranger 7 photographs and finding each and every one "a bull's-eye as far as quality is concerned."

One of the first pictures released and one which was printed in many of the world's newspapers was centered on dark Mare Nubium just southwest of the ancient "ruined" crater Guericke, a low enclosure 40 miles across with but remnants of its

walls visible above the lava floor of the mare. It was taken from a height of 470 miles at an instant when the setting sun was 23 degrees above the western horizon. Even the newspaper reproduction showed numerous small craters and pits that are missing from the best photographs taken with great telescopes. The original print showed many, many more. From there on down to the last exposure the "frames" contained such an abundance of new information about the surface structure of the moon that one investigator estimated it would take three years to analyze and assimilate all of it. Only a part of the last picture (Fig. 23) was transmitted since the apparatus was smashed to bits in the act of sending it. From a height of 1400 feet it shows craters that appear normal but are no more than three feet in diameter. It covers only a city-sized lot 125 by 165 feet, and it is believed to reveal every object in that area down to rocks 10 inches across!

Less than seven months after the phenomenal success of Ranger 7, the feat was duplicated by the flight of Ranger 8, which was blasted off from Cape Kennedy five minutes after noon on February 17, 1965. The later model cut three hours and 31 minutes off the time of its forerunner to make the trip in less than 65 hours, and it crashed no more than seven seconds ahead of schedule at a spot only 16 miles from the selected impact point! It is scarcely surprising that William Pickering, director of the Jet Propulsion Laboratory, Harris Schurmeier, Ranger project manager, and several hundred others who had designed and constructed the marvelous robots rejoiced mightily. They could be proud, indeed, of their achievement and their substantial contribution to the conquest of space.

Ranger 7 had approached the lunar surface in a high-angle trajectory above Oceanus Procellarum, but Ranger 8 sped in at a lower angle, sweeping northeastward to photograph a much larger region. Its cameras were turned on earlier, and it sent back 7137 pictures of high quality before its splintering collision. The first few frames covered about the same area that Ranger 7 had televised, but soon the rugged uplands to the east came into view excellently portrayed. The whole mountainous territory between Mare Nubium and Mare Tranquillitatis was mapped in detail for the first time before Ranger 8 hit the southwest edge of the latter plain about 100 miles east of the crater Ritter. The highland crater Delambre was portrayed in remarkable detail together with the mountains to the east which contrast with the flat edge of the mare to

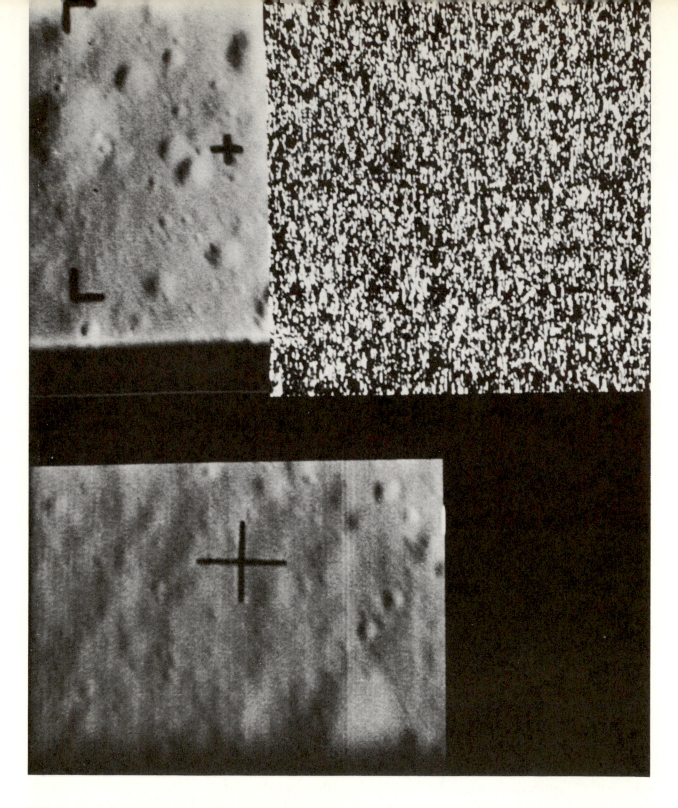

FIGURE 23.   The moon at closest range. Below: last P1
camera frame returned by Ranger 7. From an altitude of
3000 feet above the lunar surface it shows an area 110 by
160 feet. Above: last P3 camera frame. From only 1400
feet it shows an area 125 by 165 feet, and the original
negative reveals features as small as 10 inches across. Only
part of the picture was received since Ranger 7 crashed
during the transmission, as the background noise pattern
covering the right section of the frame indicates.

FIGURE 24. The crater Delambre televised by the B camera aboard Ranger 8 on February 20, 1965, at 4:51 A.M. EST, 6 minutes and 54 seconds before impact. The moon probe was 470 miles above the surface at that instant, and the picture covers an area 94 miles east—west by 76 miles north—south. Theon Junior shows well west of Delambre, and several domes may be seen on its floor. Theon Senior is bisected by the west margin, and the smooth spots near the northeast corner mark the edge of Mare Tranquillitatis.

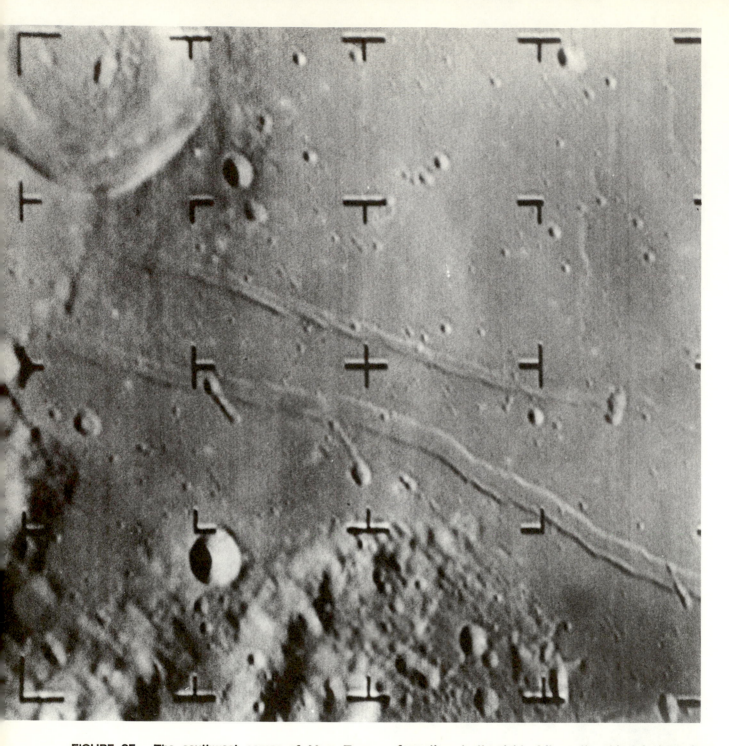

FIGURE 25. The southwest corner of Mare Tranquillitatis televised by the B camera of Ranger 8 from an altitude of 270 miles on February 20, 1965, at 4:53 A.M. EST, four minutes before impact. The large crater in the upper left corner is Sabine. The Hypatia Rill runs from the middle of the left margin to the lower right margin. It is a graben some 60 miles long and 2 miles wide formed by surface slumping between parallel fault lines. The smaller graben to the north shows characteristic discontinuity of its walls. The mountain ridge running from the lower left corner northeast toward Sabine and the two shorter parallel formations to the right of it are the oldest features in the picture according to Eugene Shoemaker and associates who date them "pre-imbrian." The three elongated craters or grooves along the Hypatia Rill generally have been attributed to blocks tossed out of Theophilus, but Gerard Kuiper and associates believe they were produced by collapse. Picture width 59.0 miles. The first manned landing on the moon occurred at a point slightly beyond the upper right margin. The crater at upper right was named Aldrin after one of the Apollo 11 astronauts.

the northeast (Fig. 24). The most detailed rill pictures ever obtained were taken of the double system southeast of Sabine (Fig. 25), and in that general area were found a number of curiously elongated small craters pointed toward the major crater Theophilus 250 miles to the southeast (Fig. 5).

The Ranger series of experiments was completed the following month with the launching of No. 9. It was lifted off at Cape Kennedy on March 21, 1965, at 4:37 P.M. and injected into a nearly parabolic orbit at a speed of 24,533 miles per hour. The initial trajectory, slightly corrected during mid-course maneuver, put it on collision course with the major crater Alphonsus 250 miles south of the lunar disk center. It was aimed for a point on the broad, flat crater floor about midway between the central peak and the northeast wall. The impact point was so chosen that the cameras would cover well the areas in which clouds of gas had been observed and where the results of volcanic activity were suspected. Ranger 9 hit the surface less than three miles from the chosen point at a velocity of 5979 miles per hour after a flight of 64 hours 31 minutes. Its six cameras surveyed 600,000 square miles and its television transmitter sent back 5814 pictures of superior quality. Rangers 7 and 8 had carried out their missions well, but the performance of Ranger 9 was truly exceptional.

So remarkable are the photographs that I reproduce here a sequence of 11 frames through the courtesy of the Jet Propulsion Laboratory of the California Institute of Technology and the National Aeronautics and Space Administration. On all of them the impact point is circled, and they were made at altitudes above the lunar surface ranging from 265 miles down to 2000 feet. The first one shows more detail than any observation ever made with a terrestrial telescope, and each successive photograph reveals much more than its predecessor. Notice the abundance of small craters that appear in areas which look smooth on earlier photographs. Notice also what happens to the strong rill on which the impact point is located. Where is it on the eighth picture of the series (Fig. 33)? The pictures are packed with new information, and I shall discuss them further in a sequel to this book. Ranger 9 also scored a "first" on commercial television. While it was en route to its destination a decision was made to use the scan converter unit developed for control of the Surveyor soft-landing lunar probes scheduled to follow the Ranger program. The converter made it possible to relay over commercial television channels some 200 of the pictures simultaneously with their reception at the Jet Propulsion Laboratory. Thus millions of people were able to watch the brilliant achievement in their homes as it actually took place. They saw the cracked and pitted crater floor come closer and closer, and superimposed on the bottom of the picture they saw for the first time the remarkable legend "LIVE FROM THE MOON."

Never before did the moon receive so much attention as it did in the 1960–70s, and never before have funds been poured out so lavishly in the purchase of astronomical information. Those 17,259 television pictures were expensive. The total cost of the Ranger program was $250,000,000. Was it worth the expenditure? For most of us that is a difficult question to answer intelligently. We are not accustomed to dealing in hundreds of millions of dollars. However, we may be able to form a relative concept of cost and value by comparing this with other vast governmental expenditures. A few weeks after the spectacular success of Ranger 7, the *XB-70 A* finally got off the ground on a maiden flight plagued with numerous malfunctions. Does anyone remember the *XB-70 A?* That was the monstrous supersonic super bomber which was to have been our salvation when the Defense Department began pouring our money into its development years ago. When at long last the first and only remaining model was rolled out of the construction hangar, it already had been branded obsolete by the Secretary of Defense. It cost *five times as much* as the Ranger program! Perhaps the Rangers did as much to protect us as did the *XB-70 A*, and I have the idea that they did a great deal more to enhance our image abroad. In addition they brought us new knowledge, and they gave to each of us some of the life-sustaining lift that comes from achieving the impossible.

Ranger 7, our first successful lunar probe, crashed on the northwest portion of Mare Nubium. The area is enclosed partially by the Guericke-Parry complex of ancient, melted craters on the east and by the Riphaeus Mountains on the west. Several other remnants of former features rising ghostlike above the mare lava also contribute to the enclosure of an oval area some 200 miles long by 120 miles wide. It was appropriate, therefore, that the International Astronomical Union, meeting in Hamburg, Germany, 25 days after the memorable flight, acted to separate the enclosed area from the rest of Mare Nubium and to give it the new name *Mare Cognitum*—the Known Sea.

FIGURE 26.   The A camera on Ranger 9 here televises the
great crater Alphonsus from an altitude of 265 miles on
March 24, 1965, at 9:05 A.M. EST, 3 minutes and 2 seconds
before impact at the center of the white circle. Note how the
smaller Alpetragius has been cut out of the southwest outer
wall of Alphonsus. Part of Davy with its wall broken by Davy
A is seen near the northwest corner of the frame. Compare
the floor of Alphonsus with the smoother Mare Nubium west
of the crater. The width of the picture is 120 miles.

FIGURE 27. Ranger 9 view of Alphonsus from 140 miles above the crater 1 minute and 35 seconds before impact at the circled point. A large rectangular section of the crater rim, indicated by the arrow, has subsided. Using the picture width of 65 miles to establish the scale, we find the section is about 20 miles long and 4 to 6 miles wide. Measuring the width of the narrow wall shadow on the left of the section and using the known altitude of the sun, we find the 100-square-mile block has slumped about 2000 feet.

FIGURE 28. As Ranger 9 hurtles toward the floor of Alphonsus, the A camera records this view from an altitude of 96 miles 1 minute and 4 seconds before impact. The crater walls can be seen only in the upper corners. The floor is especially dark near the middle of the left margin indicating that the two larger craters within the dark spot are volcanoes which showered their surroundings with ashes and cinders. Picture width is 44 miles.

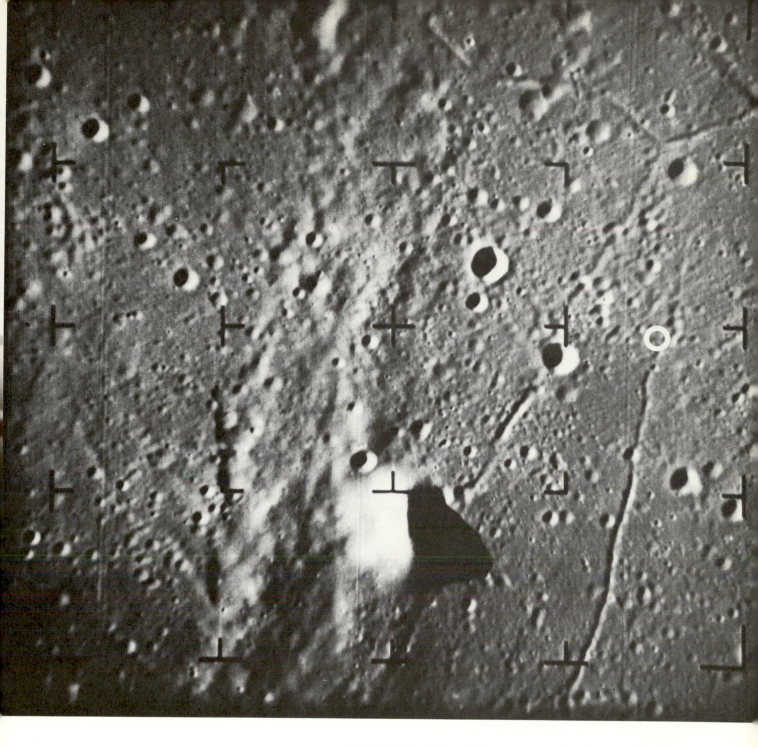

FIGURE 29. Ranger 9 is now so near the floor of Alphonsus that the crater walls are outside the field of view. According to Kuiper, the feathery mass of ridges running vertically across the picture to the left of the brilliant central peak is a complex spine built up by lava extruded through a network of cracks in the floor some of which still are visible. On the other hand, Harold Urey regards that formation as piles of debris from the Mare Imbrium-forming explosion or material torn from the north wall of Alphonsus by that explosion. The space craft altitude is 65 miles 43.9 seconds before impact, and the picture width is 30.1 miles.

FIGURE 30. The floor of Alphonsus appears increasingly disturbed as Ranger 9 speeds toward it. Here is a section 25.7 miles wide viewed from an altitude of 55 miles 36.7 seconds before impact within the circle on the right. The razorback central mountain (near lower margin) is 5.3 miles across, has a crest 2.5 miles long, and rises 3800 feet above the floor. The latest of various theories of its origin is that of Kuiper who believes it is a volcanic cone slowly accumulated by lava emission accompanied by vapor which covered it with a bright coating.

FIGURE 31. The first six craters in order of size shown here can be identified readily as small pits in Fig. 26, located within three impact-circle diameters of that circle. Note the great number of craters shown here in areas that appear smooth in Fig. 26 and later pictures. The two prominent rills are disintegrating into chains of craters and shallow depressions as resolution continues to increase. Two general classes of craters are now evident: sharp, round, deep ones and soft, shallow ones. Some of the latter are irregular in outline. Ranger 9's altitude is 35.1 miles 23.5 seconds before impact. Picture width is 16.2 miles.

FIGURE 32.   Ranger 9 is now only 12.1 miles above the floor of Alphonsus and will crash within the white circle in 8.1 seconds. The huge crater near the west (left) margin is the small but well-defined one just west of the impact circle in Fig. 26. It is 1.6 miles across, and it resembles the volcanic vents seen in Fig. 28. The area around it is not darkened, but it is blanketed according to Shoemaker and associates who conclude that the crater is a volcano. Picture width is 5.64 miles.

FIGURE 33. Large shallow depressions and tiny sharp craters characterize this Ranger 9 view of Alphonsus' rough floor from an altitude of 4.5 miles, 2.97 seconds before impact. The sharp round craters were produced by the impacts of meteoritic bodies colliding with the moon. The soft shallow depressions are believed by Kuiper and associates to be collapse features produced when parts of the surface layer dropped into subterranean chambers or empty tubes once filled with molten lava. Collapse resulted either from withdrawal of the supporting lava or from the shock of heavy meteoritic impacts perhaps hundreds of miles away. This is the last A camera picture, and its width is 2.06 miles.

FIGURE 34. The last three frames televised by the P3 camera aboard Ranger 9 show small sections of Alphonsus' heavily cratered floor. In the order numbered, the frames were taken from altitudes of 2.90 miles, 1.63 miles, and 2000 feet respectively, 1.92, 1.08, and 0.24 seconds before impact. The widths of the pictures are 1960, 1100, and 250 feet respectively. The largest crater shown in frames 1 and 2 is the small one on the bottom edge of the impact circle in Fig. 33. It is an impact crater 170 feet in diameter. The large double crater to the right in frame 3 is 34 by 42 feet across, and it may be seen in frame 2 to the right of the impact circle which it resembles in size. In addition to craters down to less than one foot across, some forty rocks have been spotted in frame 3. They are believed to be blocks blasted out of the 170-foot crater which lies just below the lower border of the frame. Two rocks about 5 feet in diameter may be seen in the directions of the arrows halfway across the entire frame. They are bright on the left and dark on the right—opposite shading to that of the craters. Ranger 9 crashed within the circle at 9:08:19.994 A.M. EST, March 24, 1965, while the last frame was being transmitted.

The Ranger program increased enormously our knowledge of small-scale lunar topography. In three representative regions, judiciously selected, we gained abundant photographic information in which increasing resolution was accompanied by decreasing area of coverage. By stretching the final frames to the limit, it could be said that man had examined about ⅔ of an acre on the moon as it would appear to the unaided eye at a distance of several city blocks. The severe limitations of weight and size, under which the Ranger vehicle had to be designed, restricted drastically the amount of power available for the experiment. Consequently, the cameras and the transmitters had to use the battery-power supply in alternate shifts instead of providing an apparently continuous picture in the mode of terrestrial television. Such retarded sequencing combined with the enormous speed of Ranger's approach ruled out the possibility of observations from less than about ¼ of a mile. How could we get closer? How could we learn more?

Obviously, the next step was to land a television system on the moon in working condition. That operation not only requires additional skills of high sophistication but it also increases the launching weight of the pay load by a factor of nearly three. For these reasons no attempt was made to slow the Rangers, and they collided with the surface at terrific speed, probably leaving little to mark each touchdown spot other than a new crater some 22 feet in diameter. Soft landing was the mission of the Surveyor program which had been in the developmental and testing stages for about five years when the operational phase of the Ranger project ended in 1965.

At this point it may be well to note that picture taking and chemical-physical analysis of the lunar surface are not the primary objectives of the new technology which we might call moon probing. The resulting data are, of course, tremendously important to the advancement of selenography and lunar geology, but we have yet to reach the level of intellectual development at which we would be willing to spend the necessary hundreds of millions of dollars without a larger and more spectacular purpose in view. That greater purpose was outlined on May 25, 1961, by President John Kennedy who declared in a message to the Congress: "I believe this nation should commit itself to achieving the goal, before this decade is out, of landing a man on the moon and returning him safely to the earth." Accordingly, the primary mission of the

Rangers was the discovery and development of techniques with which a space craft could be accelerated away from the immediate vicinity of the earth and maneuvered onto a collision course with a selected spot on the moon. Their secondary purpose was to obtain data on the small-scale structure of the surface required for effective planning of lunar landing craft. The primary mission of the Surveyors was the discovery and refinement of the additional techniques necessary for the gentle landing of a manned space craft on the airless, waterless moon. Their secondary purpose was the further investigation of surface structure through photography, physical measurements, and chemical analysis.

Soviet scientists beat our American spacemen by four months in achieving some of the secondary goals of the soft-lander program. On February 3, 1966, at 1:45 P.M. EST, they landed an instrumented package, Luna 9, on the western edge of Oceanus Procellarum 60 miles northeast of the crater Cavalerius (see Chart IV—near the west limb north of Cavalerius and west of Reiner). The space craft that carried Luna 9 was decelerated by powerful retrorockets which began firing when it was 45 miles above the moon and continued ⅘ of a minute. During that short interval the speed of approach was reduced from 5800 miles per hour to something in the neighborhood of 10 miles per hour, and just before impact the instrument capsule was ejected to fall free of the vehicle. Luna 9 weighs (or did weigh on the earth) 220 pounds, and it is in the form of a sphere two feet in diameter. After coming to rest, it erected itself by unfolding half its protective cover in four petallike sections, extended four antennas, looked about with its television eye, and began transmitting pictures of what it saw in the early-morning glare of a superbrilliant sun.

During the next three days Luna 9 sent pictures from time to time on orders from its masters. Then, to the surprise of others, it was announced that the mission had been completed and no further information would be received because the batteries of the "automatic space station" were exhausted. Eventually about one dozen frames were released officially which, when fitted together, formed a panorama of gently-undulating territory apparently consisting entirely of small-scale depressions and protrusions. Earlier, Nikolai Barabashov, chairman of the Soviet Committee for the Study of Physical Conditions on the Moon, had announced that "the upper layer of lunar soil is a sponge-like rough-textured mass scattered with individual sharp-edged fragments of various size . . . strong enough to support more or less heavy objects." Observable depressions range in diameter from less than one inch to around 100 feet, and most of the larger ones are shallow. The protrusions are angular blocks that litter the entire area, and they range in dimensions from about ⅕ inch up to perhaps two or three feet. About half the surface is cratered, and nearly all of it is covered with debris.

The Soviets deserve full credit for the first on-the-spot photographs of the moon and for a few other observations and measurements made by Luna 9. The mission was one of extreme difficulty as anticipated by American scientists and as shown by the fact that the Soviets are known to have failed in at least four and possibly seven earlier attempts to accomplish it. It was a major achievement, but was it truly a soft landing? The instruments aboard that we know about survived and performed their functions, but could such a space craft be trusted to carry a man to the moon? Despite expressed opinions to the contrary, it seems likely that one or more astronauts could be cushioned and braced inside a spherical space capsule to withstand a crash at considerably more than 10 miles per hour with subsequent rolling and bumping over rather rough terrain. The odds on making it are good although I can think of no worse place in the universe to risk personal injury or damage to vital equipment. Man could get to the moon by that means, but he would be riding on a one-way ticket. Those who plan to return must land not only softly but in a fully-equipped space craft capable of lift off and injection into an earth-bound orbit.

Surveyor 1, first in the American series of soft landers, was launched from Cape Kennedy on May 30, 1966, in the nose cone of an Atlas-Centaur vehicle, a 113-foot rocket with a more powerful second stage than the Atlas-Agenas used for the lighter Ranger space craft. Sixty-three hours of travel brought it to a point 59 miles above Oceanus Procellarum where a series of intricate maneuvers began which reoriented it to landing attitude, reduced its velocity from more than 5800 miles per hour to about three miles per hour, stabilized its descent so that the three components of the landing gear would touch ground simultaneously, and finally cut off all power at an altitude of 14 feet. At 1:17 A.M., EST, June 2, 1966, Surveyor 1 made a perfect touchdown, the jolt of its final fall being taken up by shock absorbers and crushable blocks

of aluminum honeycomb on the legs, footpads, and frame. Next it went through a routine of reporting on its excellent condition, and within half an hour it transmitted its first picture. That was a soft landing in the full sense of the term. The same sequence of operations, with appropriate modifications, was planned for landing the much heavier LEM space craft that would place two astronauts on the moon, and then we knew that the process would work. Another significant contrast between the American and Soviet moon probes is evident from a comparison of weights. That portion of Luna 9 which traveled to the moon weighed 3480 pounds, 59 per cent more than the 2194 pounds of its Surveyor 1 counterpart. Yet the instrument package of Surveyor 1 weighed 596 pounds, more than 2⅔ times as much as the 220-pound package landed by Luna 9.

Surveyor 1 also landed on Oceanus Procellarum but at a spot 435 miles southeast of Luna 9. It touched down on the broad mare-flooded floor of a large ancient unnamed "ghost" crater, the low walls of which are tangent on the south to the much smaller and younger crater Flamsteed (see Chart IV). The sun was 28 degrees above the eastern horizon when Surveyor 1 began to record its surroundings, and its first-day pictures revealed a more familiar topography than the weird scenes televised by Luna 9. Except for the characteristic craters, the views might have been snapshots taken on a terrestrial desert strewn with rocks. The panorama is that of a gently rolling surface pocked with craters ranging in diameter from about two inches to around 1500 feet (Fig. 37). Craters in the two-inch to ten-foot class appear to cover about half the region between the space craft and the apparent horizon. Distance to the latter varies from several hundred yards to one and ½ miles, and beyond the horizon in a few directions can be seen low, bright ridges, the tops of the nearest wall remnants of the great ghost crater. Scattered everywhere are rocks from a fraction of an inch up to several feet across (Figs. 35 and 36). Most of them are angular blocks while the rest are rounded as if erosional forces had long been at work upon them. The majority of the sharp blocks appear to rest essentially on the surface while many of the rounded masses seem to be partially buried. The general appearance and the variation in distribution density of the protuberances suggest that most of them are chunks of lunar rock ejected from the larger visible craters and others beyond the horizon when those craters were carved

FIGURE 35. Lunar scene on Oceanus Procellarum as viewed by Surveyor 1 on June 12, 1966. The horizon was slanted by the rotatable mirror used to reflect the image into the vertically-mounted television camera. Note the black sky and black shadows cast by the rubble. A medium-sized crater includes the largest dark spot on the surface. The crater appears as a long, narrow ellipse parallel to the horizon, centered in the picture, and extending a distance five times the length of the dark spot. It runs from the left edge of the bright area on the left to a point one spot-length beyond the spot on the right. The rim is marked in part by stones of varying size. It is late afternoon, moontime, ten days after Surveyor landed and two days before sunset.

FIGURE 36.   Landscape by Surveyor 1 shortly after its
successful landing on the moon June 2, 1966. The horizon
slants across the upper left corner. The conspicuous object
to the lower right is one of the larger rocks near the space
craft. It is about 1.5 feet long, and the early-morning sun
gives it a sharp shadow. It is definitely rounded and cut by
grooves and cavities, and it could have been produced by a
gaseous melt. About halfway from the rock to the horizon
is the elliptical outline of an apparently rimless crater ten
feet in diameter. Its bright inner right wall contrasts with the
darker surroundings, and its darkened inner left wall is cut
off by one of the bright round solar reflections (ghosts) in
the television system. The black dots regularly spaced in
horizontal and vertical lines are reticle marks for calibration.

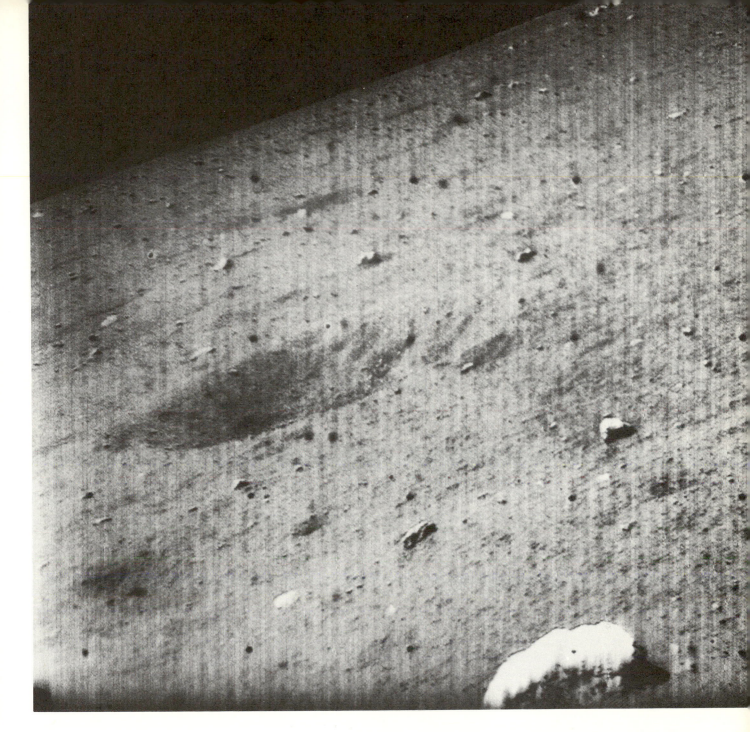

FIGURE 37.   The changing scene three days later. The same region as that of Fig. 36 was photographed by Surveyor 1 on June 5, 1966, under late-morning illumination. The left edge of the 10-foot crater shows clearly, but the right edge is almost washed out against its surroundings. It is located 35 feet from Surveyor. Note the line of small bright objects near the upper right corner close and parallel to the horizon. That is the most concentrated portion of a ridge of large blocks along the edge of a crater several hundred yards in diameter. Its continuation can be traced approximately parallel to the horizon clear across the picture. The far wall of the large crater coincides with the horizon at the upper right.

out of the moon's surface by cataclysmic forces. Since the three footpads of Surveyor 1 sank below the surface about one inch and pushed some of the material aside to form small craters with raised rims, it was concluded that the surface layer consists of a granular soil-like substance of unknown composition and fine-grained structure. (Grains as small as 1/50 of an inch in diameter have been resolved in the pictures.) Most students of the moon expected some dust on the barren plain, but an experiment designed for its detection revealed none whatsoever. In appearance and in its reaction to impact the surface material resembles fine-grained damp terrestrial soil. Lunar rocks and soil brought back by astronauts differ chiefly from earthly counterparts in that apparently no water has touched them since they were formed.

Surveyor 1 didn't weaken and die after working for a few hours to send a few dozen pictures. Like the Rangers, it was equipped with broad panels of solar cells which converted sunlight into electricity and continually charged its batteries to keep all apparatus operating at peak efficiency. Each night when the moon rose into the California sky the great radar dishes of the Goldstone Deep Space Network on the Mojave Desert were directed toward the spot where the invisible robot was known to be, to transmit to it orders from the Jet Propulsion Laboratory in Pasadena and to receive from it the precious signals that were turned into pictures—hundreds of excellent pictures every night. On the second day Surveyor took its own temperature and reported a rather warm 180 degrees Fahrenheit. Slowly the sun continued the upward climb toward its culmination point when a sizzling 240 degrees was anticipated, and Surveyor was given a midday siesta. Soon it was called back to work and the picture-taking sessions continued until the sun set on June 14, 1966 (Fig. 38). Half an hour after that event Surveyor took one more picture, a self-portrait in the light of the earth which was nearly full at the time. That was the 10,338th picture it had sent back! Immediately afterward it was put to bed for a long winter's snooze at a chilly 250 degrees below zero.

At sunrise on June 28, 1966, a button was pressed at the Jet Propulsion Laboratory to ring Surveyor's alarm clock, but there was no response from the ghost crater on Oceanus Procellarum. Scientists had hoped the automaton would survive the ordeal of cold, but they had not expected it to do so. It had performed its assignment far above and beyond the most optimistic anticipation, and there really

was little more it needed to do. Yet its persistent creators continued to needle it, and the following week it revived. Understandably its health was not good, but the plucky robot went back to work and sent about 1000 more pictures in spite of serious attacks of fever.

Other Surveyors followed the first. They touched down in various locations carefully selected for probable variety of terrain. They sent back more thousands of pictures, and they performed experiments on the surface material which I shall discuss elsewhere since they are more closely related to lunar geology than to selenography. The Rangers and the Surveyors had extended in a truly fantastic manner the observations made with terrestrial telescopes until we have acquired what is probably a fairly complete and representative atlas of lunar features ranging in dimensions all the way from a tiny fraction of an inch to hundreds of miles. We have learned a great deal since Galileo turned his telescope on the moon in 1610, and the vast bulk of our knowledge has been accumulated during your lifetime and mine. But the true scientist never has enough information. The very fact that he is a scientist makes it impossible for him ever to be satisfied. So the arguments over the nature of the lunar surface went on and on, and from time to time one heard that exclamation of frustration: "If only we could get hold of a few handfuls of that stuff!" That they did—hundreds of handfuls from late 1969 to early 1973.

The ultimate instrument of the selenographer's dream was the Lunar Orbiter. Its secondary mission was similar to that of the last four Rangers but much more extensive. It would neither land gently nor crash, but, as its name implies, it would be placed in orbit around the moon. The techniques of transfer from a collision trajectory to an orbit around our satellite constituted its primary mission since that operation had to be performed with precision by the three astronauts aboard the Apollo space craft three years later. Also it would photograph at close range all the areas tentatively selected for the Apollo touchdown in order that the most favorable might be picked and detailed charts could be prepared to assist the navigator in finding it.

Lunar Orbiter carried two cameras and four large solar cell panels to supply the necessary power, and it looked somewhat like a plump bug on snowshoes. But what, you may ask, could it do for selenography that had not been done already? Did we not then have excellent lunar charts? Yes, we did—for approximately the central half of the moon's disk,

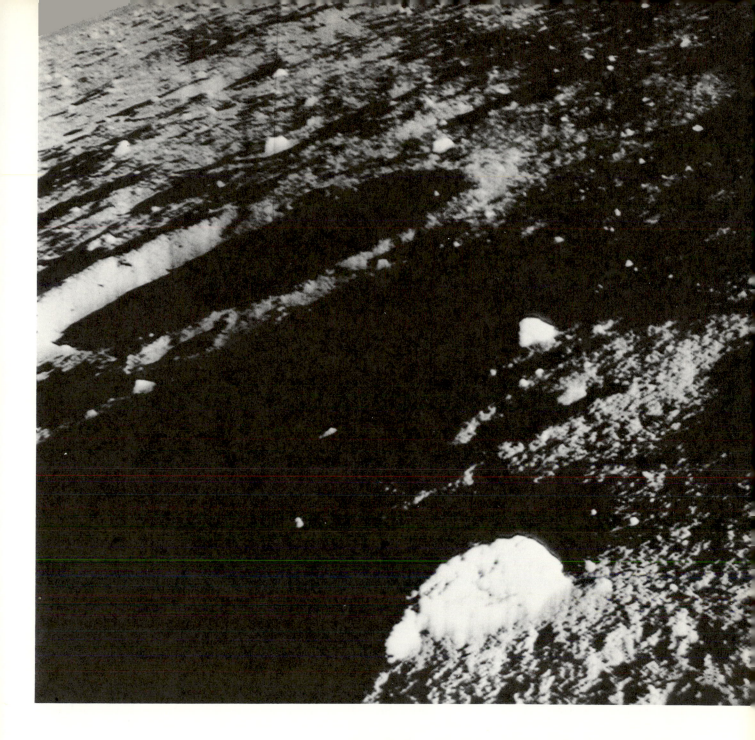

FIGURE 38. The same view as that of Fig. 37 photographed 9 days later by Surveyor 1 on June 14, 1966. As sunset approaches, depressions disappear in black shadow, protrusions shine brilliantly in full sunlight, and the landscape takes on a surrealistic aspect. The 10-foot crater resembles a bathtub full of ink, and the scattered rocks might be bergs of meringue in a sea of oatmeal. Note that the crater now exhibits a raised rim, its right outer wall sunlit and its left wall casting a shadow to the left. The block-studded rim of the large distant crater shows well as it parallels the horizon. The texture of the lower right portion of the picture, below and to the right of the large rock, closely resembles the appearance of the lunar surface in all the Soviet Luna 9 photographs that have been released.

but even there great improvement in detail could be made, as may be seen by comparing a pre-Ranger chart of the Mare Cognitum region with a post-Ranger chart of the same area. The remaining outer half of the disk, which actually comprises about ¾ of the moon's visible surface, was less well known. In the vicinity of the limb, where we view the level surface nearly edge-on, we knew very little about even the large features. We knew much less about the far side. So our best maps of the central half of the lunar disk could be improved by a uniform series of Ranger-type photographs, and our charts of the other ⅞ of the lunar surface were in real need of such observations.

The far side had been photographed to a considerable extent. On October 4, 1959, Soviet scientists celebrated the second anniversary of the Space Age by launching the first automatic space station, Lunik 3. Three days later it curved behind the moon and began taking photographs at a distance of about 40,000 miles above the surface. At least 38 negatives were made during the next 40 minutes as the probe receded. The moon was five days old at the time, and the fraction of the far side that was illuminated corresponded to the fraction of the known side we see when the age is 20 days (or 10). Consequently, about ⅔ of the unknown side was sunlit and accessible through photographic observation. Lunik 3 was the first space probe that had done anything about the moon other than hit it or miss it, and most of its predecessors had missed. Its successful operation by remote control was a major breakthrough in the technical advancement of civilization, but the pictures themselves left a great deal to be desired. Resolution was poor, and operational difficulties resulted in images of the disk that were covered with defects. The latter were separated from what appeared to be surface features by analytical methods, and an atlas was published by the USSR Academy of Sciences. In 1961 it was made available with English translation (see Barabashov in the reference section at the end of this book). In August, 1965, the Soviets announced that again they had photographed the far side, on July 20, from the unmanned space craft Zond 3. One of the three pictures released showed better detail confirming the announcement that the far side is heavily cratered but deficient in maria. According to an official spokesman, Lunik 3 and Zond 3 had photographed 92 per cent of the far side, and an improved atlas based on the later data was published in 1967.

Again I must report that the first Lunar Orbiter was blasted into operation by the Soviets. On April 1, 1966, they announced that a 540-pound automatic space station, Luna 10, had been launched the previous day and was on course to its destination. It was directed into a closed orbit two days later, an orbit in which, according to official announcement, it circumnavigated the moon every three hours within an altitude range of 220 to 620 miles. Surprisingly, Luna 10 carried no camera and contributed nothing to selenography. It was provided with sensors for radiation, magnetic, and meteoritic measures and a radio transmitter which broadcast its findings as long as its batteries supplied the necessary power. Luna 10, doubtless, still is making its rounds, but its brief period of usefulness ended long ago. It has the distinction of being the first known piece of defunct scientific apparatus indefinitely in orbit around our satellite. A few years later it had abundant companions, but it seems unlikely that the clear lunar skies ever will become so contaminated with mechanical junk and other refuse as is space in the vicinity of the earth.

Lunar Orbiter 1, the first space craft to photograph the moon from a stable orbit, ascended from Cape Kennedy on the afternoon of August 10, 1966, in the nose cone of an Atlas-Agena rocket. The 850-pound photographic studio took 92 hours for the trip (nearly half again as long as Ranger 9's flight) in order to minimize its terminal velocity and permit better control of the crucial orbit changing maneuver. At 10:23 A.M., EST, on August 14, the 100-pound-thrust motor of the space craft was fired for 10 minutes in the direction of motion. That powerful, frictionless brake slowed Orbiter enough for the moon's gravitational field to take control of its destiny and switch it into an elliptical orbit that carried it around our satellite in a period of 3 hours and 38 minutes at altitudes ranging from 117 to 1159 miles above the surface. For a week it continued in that orbit as our tracking stations "watched" it closely for slight progressive changes in its motion that would reveal small departures from uniformity in the structure of the moon.

The first series of pictures was taken by Lunar Orbiter on August 18, 1966, from an altitude of 133 miles as it sped toward the eastern limb above Mare Smythii (see Chart II). One of the photographs showed 10,000 square miles of mare plain and rugged upland to the east with a detailed representation of a nearly circular crater 18 miles in diameter in a location where terrestrial observations reveal little but short, narrow streaks parallel to the limb. Three days later Orbiter sent back several photographs of the far side. The high-resolution camera recorded 7500 square miles of craters, ridges, depressions, and plains, while the medium-resolution camera captured 130,000 square miles of the same from an altitude of 930 miles. Craters no larger than 1500 feet across showed clearly on the published prints. Also on that day the rocket motor was fired again to reduce speed and shrink the minimum altitude from 117 miles to 36 miles, and the survey of Apollo touchdown sites was begun. During the sixteenth pass on August 23 Lunar Orbiter was instructed to photograph the small waning crescent of the earth in a black day sky as it was about to set behind the bright west limb of the enormous moon. Three days later the newspapers of the world published the first photograph man had ever seen of the earth as a planetary body viewed from the moon (Fig. 41). As expected, the picture showed principally cloud patterns, but contrary to expectation, it revealed a sharp terrestrial terminator closely resembling that of the airless moon.

Lunar Orbiter took its last photographs on August 29, 1966, and began the two-week task of slowly and meticulously transmitting its accumulation of 215 pictures to the NASA Langley Research Center. As the scientific world waited to take a close look at possible Apollo touchdown areas, it was announced that analysis of Orbiter's slight deviations from its assigned circumlunar course had shown the moon to be pear-shaped like the earth was found to be a few years earlier by exactly the same method. Our satellite has a bulge $\frac{1}{4}$ mile high over its north pole, $\frac{1}{4}$ mile flattening at the south pole, a chest shrinkage of $\frac{1}{8}$ mile in the northern hemisphere, and a spread below the belt of $\frac{1}{8}$ mile in the southern hemisphere. Why haven't we noticed these deformities before? The greatest departure of the limb from a true circle which they could produce on a perfect photograph of the moon nine inches in diameter would amount to 1/1000th of an inch! Not only do natural elevations 20 times that high occur along the limb, but atmospheric disturbance, optical imperfection, and photographic distortion combine to produce greater discrepancies in the best lunar portraits we have.

Due to Orbiter's low-minimum altitude and the particular irregularities in the moon's gravitational field, it was predicted that the space craft would continue to swing closer to the surface and crash in about eight months. However, since it might interfere with future probes, it was ordered on October 29, 1966, to fire a small retrorocket and reduce speed from 2150 to 1750 miles per hour. The loss of momentum threw it into a glide which ended with a crash on the far side. Our pioneer probe of possible Apollo landing sites experienced some difficulty with its high-resolution camera, but on the whole the mission was judged a success. It provided the photographic data for preliminary screening of the nine tentative touchdown spots along the lunar equatorial belt, and later Orbiters narrowed the choice.

Lunar Orbiter 2 repeated the performance of Lunar Orbiter 1 but did it better. Blasted off from Cape Kennedy on November 6, 1966, it reached the trajectory switching point nearly four days later and was deboosted into a circumlunar orbit almost identical with the first orbit of its predecessor. Once Lunar Orbiter began taking pictures the process could not be interrupted for more than about eight continuous hours lest the film acquire permanent kinks where it stood wrapped around the various take-up spools. Moreover, a pair of exposures (high- and medium-resolution) had to be made every four hours to prevent excessive drying of the chemical solution on the developing web. Such limitations might seem to be unacceptable since the end of a maximum rest period could and often did find the spacecraft far from one of its assigned target areas. However, the four-hour requirement turned out to be a tremendous advantage from the selenologist's point of view. Most of the excellent far side photographs were taken because the film needed to be advanced between periods of photographing Apollo landing site candidates, Orbiter's primary assignment. On November 23, 1966, the cameras of Orbiter 2 required activation as it cruised across Oceanus Procellarum 150 miles south of Copernicus at an altitude of 28 miles. It was ordered to direct its cameras toward the great crater and take a pair of pictures (see Figs. 54–56). The resulting high-resolution frame revealed so much new detail of the walls and floor that a NASA spokesman evaluated it as "one of the great pictures of the century."

Lunar Orbiter 3 was launched February 4, 1967, given a small mid-course correction 38 hours out, and deboosted into circumlunar orbit after a four-day journey. Its mission was to check under different lighting 12 of the best candidate Apollo sites selected from the many photographed by its predecessors. Also, the probe was to continue measurements of the moon's gravitational field and the flux of micrometeorites and solar radiation, photograph

FIGURE 39.   Lunar Orbiter 1 gives man his first good look at the surface of the moon's far side. This frame, one of many, was taken August 24, 1966, and it shows an area some 800 by 900 miles across. The sun is 16 degrees above the western (left) horizon at the central crater, and the region shown is in the southern hemisphere beyond the east limb about ⅕ the way across the far side. This is a single exposure, developed on board, scanned, and televised to earth in strips which were put together. The large heavily-battered crater covering nearly six strips (lower center) is about 180 miles in diameter, and the floor of its largest intruder shows flooding stain. A larger crater in the upper left is less well marked. It covers strips 4 through 10 from the top, its west wall is cut by the left margin, and its east wall lies ⅓ the way across the frame. It is about 250 miles in diameter, larger than any recognized crater on the known side, and approximately the size of the basins for the circular Maria Nectaris and Humorum as measured by Ralph Baldwin. The crater is extremely old, and the only difference between it and a typical mare may have been the non-availability of molten lava some 4.5 billion years ago.

FIGURE 40.   Lunar Orbiter 1 scans part of the great dome field south of Copernicus and northeast of Apollo landing site 7. Craters are bright on the left, and domes are bright on the right. The area 23 by 28 miles across was photographed August 26, 1966, with the sun 13 degrees above the eastern horizon. It can be located on Fig. 4 about two-thirds the way from Lansberg to Gambart. The largest crater (Gambart AA in upper right) is 1.9 miles in diameter, and the dome just southwest of it is 3 miles across. Here earth-based photographs have revealed virtually nothing with dimensions smaller than one strip width. Note the depressed corridor 2 miles wide west and northwest of Gambart AA and the many grooves, crater chains, and rills. The smallest craters are about 250 feet in diameter. There is some east—west foreshortening near the right margin and considerable near the left.

FIGURE 41.   Man's first view of the earth as seen from the moon. Photograph by Lunar Orbiter 1 in its sixteenth circuit on August 23, 1966, at 11:36 A.M. EST. The moon was 7 days old (13.5 hours past first quarter) at the time, and the earth in its corresponding cycle of phases was 22 days old or a little past last quarter. The terminator cuts through western Europe and Africa, but continental outlines are very weak in contrast to the highly-reflecting cloud patterns. Orbiter is moving eastward and witnessing the beginning of "earthset" as the dark limb of the earth moves behind the bright limb of the moon. Note the unexpected sharpness of the terminator near the south horn where the cloud cover is continuous over Antarctica, and note the characteristic black sky above the sunlit moon. Craters ranging from tiny specks to a large ellipse (lower right) that stretches 100 miles or two-thirds the picture height are portrayed in magnificent relief. The conspicuous dark crater with bright multiple floor mountain, through which the right border cuts, is 30 miles in diameter, and the distance from the upper right corner to the limb near the earth is some 400 miles. Features of many shapes and ages are represented in this low-angle view. In the upper right section the remarkable detail resembles the surface of a storm-tossed ocean.

32 features of special interest, and provide practice for the personnel of the Manned Space Flight Network organized to track the Apollo spacecraft enroute to the moon. One of its best photograph pairs (Figs. 58 and 59) was the prominent far side crater Tsiolkovsky, 115 miles in diameter, with a rugged central mountain and a floor flooded by dark lava. In the medium-resolution frame the floor appeared "smooth as a mirror," but in the high-resolution view it was seen to be heavily pocked with small craters. Like the Ranger 9 series (Figs. 26–34), the pair was yet another reminder that in spite of the best advance scouting, some surprises surely awaited the Apollo astronauts in the final seconds of landing.

Five Lunar Orbiters had been ordered in the hope that two or three could be counted on not only to orbit the moon but to perform the long series of intricate operations that would supply the Apollo mission planners with large-scale photographs for the selection of most favorable landing sites. To the amazement of almost everyone, the first three had completed the job. The remaining two would be flown for further site checks and Apollo rehearsals for Mission Control personnel, but almost all of the photographic capacity could be devoted to pure science. Each Orbiter would be placed in a polar orbit to revolve perpendicular to the lunar equator and near the terminator. Each revolution would carry it over a different strip of near and far side as the moon rotated below it. Lunar Orbiter 4 departed on May 4, 1967, and a week later it began 15 days of photography. Its spectacular view of Mare Orientale (Figs. 52 and 60) will be discussed later. Lunar Orbiter 5 was launched August 1, 1967, and it performed as directed to complete a chapter in space exploration with the remarkable record of 100 per cent efficiency. Among its contributions were far side photographs under different lighting of the huge flooded crater Mare Moscoviense (Fig. 65). Stains on the crater's floor indicate several different lava flows. The rugged 25-mile postflow floor crater Titov contrasts with smaller preflow floor "ghost" craters. Another photograph of remarkable detail revealed two boulders which had rolled down the central peak of the crater Vitello (Fig. 67) on the south shore of Mare Humorum.

A MONTH WITH THE MOON

In an earlier section I traced the moon through its cycle of phases showing how you can predict approximately when and where in the sky to look for it each night of the month, guided by the phase dates given on most calendars. Now let us go through the cycle observing the more conspicuous features as they are brought into prominence from night to night by the advancing terminator. I shall make nightly reference to the seven full-page charts (pp. 83, 91, 97, 117, 147, 171, 183) discussed in the chapter preceding the lunar gazetteer. As you will note, each chart is double and covers two pages. On the right-hand page is a photograph which may be compared directly with the moon as seen through binoculars. On the left-hand page is a photographic diagram to the same scale identifying the various features that show well at that particular phase. Having located the major features each evening, you then may proceed on your own to find and examine as many of the smaller markings as you may desire. In the gazetteer I list alphabetically all the named features visible through binoculars with data and brief description.

Already I have used from time to time the compass directions north, south, east, and west to locate objects on the face of the moon. Unfortunately, there is some confusion today concerning such lunar coordinates, but that is a result of our progress into the Space Age. Rudyard Kipling in a familiar ballad wrote: "Oh, East is East, and West is West, and never the twain shall meet." Whether or not the United Nations proves him wrong on the earth remains to be seen, but the International Astronomical Union already has belied him on the moon. Not only have the twain met, but simultaneously they have been wedded together, exchanged places, or disappeared entirely. East is now West except when it is East, but to avoid ambiguity we are urged to call it Left which for some observers turns out to be Right, and vice versa. If this arrangement is not entirely clear it is because of the great compromise reached by the International Astronomical Union at Berkeley in August, 1961, when that august body decided to retain the old coordinates familiar to astronomers and at the same time adopt the new coordinates which astronautical planners demanded. Perhaps I had better begin at the beginning.

Picture the full moon at its greatest altitude when it stands high above the southern horizon bisected by the celestial meridian. For generations it has been customary to call the lunar limb toward the west half of the sky the western limb and the limb toward the east half of the sky the eastern limb. North was up for observers in the

Northern Hemisphere, and south was down. That was logical and simple. Of course it meant that the terminator moved across the moon's disk from west to east, but that didn't bother astronomers as they observed from the earth, and no one else cared. It also meant that if man ever landed on the moon he would find the sun rising in the "west" and setting in the "east." That didn't bother astronomers either—not until the Space Age was well under way. The likelihood that men actually would land on the moon, coupled with the fact that both earth and moon rotate on their axes in the same direction, indicated that the direction toward which the terminator moves across the moon should be called "west," just as it is on the earth. Accordingly, the new coordinates are identical with those used on terrestrial maps; north is at the top, south is at the bottom, east is to the right, and west is to the left. Since we plan to explore the moon we use the new coordinates, and we use them on every chart and photograph in this book. On all identification charts that accompany photographs the directions of the compass are shown since a few of the photographs have been cropped in such a way that north is not toward the top. When consulting other books remember that older books use the old system, and a few use both.

It would facilitate your acquaintance with the moon if I could say to you, for example, that whenever the moon is seven days old you will find the important crater Albategnius bisected by the terminator and located ⅓ the distance from the south limb to the north limb. This is sometimes true, but, I regret to report, not always. The actual instant at which the moon reaches the exact age of seven days may not take place until some hours after your observation, and Albategnius may be largely in darkness west of the terminator when you look for it. On the other hand, the moon may have reached its seven-day age some 18 hours prior to your observation, and Albategnius may be more than its diameter inside or east of the terminator. Moreover, the libration in latitude can shrink or stretch that "⅓ the distance from the south limb" considerably by tipping the lunar north pole either toward or away from us. You might suspect that the libration in longitude would further complicate the picture, and you would be correct. However, it doesn't do what you probably think it does. The predominant factor in the movement of the terminator across the face of the moon is the axial rotation of the moon which proceeds at a uniform rate, at least to a high degree of approximation.

The libration in longitude can cause large variations in the thickness of the crescent presented by the two-day moon, for example, but it does not alter the position of the terminator with respect to lunar surface features. There are other factors involved, but their effect is relatively small, and we need not discuss them here.

The libration in longitude, as we have seen, results from the non-uniform revolution of the moon in its elliptical orbit—a continuous slowing down from perigee to apogee and a continuous speeding up from apogee to perigee—as required by Kepler's second law. Such exasperating indifference to man's way of scheduling events thus accounts for another instance of the moon's appalling disregard for punctuality. It may arrive at the new moon position, between the earth and the sun, either early or late, as measured by our clocks. In extreme cases it may be off schedule by as much as ⅔ of a day, but generally we find its "error" no greater than half a day, and often it is but a few hours. Now, since our satellite arrives at the new-moon position at varying intervals, we find corresponding variations in the position of the terminator at new moon.

We are interested primarily in the position of the terminator with respect to the lunar surface features, and the terminator advances over the lunar surface at the essentially uniform rate of the moon's rotation (about 12.2 degrees per day). Just as the combination of uniform rotation and non-uniform revolution causes the libration in longitude, that combination also causes the terminator to reach the position it should occupy at new moon sometimes before and sometimes after the instant of new moon. This means that we must expect slight differences in the location of the terminator at the moment of new moon from one lunation to another. These differences account for part of our uncertainty as to whether or not Albategnius will appear exactly on the terminator when the moon is seven days old. The rest of the uncertainty has to do with the time of day at which the lunation begins.

The event of new moon may occur at any time of day from, say, one second after midnight to one second before next midnight. If, for example, it should occur at 7:43 A.M. on July 3, then the moon will be two days old on July 5 at 7:43 A.M., ten days old on July 13 at 7:43 A.M., etc. Since the average length of the lunation is 29.530589 days (29 days 12 hours, 44 minutes, and 2.8 seconds), the moon is likely to reach the new phase next at an entirely different hour of the day, and that new hour will determine the exact time of its daily "anniversaries" during the next lunation. The reason why this event can take place at any time of day is, of course, that the fundamental cycles involved—the revolution period of the moon and the rotation period of the earth—are incommensurable. In such instances, the most probable time of recurrence is the midpoint of the measuring cycle used—the day in this case. If you have an almanac that gives the hour and minute of new moon as well as the day of the month, you readily can demonstrate this fact. Disregard the dates, and average the times of day for the event through the whole year (remembering to add 12 hours to all P.M. times). It is highly probable that you will find an average time between 10:30 A.M. and 1:30 P.M. Such an average taken over an interval of five to 10 years would fall within a few minutes of noon.

If the last four paragraphs scintillate with all the clarity of income-tax instructions, their purpose merely is to establish some basis for deciding where we should place the terminator of our standard 10-day moon, for example. If the moon is new on July 3 at some hour unknown to us, the most probable time at which it will be 10 days old will be July 13 at noon. Such an assumption can never be off more than ½ day, and there is a 50–50 chance that the error will be no more than ¼ of a day.

Now, what can we do about the terminator reaching the new-moon position at a time other than the instant of new moon? Since it can do so either before or after new moon, and since the extreme range is the same in either case, the *average* difference between the two events over the years will be zero. While the most probable value of the difference for any given lunation is thus zero, we must be aware of the effect and consider how it can affect the position of the terminator we observe. For most lunations, this discrepancy will amount to less than half a day. Furthermore, there is a 50–50 chance that its correction will have the opposite sign from the correction required by the occurrence of new moon at some hour other than noon. For example, new moon might occur at 7:43 A.M., and the terminator might not reach the new-moon position until five hours nine minutes after new moon. In that case the event which is important to us—the arrival of the terminator at the new-moon position—would take place at 12:52 P.M., the two discrepancies largely having offset each other.

While one correction happily can be expected to differ in sign from the other correction for half the lunations to come, the numerical values of the two corrections cannot be expected always to come out as nearly equal as those in the example. So we must anticipate frequently a quarter-day, sometimes a half-day, and rarely a whole day net discrepancy between the position of the standard terminator, which I shall describe for each day of the moon, and the actual terminator which you will find when you observe.

Of course, we could avoid all this fussy technical discussion by just setting up a lunar almanac for the next 10 years or so. In it I might list the date of each new moon and the time at which the terminator would reach the new-moon position. The time used would be Universal Time, and I would explain to you how to convert that to your particular brand of standard time and what to do about daylight-saving time if and when you have it. Standardized observation times would have to be specified for each night of the moon, and corrections to those hours would have to be given for each lunation, and by the time it was all worked out and set up with tables and rules, we would be more confused than we are now. It is much better for us to make the most probable assumption—that the moon changes its terminator age every day at noon—and to recognize that some small adjustment or allowance usually will have to be made to fit the actual appearance of the moon as we observe it.

One final point remains to be mentioned, then we shall have finished all this academic twiddle-twaddle, and we can go out to observe the moon. We shall hope that the moon will have its terminator as near as possible to the proper place each day at noon. However, we are not going to observe the moon at noon. The two-day moon we shall look for shortly after sunset. On subsequent evenings we shall observe it later to take advantage of the best conditions, and by the time it is nearly new again we shall be straining for a glimpse of it just before sunrise. So our final, and I do mean final, adjustment of the terminator will be made for the interval from noon to the most favorable time of night for observation of the moon at the age given.

Here ends our first (and last) lesson in statistics. You now see how I selected terminator positions for each day of A Month with the Moon. If it didn't make sense, don't be concerned. Just accept my assurance that while the moon you observe may differ in appearance at the terminator from the moon I shall describe, the difference generally will be small and should cause no difficulty in locating and identifying the features described.

As I endeavor to select the more noteworthy objects for successive evenings' observation I am reminded of the student in an astronomy class who, for another reason, experienced selection difficulties. Caught unprepared by an unannounced quiz, he was asked to describe the most interesting star in the constellation Cetus. During the 10 minutes allotted he searched the dusty shelves of his empty cupboard and then hastily wrote: "All the stars in Cetus are so interesting that I can't choose among them." I am even more troubled by the memory of an illustrated lecture which I attended many years ago on "The Castles of the Old World." It was delivered by the owner of the feed-and-grain store who had "made the trip" abroad and had taken the pictures, such as they were. The pictures were stills, but the sputtering arc light in the "magic lantern" gave them a certain degree of motion. As each slid onto the bed-sheet screen, it was introduced with: "Now this here is the ancient old" whatever-it-happened-to-be. So I ask your indulgence if you find a redundancy of such adjectives as "conspicuous," "prominent," and "magnificent." Perhaps after you have become acquainted with the face of the moon, you, too, will be inclined to describe its features in superlatives.

Pick an observing night, and determine the moon's age from the calendar (number of days since the date of previous new moon). Turn to the section for that age in the pages that follow, and put a paper clip on the page. Next turn to the chart indicated for that age, and put a paper clip on that page. If the major features that are supposed to be close to the terminator cannot be found, they probably are lost in the darkness beyond and will not appear tonight. In that case shift to the description for one day earlier if the moon is waxing (before full) or one day later if the moon is waning (after full). If you should note conspicuous features along the terminator that are not mentioned, you likely will find them described in the section for one day later if the moon is waxing, or one day earlier if the moon is waning.

You are about to embark upon a voyage of discovery and exploration that can bring you much of the thrill and satisfaction experienced by the adventurers of the sixteenth and seventeenth centuries who crossed the Atlantic Ocean to open the New World. They encountered hardships, however, and so will you. There may be the hardship of frosty nights bringing cold feet and numb fingers. If such misery seems too great a sacrifice, perhaps you can draw new strength from the zeal of those thousands of hardy and dedicated souls who in midwinter sit outdoors all afternoon to watch a football game. You will encounter the severe hardship of responding to the alarm clock at 3 or 4 A.M., leaving the comfort of a warm bed, and driving yourself out in the cold, damp solitude of predawn to observe the waning crescent. But the rewards will prove to be worth the hardships as you inspect the lunar landscape and go over the ground upon which the pioneers of the Space Age have walked, experimented and gathered samples.

Without dampening your enthusiasm, let me guard against possible disappointment by injecting a word of caution. Don't expect too much too soon. The major craters and the great mountain ranges should be yours from the beginning, but to see the small ones you will need:

1. Binoculars of highest quality with sturdy mounting.
2. Experience as an observer.
3. Excellent atmospheric conditions.

You can acquire the first two requisites, but you might have to move far away to be sure of the third. If you have difficulty resolving craters smaller than, say, 30 miles in diameter, omit them at first. Later, as you gain No. 2, and on those nights when you find that you have No. 3, try for the smaller objects, and you will find most of them. If, after you have become acquainted with the moon, you want to do something about No. 1, let me make a suggestion. Don't buy expensive binoculars to replace those you have been using. Instead, invest in a three-inch refractor or a six-inch reflector. With such a telescope you will be in position to study with ease and in detail every object mentioned in this book and a great many more.

ONE-DAY MOON

(Chart I)

The one-day moon, if seen at all, is a very slender crescent indeed. Moreover, its altitude is low, its light is feeble, and the sky background is bright. All these factors conspire to thwart the observer. The most favorable time for viewing it in the latitudes of the United States is in March or April, for with the sun near the vernal equinox the young moon stands higher in the sky at sunset than at any other time of year. Unless we had a total eclipse of the sun to inaugurate the present lunation, the moon recently has passed either north or south of the sun in the sky. Consequently, the line joining the horns or cusps of the crescent still may be inclined at a considerable angle to the north-south line on the moon's disk. If the lunation began near noon yesterday the angle should not exceed about 15 degrees, but if the moon is much younger than 30 hours the angle may be considerably greater. Since that angle is equal to the displacement of the midpoint of the crescent from the east point on the moon's disk, there is some uncertainty as to what portion of the limb the illuminated crescent will cover.

My advice to you is to wait until tomorrow evening before trying to identify any lunar surface features even if you have succeeded in locating the one-day moon, a major triumph for anyone on his first observation evening. After you have observed through a lunation and have become experienced in locating surface objects, you may wish to try your skill on the one-day moon.

Get ready to observe before sunset, and be sure you get out in time to note precisely the point on the horizon where the sun sets. Don't point your binoculars toward the sun before atmospheric scattering and absorption have reduced it to a dull red disk at the horizon. Your binoculars probably have a field diameter of seven or eight degrees. You can check their area coverage almost any night by observing the Pointers of the Big Dipper which are a little more than five degrees apart. If the field diameter is about seven degrees, the average one-day moon should be found two field diameters east of the sun. From the sunset point on the horizon lay off two field diameters up and to the left. Then through that point sweep your binoculars in a circular arc centered on the sunset point from a low to a high altitude. In a few minutes you should pick up the faint, thin sliver of a crescent. Your first impression will be that it is completely blank. However, the sky still is bright and the moon isn't. The relative brightness will change rapidly in the next few minutes, and you will begin to notice surface detail.

The largest and best-defined crater visible this evening is HUMBOLDT, 125 miles in diameter and 15,400 feet deep. Its dark inner east wall shows as a short black streak parallel to the terminator and not far from it. Its apparent location depends upon the cusp line inclination, and so it may be found anywhere from the midpoint of the crescent to ⅔ the way from the midpoint to the south cusp. As more detail appears we can check the identification of Humboldt by looking for another black streak, thicker but only ⅔ as long and located between it and the terminator. That is PHILLIPS, 80 miles in diameter and 10,500 feet deep. Just north of Humboldt and appearing almost as a continuation of it is a stronger black streak marking the east wall of HECATAEUS, 87 miles in diameter, 15,700 feet deep, and joining another crater on the north for a total length of 115 miles. Just north of Hecataeus a shorter, stronger black dash marks BEHAIM, 33 miles in diameter and 11,000 feet deep. A little farther north may be seen a pair of black specks. The farther southern one is ANSGARIUS, 58 miles in diameter and 10,000 feet deep. The nearer, weaker, northern one is LAPEYROUSE, 46 miles in diameter and 10,000 feet deep.

In a telescope the craters we have been looking at can be seen as very narrow, cigar-shaped ellipses with some irregularities here and there. The reason that we see only black specks and streaks through binoculars is that we are picking up only their strongest parts—black-shadowed inner east walls. Their bright inner west walls are tipped away from us, and their illuminated floors blend with the light surroundings.

If now we jump north from Lapeyrouse about twice the Humboldt-Lapeyrouse distance we should find another strong black dash which is PLUTARCH, 41 miles in diameter and 9200 feet deep. North of Plutarch, a distance equal to that between Humboldt and Behaim, is the mountain-walled plain GAUSS, 112 miles in diameter and 13,000 feet deep. Its dark streak is not so strong as the others, and you will have to look hard to see it. Just before the moon disappears behind the trees or into the thickening haze we get the impression that the whole crescent is pocked with craters and by no means blank.

TWO-DAY MOON

(Chart I)

The largest feature of the two-day moon is the east part of smooth MARE CRISIUM which is cut almost

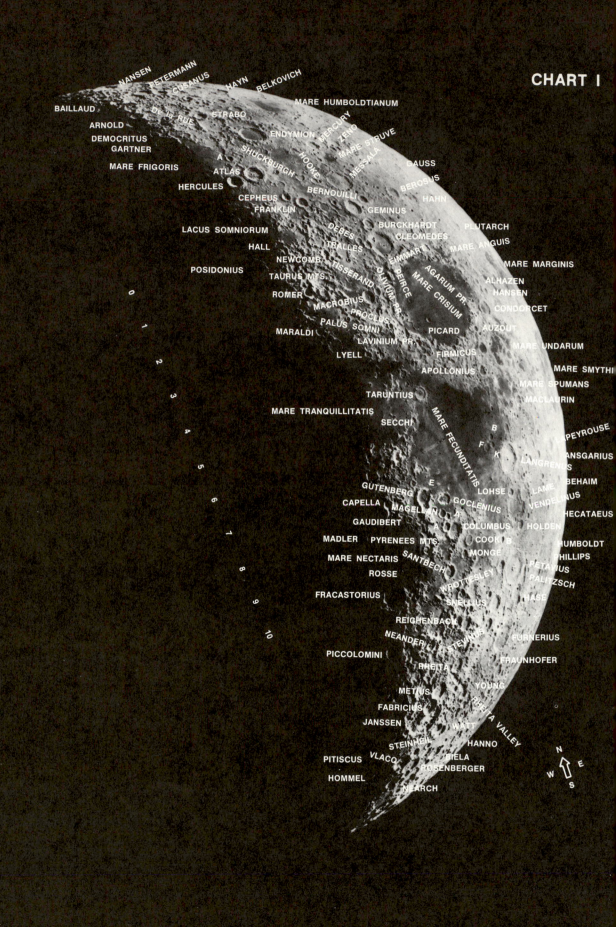

CHART I

NANSEN
PETERMANN
BAILLAUD
CUSANUS
HAYN
BELKOVICH
ARNOLD
DE LA RUE
STRABO
MARE HUMBOLDTIANUM
DEMOCRITUS
ENDYMION
MERCURY
GARTNER
ZENO
MARE FRIGORIS
SHUCKBURGH
HOOKE
MARE S. STRUVE
A
MESSALA
GAUSS
ATLAS
BERNOUILLI
BEROSIS
HERCULES
HAHN
CEPHEUS
GEMINUS
FRANKLIN
BURCKHARDT
PLUTARCH
CLEOMEDES
LACUS SOMNIORUM
DEBES
MARE ANGUIS
HALL
TRALLES
NEWCOMB
EMMART
MARE MARGINIS
TISSERAND
AGARUM PR.
POSIDONIUS
TAURUS MTS.
OLIVIUM PR.
ALHAZEN
PEIRCE
HANSEN
ROMER
MARE CRISIUM
MACROBIUS
CONDORCET
PROCLUS
AUZOUT
MARALDI
PALUS SOMNI
PICARD
LAVINIUM PR.
MARE UNDARUM
LYELL
FIRMICUS
MARE SMYTHI
APOLLONIUS
MARE SPUMANS
TARUNTIUS
MACLAURIN
MARE TRANQUILLITATIS
SECCHI
B
LAPEYROUSE
F
ANSGARIUS
K
LANGRENUS
BEHAIM
GUTENBERG
E
LAME
LOHSE
VENDELINUS
CAPELLA
GOCLENIUS
MAGELLAN
A
HECATAEUS
GAUDIBERT
COLUMBUS
HOLDEN
A
COOK
B
MADLER
PYRENEES MTS.
MONGE
HUMBOLDT
PHILLIPS
MARE NECTARIS
SANTBECH
PETAVIUS
ROSSE
WROTTESLEY
PALITZSCH
FRACASTORIUS
VASE
SNELLIUS
REICHENBACH
NEANDER
FURNERIUS
STEVINUS
PICCOLOMINI
FRAUNHOFER
RHEITA
YOUNG
METIUS
RHEITA VALLEY
FABRICIUS
JANSSEN
WATT
HANNO
STEINHEIL
PITISCUS
VLACO
BIELA
ROSENBERGER
HOMMEL
NEARCH

MARE FECUNDITATIS

0
1
2
3
4
5
6
7
8
9
10

N
W E
S

down the middle by the terminator. It is a large semiellipse just north of the center of the crescent, and its length is about equal to the width of the crescent there. Close examination shows that it is not quite so bright as the rest of the crescent, which at first appears to be uniformly brilliant. As the moon grows older the mare will darken, and the rough territory which borders it on both north and south will brighten, presenting a strong contrast in both shading and texture, but this evening we might miss it entirely if we did not know where to look.

Mare Crisium is unique in its lack of connection with the other maria. We view it far around the curved surface of the moon where circles are foreshortened into ellipses. Consequently, when it is revealed fully tomorrow evening we shall not be surprised at its oval outline. Yet if we compare it in shape with some of the craters located about the same distance from the limb, we shall see that it is not flattened as much as they are. From such an observation we infer that Mare Crisium is not circular but truly elliptical with its long axis running east-west rather than north-south as it appears to us. Its actual dimensions are 270 miles (north-south) by 350 miles (east-west), giving it an area equal to that of the state of Washington. Its apparent area is only half as great or equal to that of Maine. Since its isolation and low reflectivity make it an easy landmark any time between the lunar ages of three and 17 days I shall use its *apparent* dimensions as convenient yardsticks for indicating distances between features. Thus "one Crisium length" will represent 270 miles, and "one Crisium width" will denote about 180 miles.

A conspicuous crater this evening is LANGRENUS, which is 85 miles in diameter (Fig. 47). It also is bisected by the terminator, and it is located south of the Mare Crisium south shore a distance a little greater than one Crisium length. The walls, which rise 6500 feet above the surrounding plain and drop 16,200 feet to the black floor, appear exceptionally steep and rugged in this light, and their outline is completed on the west by a bright crest extending out over the dark side. Perhaps later you may see the sun rise on the central mountain which is small for such a large crater and a difficult object through binoculars at this phase. Continuing south along the terminator we come to almost equally prominent, larger PETAVIUS with dimensions of 99 by 110 miles. Its distance from the south border of Mare Crisium is double that of Langre-

nus, and it also is cut in two by the terminator this evening. At its center Petavius has a whole group of mountains which easily are seen rising from the black floor some 8200 feet. Here we view one of the older craters but one which nevertheless is 13,800 feet deep. Its bright northwest crest extends slightly beyond the terminator, but the rest of the west wall is dark.

Between Langrenus and Petavius is the large, old, lava-flooded crater VENDELINUS, 92 by 100 miles across and 14,700 feet deep (Fig. 47). Its floor also is in black shadow at the terminator, but it is completely bounded on the west by the bright wall crest upon which the sun already has risen. We note that the northeast wall has been moved a dozen miles inward by the younger, well-outlined crater LAME, 55 miles in diameter and 11,800 feet deep (Fig. 47). Farther south, about ⅔ the Langrenus-Petavius distance beyond the latter, the terminator cuts through FURNERIUS, another major crater 81 miles in diameter and 11,000 feet deep. Like Vendelinus, it has no central mountain or peak, but both have many intruding craters on floor and walls, some of which may be seen through binoculars with favorable lighting. In a few nights it may be noted that a spot just northwest of Furnerius is the center of a bright, broad ray system which extends outward more than 150 miles in many directions and becomes so bright that crater outlines are obliterated in its glare.

Furnerius belongs to a subgroup of large craters known as *mountain-walled plains*. They are rather shallow for their size, and usually they have smooth floors, no central mountain, and low or no outside walls. They are often polygonal in outline rather than circular or elliptical, and they appear to have been formed simply by dropping their floors rather than having thrown up their walls by explosive action of some sort. Some authorities believe they originated in that manner.

If now we go north from Mare Crisium a distance a little less than that of Langrenus to the south we come to another large crater nearly bisected by the terminator. That is MESSALA, an old mountain-walled plain the walls of which are deformed by smaller craters in several places. It is a weak feature compared to those we have just observed, but it is well outlined by a narrow black east wall and a bright west wall. The south half of the floor is still in darkness, but the north section is illuminated dimly. It is 72 miles across, and its north wall, which rises 8900 feet above the floor, has a linear section 30 miles long. The curvature of the south wall is reduced, and the crater has the apparent outline of an ellipse with the ends cut off.

It is a curious coincidence that all but one of the features we have observed this evening are approximately bisected by the terminator and that all five major craters appear as scallops or black "bites" out of the thin crescent. Those black spots are not typical lunar views. We are going to see a great deal more and see it much better as the crescent grows. However, since craters near the terminator are usually the easiest to locate, these black half views actually may be welcome as we orient ourselves on our first regular observing night. If the new moon of the current lunation had been "born" in the early hours of the morning two days ago, the terminator now might coincide with the west walls of the five craters, and they would show much better. In any case, they will be lighted fully by the sun tomorrow evening.

Were the moon so turned that we could see these large craters, particularly Langrenus and Petavius, near the center of the disk they would be most impressive sights. Even from their outfield positions they show well. If you look closely at the crescent you are likely to see quite a few craters in addition to those I have pointed out, but most of them are smaller and likewise black. Should you be eager to press on you can identify most of them by reference to Chart I, but if you have located Mare Crisium and the six craters listed you have done well. Later, as you become accustomed to binocular observation of the moon, there will be plenty of objects to find.

THREE-DAY MOON

(Chart I)

Tonight Mare Crisium is disclosed fully and identified easily, but as yet it cannot be called conspicuous. Its west half definitely is darkened, but its east edge is almost as bright as the brilliant highland that borders it. The terminator passes through the mountains that form its west boundary. Prominent in relief, they appear very high tonight, and their east slopes shine as brilliantly as if they had been brushed with gold paint. Beyond the Crisium uplands to the south lies the larger, more irregular, and less well-defined plain known as MARE FECUNDITATIS, a rather pretentious description for an empty basin that is likely to prove perfectly sterile. A few nights hence it will have darkened enough to stand out strongly, but tonight it is revealed more by its smoothness than by the slight shading which it exhibits.

One of the conspicuous features of tonight's moon is the large crater CLEOMEDES which is seen to the north of Mare Crisium and separated from it by about 40 miles of mountainous territory. Its black inner east wall and bright west wall crest give it high contrast. It has dimensions of 81 by 92 miles, a depth of 14,300 feet, and it belongs to "Class 5."

Ralph Baldwin in his major contributions to lunar studies, *The Face of the Moon* (1949) and *The Measure of the Moon* (1963), has classified more than 300 craters according to their relative age. Those of his Class 1 are the newest and sharpest formations. They are *postmare*, which means that none of them is more than 2 to 3 billion years old (his age range for the maria), and some are very much younger—perhaps only a few million years old. The youngest of the *premare* craters are of Class 2. Still older ones belong to Class 3, and a few very ancient enclosures have been placed in Class 4. The latter may be almost as old as the moon itself (4½ to 5 billion years). Baldwin uses Class 5 to identify craters which have undergone various degrees of flooding during the period of the lava flows. They are among the older depressions, and they have secondary floors of congealed lava (or perhaps ashes) which make them shallower than they were originally. Where available, Baldwin's classifications are given for the craters discussed here.

Baldwin also has measured with precision the diameters and depths of the craters he has classified. However, later publications, such as the U. S. Air Force's Lunar Charts, Army Map Service's Lunar Maps, and Kuiper and associates' *Rectified Lunar Atlas* (1963), show that many lunar craters depart considerably from the circular outline previously assumed for most of them. In fact, those that look circular after correction for foreshortening seem to be in the minority. Crater sizes given in this book have been measured in the publications just cited. Where there is not a large departure from circular outline an average width has been recorded as a "diameter." In other cases the term "across" has been used or both minor and major widths have been recorded to indicate the degree of irregularity or distortion. Crater depths and mountain heights given here have been determined from the data recorded on the U. S. Air Force's Lunar Charts or the Army Map Service's Lunar Maps.

Another prominent feature is Class 1 Langrenus on the east shore of Mare Fecunditatis. It is a splendid object with its bright floor, black inner east wall, and small central mountain which you may detect as a black speck. While Langrenus has retained fully its qualities of eminence, the same cannot be said for its rivals of last evening. Petavius, Vendelinus, and Furnerius (Classes 5, 5, 3) can be seen, but they have faded considerably and are no longer serious competitors. This is but one example of an endless list of more or less startling changes in appearance which the student of the moon can record as he compares its features from night to night. Another faded crater is Messala. Its low wall is difficult to detect, and within a few nights it will lose all its shadows and disappear from the scene.

Nearly two Crisium lengths northwest of Cleomedes, at the terminator, may be found the prominent crater ENDYMION, of Class 5, with a diameter of 77 miles and a depth of 16,100 feet. Its unusually dark, smooth floor tonight appears no darker than its surroundings, but the bright west wall and the broad shadow of the east wall outline the crater well. Also northwest of Cleomedes, at a distance of about twice that crater's length, is the somewhat smaller but sharply outlined Class 1 crater GEMINUS, 53 miles in diameter and 8200 feet deep. It stands out distinctly in contrast with the softer outlines of Cleomedes to the south and faded Messala to the north. Its larger neighbors are old craters while it is a relative newcomer. Its small central mountain group is beyond the reach of our binoculars.

A little southwest of the spot occupied by the once-prominent Petavius we see a pair of smaller but easily-spotted craters, SNELLIUS and STEVINUS. Located much closer to the terminator than the limb, and on a line parallel to the latter, they show prominently as two ellipses, $\frac{1}{3}$ and $\frac{2}{3}$ black respectively, with bright west wall crests. Their diameters are 52 and 48 miles respectively, and their depths are 11,700 and 15,400 feet. Watch them for several nights, and note their clear, sharp outlines characteristic of Class 1 craters. Between Snellius and the limb, on the south edge of Petavius, is an odd fish-shaped combination of confluent craters

about 70 miles long and 7600 feet deep. Although much weaker than the last two features, it is easily seen as a short, fat, crooked black streak, the north end of which is the Class 2 crater HASE, 50 miles across and 9000 feet deep.

FOUR-DAY MOON

(Chart I)

Mare Crisium now has darkened enough to be recognized as one of the "dark" lunar areas. The larger Mare Fecunditatis to the south also is shown fully, and it is almost as dark as its neighbor. It is a little larger in area than California. At the terminator near the center of the crescent we see a shaded triangular area. That is the eastern portion of MARE TRANQUILLITATIS, which will grow for the next two nights. South of its south shore about one Crisium width the terminator cuts into a similar but smaller smooth, shaded region. This is MARE NECTARIS, which will be disclosed completely in another evening. It then will be seen to be a little smaller than Mare Crisium (about the size of South Carolina) and approximately square in outline.

Langrenus is still one of the more prominent craters in view tonight, but Petavius is almost "washed out" and lost in the shadowless area that surrounds it. Its disappearance is aided by the brightening of the area just north of it. That is caused by PETAVIUS B, 21 by 23 miles across and 11,800 feet deep, a small ray crater on the south shore of Mare Fecunditatis. The crater, $\frac{1}{3}$ black, may be spotted at the north edge of the bright splash pattern. Cleomedes, Geminus, and Endymion still show well, but they are now weaker, having suffered considerable loss of face in the last 24 hours. Many more such changes may be noted by the systematic observer, especially if he makes a record of relative prominence of features from night to night.

Where the north tip of Mare Fecunditatis joins the east extremity of Mare Tranquillitatis rises the ancient Class 5 crater TARUNTIUS, 36 miles in diameter and 3000 feet deep. Like a castle tower guarding the "straits," it stands at the end of a mountain arm thrust out between the maria 200 miles from the highlands to the southwest. Tonight it is a delicate bright ring, but around full moon a better simile for Taruntius might be a lighthouse since its rays then may be traced out over the maria for 100 miles.

Along the east boundary of Mare Nectaris runs an irregular black line which is seen easily and observed to be double in places. This marks the west face of the PYRENEES MOUNTAINS, which extend some 200 miles in a north-south direction and rise 10,800 feet above the Nectaris plain. More of a *scarp* or extended cliff than a mountain range, the ridge ends on the north in what, at this phase, resembles a gigantic bright claw embracing the crater GUTENBERG. That old crater, 45 miles in diameter, 7500 feet deep, and partly filled with lava, has been distorted severely. Its northeast wall is broken by GUTENBERG E, 18 miles in diameter and 2600 feet deep, which also was formed before the flood. If you can resolve its low-contrast outline you are doing far too well. The southern part of the main crater looks like an overlapping crater, but actually it is an enclosure about 30 miles square, 3600 feet deep, and largely filled with low mountain ridges running north and south. About one Crisium width south of Gutenberg is SANTBECH. A young Class 1 crater 40 miles in diameter and 13,100 feet deep, its half-black interior and brilliant inner west wall give it high contrast. The RHEITA VALLEY, previously described, is seen easily tonight about halfway from Mare Fecunditatis to the south horn of the crescent and about halfway from terminator to limb. It resembles a dark, curved-handle sword.

Just west of Rheita Valley are two large, conspicuous craters tangent and in a line parallel to the limb. The north one is METIUS, and its close neighbor to the south is FABRICIUS. Their diameters are 53 and 48 miles, and their depths are 18,600 and 14,400 feet, respectively. Although well formed, they lack something of the sharpness of craters such as Santbech, and they are good examples of the older Class 2. Their period of formation is the same, but a close examination with greater magnifying power shows that Fabricius is actually younger since its wall intrudes upon that of Metius. Further inspection of the region reveals the prominent walls of a very large ancient enclosure which includes Fabricius but excludes Metius. From the newer common wall of these two it extends southwest fully 122 miles, a distance greater than the combined length of the two craters. Its south wall is bright, its bright west crest is beyond the terminator, and its southeast wall is dark. Broken in more than a dozen places, the 9300-foot wall is definitely not circular but hexagonal, and the floor is dis-

figured by many craters and cracks. This is JANSSEN, a mountain-walled plain and possibly one of the oldest features left on the moon.

South of Fabricius, roughly 1½ times its length, lies a pair of similar craters except that they appear joined sidewise instead of lengthwise. The narrower one on the limb side is WATT, of Class 1, 41 miles long and 6500 feet deep. The broader one, which intrudes upon Watt, is STEINHEIL. It has the same length as its Siamese twin, but it is 400 feet deeper and more of its floor is black. South of Steinheil, a distance equal to that of Fabricius to the north, is the crater BIELA. Always seen near the limb, Biela is a young crater 48 miles in diameter and 6600 feet deep. Perhaps you can detect what looks like a tiny black bead protruding from the east wall. That is BIELA C, 17 miles in diameter and 2300 feet deep. At the terminator west of Biela, and about equidistant from both Biela and Steinheil, may be seen a light-bordered black ellipse which is all that shows tonight of the slightly larger and sharper Class 1 VLACQ. It is 59 miles in diameter and 9800 feet deep. About one of its lengths southwest of Vlacq along the terminator is NEARCH, another Class 1 crater, 47 miles in diameter and 6200 feet deep (Fig. 49). It looks pear-shaped because its south wall has been broken by the large intruder NEARCH A, 26 miles in diameter and 2000 feet deep. The interiors of both craters are merged in common shadow tonight. You are now far into the crater-covered "continent" of the South where there are many additional, easily-seen features which may be identified by reference to Chart I. But let us not neglect the northern hemisphere, where a prominent pair of craters awaits us.

Approximately two Crisium lengths northwest of the Mare Crisium north shore, at the terminator, we find the pair. The smaller west one is HERCULES, and the larger east one is ATLAS. Hercules, of Class 1, is 45 miles in diameter and 12,500 feet deep. The older Class 5 Atlas is 54 miles in diameter and 10,000 feet deep. In the terminator zone a small difference in depth sometimes produces a striking difference in illumination. Hercules, with the terminator near its brilliant west wall, is so deep that its interior is entirely hidden in black shadow. Atlas, under only slightly higher sun, reveals most of its gray floor bounded by the black shadow of its east wall and a bright west crest. On a line from Atlas back to Mare Crisium and about ⅓ the way may be seen a pair of considerably smaller craters—CEPHEUS, 27 miles in diameter, and FRANKLIN, 34 miles in diameter. They are of Class 1 and 2 respectively, and the smaller Cepheus is 10,500 feet deep while the larger Franklin is 10,700 feet deep.

About one Crisium length southwest of Cepheus and Franklin the terminator cuts through a broad region of upland hills called the TAURUS MOUNTAINS. This mass of craters, hills, and ridges shows well tonight, and several bright ridges can be seen running northwest onto the dark side. Near the brightest and roughest part of the region the terminator approximately bisects the Class 1 ROMER, 25 miles in diameter and 11,100 feet deep, an irregular ring for which diameters up to 35 miles have been recorded. Although dark tonight with a light west crest, it shines brightly under high sun, and at full moon it becomes the center of one of the weaker ray systems. Tomorrow night it will appear as a black spot amid bright surroundings. Between the Taurus Mountains and Mare Crisium, and closer to the latter, is an easily seen Class 1 crater 42 miles in diameter and 11,800 feet deep. It is MACROBIUS with a gray floor and a black inner east wall which contrasts conspicuously with its unusually brilliant inner west wall. The mountainous west border of Mare Crisium soon will be one of the brightest spots on the lunar surface. Already a bright area may be observed there, and within it lies the small Class 1 PROCLUS, the interior of which is ⅔ black but ⅓ exceptionally brilliant. It is 18 miles in diameter and 11,900 feet deep, and its walls have a reflecting power of 16 per cent—an unusually high value for the moon. From Proclus east to the mare the whole region will brighten. On the next few nights two rays from the crater will be seen to widen and lengthen until they form conspicuous fans of light extending fully 200 miles to the south and north, respectively. The region is well worth a look every evening as it develops.

By now the sky should be dark enough to permit the moon to shine through with undiminished brilliance, and a few fainter objects of interest may be resolved. Look again at Mare Crisium. A line from Langrenus to Cleomedes cuts off the west quarter of the plain where it also passes over two black specks that give the mare the appearance of a head with eyes gazing languidly at you. The south crater is Class 1 PICARD, 16 miles in diameter and 7600 feet deep. The north one is Class 1 PEIRCE, 12 miles in diameter and 5600 feet deep. East of Proclus, at the west shore of Mare Crisium, stands the great gate formed by a gap between two needle-like mountain ridges. The south one is LAVINIUM PROMONTORY, and the north one

is OLIVIUM PROMONTORY. They rise to heights above the mare plain of 5600 and 6900 feet, respectively. Can you resolve them? Just east of the gate a light patch on the darkened plain adds a snout to the head of a fat, sleepy, black bear.

FIVE-DAY MOON

(Chart II)

This evening Mare Nectaris is revealed completely. It lies just east of the terminator and about ⅖ the way from the south to the north cusp. North of it, at the center of the crescent, Mare Tranquillitatis continues to grow and now appears to have an appendage to the north. The latter is the east edge of elliptical MARE SERENITATIS, which is believed to be one of the older if not the oldest of the maria. An outstanding feature of the five-day moon is the magnificent Class 1 crater THEOPHILUS (Fig. 5). If it hasn't already caught your eye, Theophilus is at the terminator, and it marks the confluence of Mare Nectaris and the southern extension of Mare Tranquillitatis. Its distance from the south cusp is between ⅓ and ½ the terminator length. With diameter of 65 miles and depth of 22,300 feet, it has an unusually large multiple central mountain which may be seen through binoculars as soon as the sun brightens enough of it. At no place is the interior of this large crater flat. Except for the interruption of the central peak, the internal curvature is parabolic. That fact, incidentally, indicates the difficulty and the likelihood of error involved in the determination of mountain heights from measurements of shadow length along crater floors which often are far from flat. The interior of the majestic crater is entirely black tonight except for a pin point of light that marks the summit of the central peak. Yet the formidable gray outer east wall, which Baldwin finds to be 16 miles thick, and the brilliant inner west wall form a striking picture of this major feature. A bright, broad ridge runs south from the wall and outlines an odd dark depression about ⅓ the size of Theophilus and about one Theophilus diameter removed from it. Although given the crater designation CYRILLUS F, the enclosure seems to be more of a square valley bounded by mountain ridges. Actually it is rectangular, 20 by 30 miles on its sides and 8200 feet deep. Theophilus is the most conspicuous of three large, connected craters, but the other two probably are concealed this evening in the darkness to the southwest. I shall describe the companions here because the three belong together. Moreover, if the current lunation is running four hours or more earlier than the average cycle, all three are visible right now.

Joining Theophilus on the southwest and intruded upon by it is the somewhat smaller and rather square-appearing enclosure CYRILLUS (Fig. 5). That Cyrillus is a great deal older than Theophilus will be apparent tomorrow night when its broken and dilapidated walls are compared with the sharp and clearly defined walls of Theophilus. It is 62 miles in diameter and 11,800 feet deep, and Baldwin has placed it in his very old Class 3. Its central peak is smaller and less conspicuous than that of Theophilus, but binoculars should reveal it later tonight shining like a star against the black background of the floor. Running south through and beyond the broken wall of Cyrillus is a black valley, bordered by contrasting mountain ridges, about 20 miles long and nearly that wide. The passage leads into the still older crater CATHARINA, of the extremely ancient Class 4, 64 miles across and 9200 feet deep (Fig. 5). Its massive, almost linear outer east wall can be seen tonight, and its battered floor, which has no mountain, may be inspected under a little higher sun. That striking group of three major craters representing three different epochs of formation has much to tell us if we can decode its story. Although its details are beyond the reach of our little instrument, it presents a splendid spectacle through binoculars, and it is well worth observing every night as the changing altitude of the sun brings surprising changes in its appearance.

South of Theophilus one Crisium length we see a bright, slightly wavy line extending 100 miles northwest to the terminator and 50 miles beyond it. That unusual feature is the sunlit face of the ALTAI SCARP, the rest of which will be seen tomorrow night (Fig. 48). A gigantic, irregular cliff, it averages one mile in height but here and there rises to 2½ miles. While it is now very bright, it will appear black when viewed during the waning gibbous phase, a relationship that tells us its lower level is toward the sun tonight, that is, on the Mare Nectaris side. Its surprising parallelism to the mare shore is another important clue to the story of the moon's surface. At the southeast end of the Altai Scarp may be seen the fine Class 1 crater PICCO-LOMINI (Fig. 48). It is 54 miles in diameter and

11,800 feet deep, and it stands out prominently with its bright floor framed by a brilliant inner west wall and a black inner east wall. Its bright central peak should be visible through binoculars.

Just south of the Altai Scarp a huge crater 100 miles in diameter and 9200 feet deep stands out conspicuously with the terminator along its west wall. It is visible only in bright outline since the interior is black except where the bright crest of a small intruding crater rises out of the darkness of the western floor. The larger crater is so ancient and so close to obliteration that it can be seen only when at the terminator as it is tonight. It has no name, but the smaller intruder is Class 1 LINDENAU, 29 by 33 miles across and 8900 feet deep (Fig. 48).

About two Crisium widths south of Piccolomini and one Crisium width southwest of the Metius-Fabricius-Janssen intersection we find again the sharp crater Vlacq. It stands out strongly with its bright and black border, and with luck you might be able to resolve its small central mountain. Tangent to Vlacq on the west and extending west almost to the terminator is the large distorted enclosure HOMMEL (Fig. 49). The main crater, of Class 3, is elliptical, 90 miles long, 75 miles wide, and 9200 feet deep. It contains several craters, the two largest of which are HOMMEL A and HOMMEL C, both 33 miles in diameter and 5900 feet deep. They seem to be fitted almost exactly at the east and west edges of the broad floor. Tonight they are difficult to detect since the east third of the main floor is dark, but look again for them under higher sun. See then if Hommel doesn't resemble a mold for the traditional two eggs fried sunny side up.

Almost tangent to Hommel on the north is the smaller but more prominent crater PITISCUS, 50 miles in diameter, 10,200 feet deep, and of Class 2 (Fig. 49). Two more large craters attract our attention tonight in the far south at the terminator. About one Hommel-Vlacq length southwest of the former, the two are partially revealed in the dim zone. The smaller one to the east is MUTUS, of Class 2, with a diameter of 50 miles and a depth of 9500 feet (Fig. 49). Its black interior is separated from the dark side by a bright but narrow west wall crest. Tangent to Mutus on the southwest and appearing as a long, narrow, elliptical bite out of the disk is the larger Class 1 MANZINUS, 55 by 63 miles across and 9800 feet deep. By tomorrow night it can be seen in full, except for the black interior on which the sun rises very slowly, and the two craters then will appear elevated above their nu-

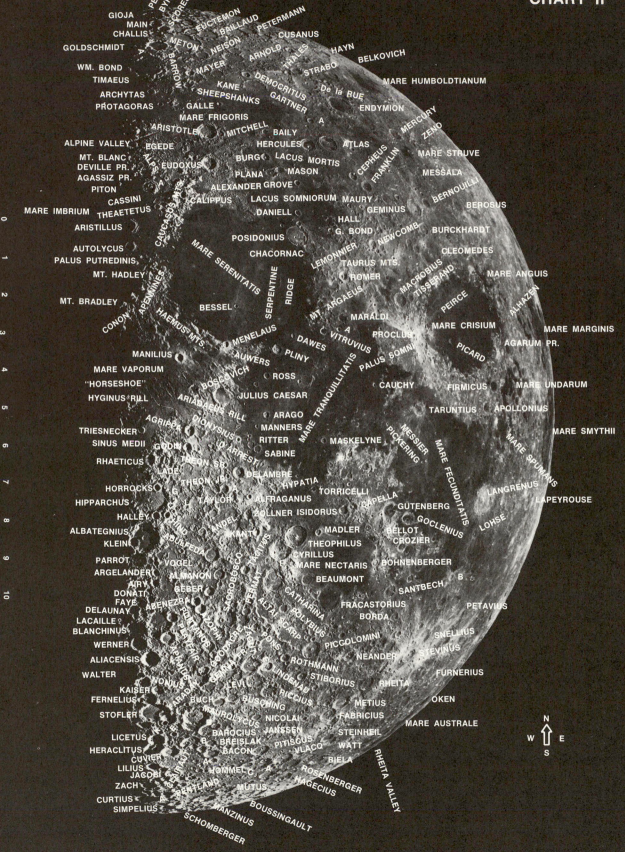

CHART II

merous lesser neighbors. The older Mutus has several floor craters, two of which might be resolved under higher sun since they have diameters of 14 and 10 miles.

If you got lost on this safari through the tangled crater jungles of the southern hemisphere, don't be discouraged. That is a common experience, one might even say the usual experience, the first time —or two, or three—around. If you made it, congratulations! The southern hemisphere is the most difficult part of the moon to navigate, and the closer one approaches the south limb, the tougher it gets. I have been lost there many times, and sometimes when I compare what I see close to the limb with what I find on certain hand-drawn charts, I have the strong suspicion that some of the trail blazers got lost there, too. The important point is not to give up if the going seems rough. There are plenty of conspicuous, important, and interesting lunar features which you can find and identify. Even if you find only a few of the more prominent ones on the first trial, you have your foot in the door a long way. Using those few landmarks and the charts, you can move from them to adjacent features and gradually work your way across the whole face of the moon. To prove that this is possible, let's move northward into less difficult territory.

On the south shore of Mare Nectaris may be seen a horseshoe-shaped "bay" delicately outlined in white and nearly washed out because of lack of shadows. This is all that remains of FRACASTORIUS, once a great crater 73 miles in diameter (Fig. 48). Evidently it was melted down by the lava flow which formed the adjacent mare. The old walls still rise to a maximum of 5800 feet above the flat new floor, but much of the north wall is destroyed. Where it stood the mare reveals only a series of low hills, ridges, and craterlets visible, at times, to the observer using more optical power than we have in hand tonight. Yet through binoculars Fracastorius often gives the illusion of a complete ring. Does it look that way to you? In the words of Percy Wilkins and Patrick Moore (The Moon, 1961): "This is one of the finest examples of a partially ruined ring on the lunar surface." Leaving it, we move to the highlands on the north edge of Mare Nectaris where two smaller craters show well, half dark and half bright. The west member of the pair is the old Class 3 ISIDORUS, 25 miles in diameter and 9900 feet deep. Its east wall has been intruded upon by that of the more prominent CAPELLA, 31 miles across and 10,500 feet deep. Both are very old, similar in dimensions, rather irregular in outline, and considerably damaged by later events. Yet Isidorus has no central mountain while Capella has an unusually large pyramidal one.

Just beyond the terminator two Crisium lengths north of Theophilus and at the confluence of Mare Tranquillitatis and Mare Serenitatis stands PLINY, a Class 1 crater 27 miles in diameter and 10,500 feet deep. Its narrow but bright crest, a dim circlet of light, makes it visible before sunrise on the dark plain even though its interior is black. A few nights hence it will show much better owing to the brightness of its floor compared with the dimness of the surrounding plain. Its small central peak has been demolished to a considerable extent by the formation of an unusually large summit crater.

On the northeast shore of Mare Serenitatis, near the terminator, and at the northwest corner of the Taurus Mountains, we find the large, ancient, mountain-walled plain POSIDONIUS (Fig. 11). Here is another partially melted and filled Class 5 crater, 52 by 61 miles across but only 8500 feet deep. The inner east wall is a thin black line, the west wall casts narrow black shadows on the mare plain, and the broad, flat floor is well illuminated even though the altitude of the sun is quite low. Posidonius resembles a light, elliptical pancake on the dark mare plain. A small telescope shows some of the major features of the variegated floor, and each increase of aperture reveals additional smaller details of the complex enclosure. Under higher sun the ring becomes quite bright and conspicuous. Look for it again when the moon is near full. Running south from Posidonius 175 miles along the edge of the mare, a dark ridge often may be seen through binoculars at about this phase. Parallel to it and some 50 miles west of it is the more conspicuous SERPENTINE RIDGE, an irregular light line that meanders 250 miles across western Mare Serenitatis approximately coincident with the terminator tonight (Fig. 11). While the ridge rises in places as high as 1000 feet above the level of the plain, that elevation is very small in comparison with its seven-mile width. These are good examples of lunar wrinkles or ridges, but they are easier to resolve when the terminator is a few degrees west of its expected position tonight.

Close to the terminator and one Crisium length northwest of Posidonius is the sharp little Class 1 crater BURG. It may be located also by noting again the conspicuous craters Atlas and Hercules and extending their center-to-center line west about twice their separation. Burg is 24 miles in diameter and 12,000 feet deep, much deeper than the far larger

Posidonius. It shows well with black interior and bright west wall crest, and it stands alone near the center of a small dark plain called LACUS MORTIS. The abrupt west boundary of Lacus Mortis is visible just beyond the terminator. It is a straight bright line running an apparent 52 (actual 75) miles north and south. This mountain wall, rising 6300 feet at some points, is the most prominent of six straight lines which bound the little plain. Moreover, the western half of the plain is a pentagon with Burg at one corner—a pentagon divided into five triangles by additional ridges and rills running from each corner to the center. The artificial appearance of the region has been noted and described many times, but, again, these details require higher optical power. Lacus Mortis appears to be the remains of an ancient mountain-walled plain 110 miles across which suffered flooding and the partial melting of its walls during some of the lava flows of the region. It differs from the famous Ptolemy, which we shall find three nights hence, mainly in the large postmare crater (Burg) which rises from its floor.

Before finishing up for the night take a look at the dark side of the lunar disk. The earthshine illuminated surface is conspicuous against a dark sky even though its intensity is fading from night to night. Does the bright portion of the moon seem to be part of a disk of larger diameter than that of the dimly lit portion? Why?

SIX-DAY MOON

(Chart II)

Tonight both Mare Tranquillitatis and Mare Serenitatis are revealed fully, with the exception of the extreme west edge of the latter. Those two are the darkest although the rest of the maria now are darkened sufficiently to stand out in contrast to the bright upland areas. Scattered 130 miles southwest of Pliny near the west edge of Mare Tranquillitatis are perhaps a few pulverized remains of the space probe Ranger 6 which crashed February 2, 1964, its six television eyes open but seeing nothing because of some failure of its electronic nervous system. Near the southwest edge of that vast plain lie the glorious fragments of completely successful Ranger 8 which sent back 7137 excellent pictures

on February 20, 1965, during the last 23 minutes of its flight. Another notable success in this region was the soft landing of Surveyor 5 on September 10, 1967, following a flight which for two days appeared doomed to failure. Not only did the space craft survive temperatures up to 283 degrees Fahrenheit, but it made space history by analyzing with its alpha particle spectrograph the soil cover of Mare Tranquillitatis. The material resembles terrestrial basalt, and its composition is oxygen 58 per cent, silicon 18, aluminum 6, magnesium 3, plus measurable amounts of iron, nickel, calcium, carbon, and sodium. Surveyor 5 also found time to register and televise 18,006 pictures.

Mare Serenitatis is seen to be sharply bordered on the southwest by the HAEMUS MOUNTAINS which end 100 miles beyond the terminator, about where they disappear tonight and where their foothills merge with those of the greater Apennine Range still in darkness to the west. Although the Haemus Mountains are more impressive than any so far encountered, the range appears to have been destroyed in part. Close inspection with powerful telescopes shows that many of the "peaks" are but truncated stumps, and the whole region looks as if it had been raked by a gigantic harrow. Thus it seems likely that at some time in the ageless past there stood here numerous mountains rising considerably above the 10,000-foot elevations measured in modern times. Near what tonight appears to be the middle of the range may be seen the small but prominent Class 1 crater MENELAUS, 16 by 19 miles across and 8700 feet deep. The sunlit west inner wall shines brilliantly, contrasting with the black floor. Later we shall see that it exhibits near the full phase a ray system reaching outward more than 100 miles.

From the base of the Haemus Mountains the surface of Mare Serenitatis curves downward rapidly, according to Baldwin who has measured lunar topography with the highest precision. Judging from his contour map, the mare elevation drops off something like 5700 feet in a distance of 10 to 15 miles from shore. It then maintains essentially that same elevation throughout the rest of its area, an area as large as New Mexico. The northwest boundary of Mare Serenitatis is marked conspicuously by the CAUCASUS MOUNTAINS through which the terminator passes tonight (Fig. 8). Most of the range is beyond the terminator, but the sun is shining on the peaks which may be seen gleaming out of the blackness of the dark side. Here we view real mountains, some of which rise to heights of

17,000 feet above the surrounding plains. The Haemus Range on the southwest and the Caucasus Range on the northwest seem to reach around the mare like two great arms.

In the uplands to the north, pointed out by the crest of the Caucasus, are two prominent Class 1 craters you can't miss. The smaller southern one is EUDOXUS, 40 by 43 miles across and 14,300 feet deep (Fig. 8). The other is ARISTOTLE, 55 miles in diameter and 12,000 feet deep (Fig. 8). Both are notable for their unusually massive walls. Those of Aristotle are definitely hexagonal rather than circular in outline, and they are buttressed by a system of radial ridges that may be followed, under excellent conditions, out from the crater a distance equal to its diameter in many directions. Eudoxus also exhibits some of these properties but to a less degree. Just beyond Aristotle to the north is a broad, narrow, smooth area which sometimes shows a little of the dark shading that is supposed to characterize the maria. That is the eastern portion of MARE FRIGORIS, the longest and the narrowest of the major lunar plains. Its total length is almost equivalent to the combined widths of the three maria between the Caucasus Mountains and the bright ray crater Langrenus near the east limb. Yet it narrows in places to no more than the width of Aristotle. Tonight Mare Frigoris is almost as bright as the highlands which bound it, and only its relative smoothness sets it apart.

An outstanding feature of tonight's moon is the huge crater MAUROLYCUS at the terminator two Crisium lengths southwest of Theophilus (Fig. 51). Since the terminator cuts through the crater, the interior is black but is bordered on the west by the sunlit crest, a delicate semicircle of light arching out over the dark side. Seventy-three miles in diameter and 16,700 feet deep, Class 2 Maurolycus dominates the vast crater-strewn continent of the south and quickly catches the eye. The multiple central mountain soon may be seen through binoculars as the sun reaches it, and the area is well worth repeated observation as it changes from night to night. Tangent to Maurolycus on the southeast is another Class 2 crater, BAROCIUS, 53 miles in diameter and 11,300 feet deep (Fig. 51). It appears unusually elongated to the northeast because its wall at that point has been destroyed by Class 1 BAROCIUS B, 25 miles in diameter and 4400 feet deep. The distortion is enhanced by the presence of another small crater wedged between Barocius B and Maurolycus and probably older than any of

its larger neighbors. Tonight the bright walls of the two large craters trace out a well-formed but backward letter "S."

Just south of Barocius is a curious group of four close craters—the largest in the middle, two smaller ones almost tangent to it on the north, and a small one tangent on the south. When seen inverted (modern astronomical telescopes usually invert the image), it makes a pretty good armless snowman or gingerbread cookie man in outline. The central crater is Class 1 BACON, 43 miles in diameter and 12,000 feet deep (Fig. 51). The next largest, due north and forming the snowman's right leg, is BREISLAK, which is 31 miles in diameter and 4900 feet deep (Fig. 51). The other two, like most small craters, have no names other than the designations BACON B (left leg) and BACON A (head). Their diameters are 28 and 25 miles, respectively, and the depth of each is 4600 feet.

North of Maurolycus, a distance equal to that of Bacon to the south, lies the old and battered Class 3 GEMMA FRISIUS, 56 miles in diameter and 17,100 feet deep (Fig. 51). Although its walls are broken in at least two dozen places there is still enough left to make it and its northern intruder a conspicuous object on the moon tonight. The intruder is GOODACRE, 28 miles in diameter and 10,500 feet deep, and it is old enough to have taken quite a beating itself. The interiors of both craters are joined in common blackness while the southwest wall of Gemma Frisius and both wall crests of Goodacre are bright. Our impression is that of a flask-shaped double crater similar to Barocius but more prominent.

East of Gemma Frisius, a distance equal to that of Maurolycus to the south, is another pair of ancient craters tangent in a north-south line. They appear about the same size, and both have numerous floor and wall craters. The north one, ZAGUT, is 50 by 55 miles across and 9800 feet deep, and it has a much shallower 23-mile-diameter crater, ZAGUT E, in its east wall and floor (Fig. 48). The south one, RABBI LEVI, is 43 by 50 miles across and 12,000 feet deep, and its floor is crossed by a line of sharp seven-mile craters as if it had been raked by the fire of a gigantic machine gun. Continuing east and a little south by the same unit distance just used, we come to the smaller but well-formed Class 1 STIBORIUS, 27 by 31 miles across and 11,200 feet deep. When near the sunset terminator, Stiborius readily is seen to lie within an almost obliterated crater twice its size and sharing with it a common east wall. This is evidently another extremely old feature, perhaps even older than the vast mountain-walled plain Janssen 100 miles to the south.

About halfway between Catharina and Gemma Frisius may be seen another prominent ancient crater—SACROBOSCO, 64 miles in diameter, 9200 feet deep, and assigned to Class 3. Among others, it has three relatively new floor craters 8 to 11 miles in diameter. However, it is of interest mainly because of its appearance under low sun when it combines with a similar but even older crater remnant to the north and resembles the darkened outline of a great sunken race track.

Near the terminator, ½ Crisium length south of Menelaus and the Haemus Mountains, and just west of Mare Tranquillitatis' shore, stands the large, "ruined" crater JULIUS CAESAR (Fig. 10). It is almost equidistant from Theophilus and Posidonius. About 55 miles long and 45 miles wide, it exhibits a low, broken west wall some 4000 feet high which shows well under the rising sun. On the south and east the walls have been removed completely in places, and nowhere do they attain an elevation as great as 2000 feet. A long black gash may be seen bordering the northeast wall remnant and running from northwest to southeast. Perhaps you can detect one or more similar but smaller grooves along the shore line to the north. It seems likely that Julius Caesar was once a far more conspicuous crater than it is today and that the great forces which produced the grooves also largely destroyed its walls. Here we find additional clues to some of the astounding events in the moon's violent past. On a line from Julius Caesar to Catharina and nearly halfway there is seen the smaller but more prominent crater DELAMBRE, 32 miles in diameter and 11,500 feet deep (Fig. 24). It stands out well, a Class 2 enclosure, in the rough foothill region, and its bright inner west wall contrasts with the black-shadowed east floor. Its walls are unusually massive, being seven miles thick at the base.

Several craters previously observed also show well tonight, some of them better than when introduced, since they are illuminated fully. Proclus in the mountains west of Mare Crisium shines brilliantly, and its fanlike north and south rays are easily seen. Theophilus and its older neighbors Cyrillus and Catharina are strong features as are also Vlacq, Pitiscus, and Hommel. Southwest of the last group look for Manzinus and Mutus. Pliny shines brilliantly on the plain east of the Haemus Mountains.

SEVEN-DAY MOON

(Chart II or III)

An outstanding feature tonight is the great hexagonal mountain-walled plain ALBATEGNIUS, close to the terminator and about ⅓ the way from the south to the north limb. The black shadow of the east wall still covers the edge of the floor, and you may be able to see the small white central peak through binoculars as it stands out against the dark stained floor. The inner west wall is bright. Eighty-one miles in diameter and 14,400 feet deep, the crater reveals a noncircular outline which is unmistakable, and its broad, flat floor of dark material indicates that it has been filled partially with lava. Tonight it stands out in bold relief, but, like other excavations of its type, it will fade away as the sun rises higher and virtually disappear around full-moon time. Albategnius is an old formation that has acquired numerous wall craters, the largest of which is seen extending from the southwest wall almost to the central mountain. That is KLEIN. It is 30 miles in diameter and 6600 feet deep, and it has a low wall on the northeast which indicates that it likewise has experienced the lava-filling process. In addition to its prominence when near the terminator, Albategnius has the distinction of being the first lunar object illuminated and detected by red laser radiation from the earth. That historic experiment was performed on May 9, 1962, by Louis Smullin and Giorgio Fiocco of the Massachusetts Institute of Technology.

Just north of Albategnius is the similar, larger, but far older and more dilapidated enclosure HIPPARCHUS. That vast hexagonal mountain-walled plain, which shows well only briefly as the terminator passes across it, is seen readily, but by tomorrow night it will have faded considerably. It is 83 by 89 miles across but only 7500 feet deep owing to partial flooding and other destructive forces that have been at work upon it. Numerous smaller craters have eroded the wall remnants and pitted the floor, and the ancient crater is well on its way to obliteration. The largest interior crater is seen easily near the north edge of the sunlit floor. It is HORROCKS, 17 by 19 miles across, of Class 1, and 9200 feet deep—more than 1500 feet deeper than Hipparchus. Its wall is bright, and its floor is hidden in black shadow. The slightly larger HALLEY breaks the south wall of Hipparchus. A Class 2

CHART III

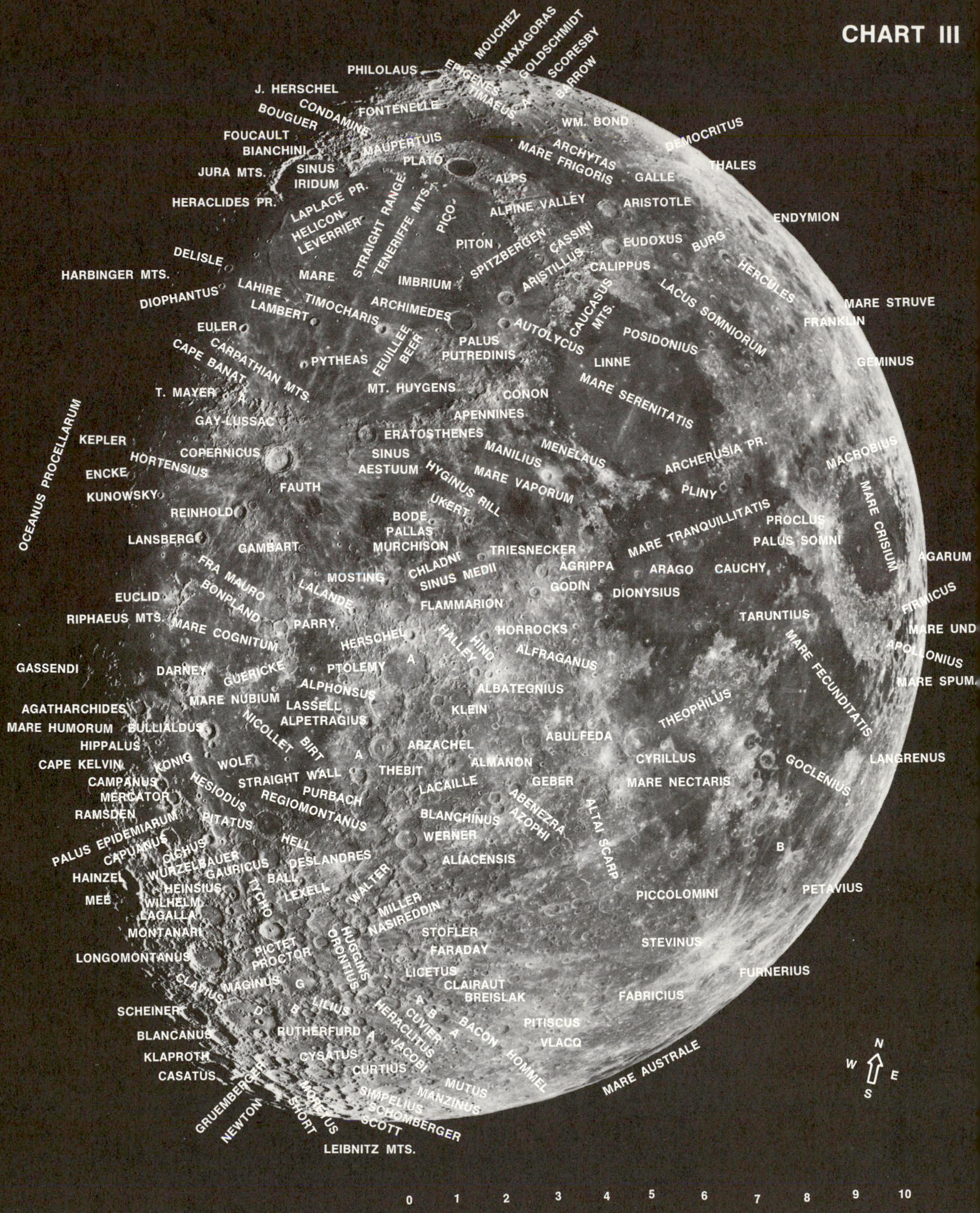

MOUCHEZ
ANAXAGORAS
GOLDSCHMIDT
SCORESBY
BARROW
PHILOLAUS
EPIGENES
TIMAEUS
A
J. HERSCHEL
CONDAMINE
FONTENELLE
BOUGUER
FOUCAULT
BIANCHINI
MAUPERTUIS
WM. BOND
DEMOCRITUS
ARCHYTAS
MARE FRIGORIS
PLATO
JURA MTS.
SINUS
IRIDUM
GALLE
THALES
ALPS
HERACLIDES PR.
LAPLACE PR.
STRAIGHT RANGE
TENERIFFE MTS.
ALPINE VALLEY
ARISTOTLE
HELICON
LEVERRIER
PICO
ENDYMION
PITON
EUDOXUS
BURG
SPITZBERGEN
CASSINI
DELISLE
ARISTILLUS
CALIPPUS
HERCULES
HARBINGER MTS.
MARE
IMBRIUM
LAHIRE
CAUCASUS MTS.
LACUS SOMNIORUM
DIOPHANTUS
TIMOCHARIS
ARCHIMEDES
LAMBERT
MARE STRUVE
FRANKLIN
EULER
FEUILLEE
BEER
AUTOLYCUS
POSIDONIUS
CARPATHIAN MTS.
PALUS
PUTREDINIS
CAPE BANAT
PYTHEAS
LINNE
GEMINUS
MT. HUYGENS
T. MAYER
CONON
MARE SERENITATIS
OCEANUS PROCELLARUM
GAY-LUSSAC
APENNINES
ERATOSTHENES
KEPLER
COPERNICUS
MANILIUS
MENELAUS
ARCHERUSIA PR.
MACROBIUS
HORTENSIUS
SINUS
AESTUUM
ENCKE
HYGINUS RILL
MARE VAPORUM
FAUTH
KUNOWSKY
PLINY
MARE CRISIUM
UKERT
REINHOLD
BODE
PALLAS
PROCLUS
MURCHISON
MARE TRANQUILLITATIS
PALUS SOMNI
LANSBERG
GAMBART
TRIESNECKER
AGRIPPA
ARAGO
CAUCHY
AGARUM
FRA MAURO
CHLADNI
MOSTING
SINUS MEDII
GODIN
DIONYSIUS
BONPLAND
LALANDE
EUCLID
FIRMICUS
FLAMMARION
TARUNTIUS
RIPHAEUS MTS.
PARRY
MARE UND
MARE COGNITUM
HORROCKS
HERSCHEL
HIND
HALLEY
MARE FECUNDITATIS
ALFRAGANUS
APOLLONIUS
GASSENDI
DARNEY
A
GUERICKE
PTOLEMY
MARE SPUM.
ALPHONSUS
ALBATEGNIUS
AGATHARCHIDES
MARE NUBIUM
LASSELL
KLEIN
THEOPHILUS
ALPETRAGIUS
MARE HUMORUM
BULLIALDUS
NICOLLET
ABULFEDA
HIPPALUS
BIRT
ARZACHEL
A
CYRILLUS
LANGRENUS
CAPE KELVIN
KONIG
WOLF
ALMANON
GOCLENIUS
THEBIT
STRAIGHT WALL
CAMPANUS
GEBER
MARE NECTARIS
HESIODUS
LACAILLE
MERCATOR
PURBACH
ABENEZRA
ALTAI SCARP
RAMSDEN
REGIOMONTANUS
BLANCHINUS
AZOPHI
PITATUS
PALUS EPIDEMIARUM
HELL
WERNER
CICHUS
B
CAPUANUS
DESLANDRES
ALIACENSIS
WURZELBAUER
GAURICUS
HAINZEL
BALL
WALTER
HEINSIUS
TYCHO
LEXELL
PICCOLOMINI
PETAVIUS
MEE
WILHELM
LAGALLA
MILLER
NASIREDDIN
STOFLER
MONTANARI
PICTET
FARADAY
STEVINUS
LONGOMONTANUS
PROCTOR
HUGGINS
ORONTIUS
LICETUS
FURNERIUS
CLAVIUS
MAGINUS
G
CLAIRAUT
BREISLAK
FABRICIUS
SCHEINER
D
B
LILIUS
A
B
CUVIER
HERACLITUS
PITISCUS
BLANCANUS
RUTHERFURD
BACON
A
VLACQ
KLAPROTH
CYSATUS
JACOBI
HOMMEL
CASATUS
CURTIUS
MUTUS
MARE AUSTRALE
GRUEMBERGER
SIMPELIUS
MANZINUS
NEWTON
MORETUS
SHORT
SCHOMBERGER
SCOTT
LEIBNITZ MTS.
N
W
E
S
0
1
2
3
4
5
6
7
8
9
10

crater, it is 22 miles in diameter and 9000 feet deep. Just east of Halley in the outer wall of Hipparchus is the Class 1 HIND, 17 miles in diameter and 9200 feet deep, which the binoculars should reveal without undue effort. If you want to continue this game of diminishing diameters, look immediately northeast of Hind where you may be able to see the bright west wall of the sharp new little crater HIPPARCHUS C, 11 miles in diameter and 8700 feet deep. If successful, move on in the same direction the same distance to bright HIPPARCHUS L, 8 miles in diameter and 7100 feet deep. Return to this spot near full moon and you are sure to find Hipparchus C since it then will appear very bright as the center of a minor ray system.

One Crisium length north of the north wall of Hipparchus we see MARE VAPORUM, a smooth dark plain somewhat smaller than the others, and an area in which Baldwin has identified numerous different lava flows (Fig. 7). Some are marked by dark stains, easily seen with binoculars. The most prominent is found in the southeast where it takes the form of a "U" or horseshoe mark about 75 miles long and 50 miles wide. Through a small telescope it resembles a vast strip mining region bounded on the west and south by the Hyginus Rill. One Crisium width southwest of Mare Vaporum and cut in two by the terminator lies an even smaller smooth plain known as SINUS MEDII. It covers some 13,000 square miles, the combined area of Massachusetts and Connecticut, and, as might be inferred from the name, it contains the adopted center of the lunar disk—the origin point from which latitude and longitude are measured. No visible feature marks the central point, but it is located near the midpoint of the south shore. The plain is also of interest as the spot selected in 1930 by Edison Pettit and Seth Nicholson for the measurement of surface temperature at full moon. Their precise observations yielded a value somewhat above that of boiling water. While the figure was lower than those of certain earlier observers, it surprisingly confirmed within 1 per cent the pioneer measures of Lord Rosse begun in 1868!

A century later measurements were in progress on Sinus Medii which would have astounded not only Rosse but Pettit and Nicholson as well. Surveyor 6 set down gently near the center of the small plain on November 9, 1967, dug into the soil, and analyzed the excavated samples with its "jewel box" (alpha particle spectrograph). It confirmed the physical and chemical measures of Surveyor 5 located 400 miles to the east. A week later, on orders from the Jet Propulsion Laboratory, Surveyor 6 made the first intentional take-off from the moon by firing the jet thrusters on its three legs, rising 10 feet, and landing 8 feet away after a flight of 6½ seconds. That historic maneuver was followed by the televising of 4000 pictures for stereoscopic combination with the 12,000 taken the first week.

At the terminator west of the Caucasus are two splendid Class 1 craters. The larger one on the north is ARISTILLUS, 36 miles in diameter and 10,500 feet deep (Fig. 42). Tonight it appears as a bright circle, broad on the east and thin on the west, with its interior completely hidden in black shadow, but with higher sun you may be able to detect some of the many radial ridges that appear to reinforce its eight-mile-thick walls. Under excellent conditions you also might see its peculiar multiple-peaked central mountain. After a few more nights its bright rays extending outward some 200 miles will be evident. The smaller crater on the south is AUTOLYCUS, 25 miles in diameter, with walls nearly five miles thick rising 11,100 feet above the concave floor (Fig. 42). It lacks the wall buttresses and central mountain of its northern companion, but it has its own ray system which Baldwin has traced 135 miles from its walls. Between Autolycus and the mountains to the south is a small, dark, smooth area 75 miles long and 45 miles wide bounded by ridges or mountains and shaped roughly like a grand piano (mouse's eye view from the floor beneath). The small plain, which is dim and hard to trace tonight, has been given the highly imaginative if somewhat unesthetic name PALUS PUTREDINUS (Fig. 42). Not far from the center of the "Rotten Swamp" a tiny black dot was reported by several European astronomers on the night of September 13, 1959, a few minutes after the predicted time for the crash landing of the Soviet space probe Lunik 2 in that general area. The dot expanded and faded for five minutes, being about 25 miles in diameter when last seen.

The plain surrounding Aristillus, particularly that to the north, has long been known as PALUS NEBULARUM, but the special designation was dropped since the area is but an open portion of Mare Imbrium, a plain so large that the terminator requires four days to cross it. On the Palus Nebularum north of Aristillus, a distance twice that of Autolycus to the south, stands a conspicuous and unusual feature. It is PITON, one of the very few isolated mountains found on the moon (Fig. 9). Rising 8200 feet from a rather bright plain, it shines brilliantly as a tiny triangle at the terminator.

East of Piton, at the edge of the palus, is a crater fully as large as the prominent Aristillus (36 miles in diameter) but difficult to see since its interior has been filled and its walls have been melted down almost to ridges which rise no more than 3500 feet above their surroundings. That is the very old Class 5 crater CASSINI which may once have had a central mountain (Fig. 9). Where such a peak may have risen we now find the Class 1 crater CASSINI A, 10 by 12 miles across and 5100 feet deep. Sometimes the latter actually shows better than its host although it covers only $\frac{1}{11}$ as much area, but tonight you will do well to resolve it. Using a large telescope in 1952, Wilkins and Moore discovered at the bottom of Cassini A a smaller, shallower crater which in turn had a tiny pit at its center. The new feature strongly resembled the "Washbowl" which they promptly named it.

From dim Cassini northwestward beyond the terminator stretch the impressive ALPS (Fig. 9). The range rises gradually from the east through an extended series of foothills bordering Mare Frigoris, and the highest peaks are found in the western section where elevations of 6000 to 14,000 feet have been measured. If atmospheric conditions are favorable tonight we have our best opportunity to view the curious ALPINE VALLEY (Fig. 9), discussed in an earlier section. It cuts into the foothills west of Aristotle and proceeds along a straight line, which, if extended backward, would cross Mare Frigoris at its midpoint north of Aristotle. Look for a very thin black line. It is not an easy object through binoculars, but the satisfaction of seeing it makes the search worth while.

Now that we have $\frac{1}{2}$ the moon's face under observation, you may have noticed that the north polar region (area north of Mare Frigoris) resembles the much more extensive crater-torn continent between the center of the disk and the south limb. The most prominent of the ancient northern craters in view tonight is BARROW, which lies due north of the Alpine Valley, close to the terminator, and roughly halfway from the poorly defined north shore of Mare Frigoris to the limb. I must emphasize the roughness of this direction because when one is observing that far north (or south), the libration in latitude can change greatly the apparent location of the object with respect to the limb. Barrow appears as a narrow black ellipse with a bright border, 60 miles across and 7900 feet deep. It is actually egg-shaped, 63 miles long and 53 miles wide. In its southwest wall it has a sharp little crater, BARROW A, 20 miles in diameter

and 4300 feet deep, which may be too close to the terminator to be seen. Under higher magnification the walls of the larger crater appear to be unusually irregular and jagged along the crest. In fact, Barrow resembles the hole torn in a tin can by a rifle bullet as viewed from the emerging side.

In the uplands east of Sinus Medii, about halfway between Hipparchus and Julius Caesar, stands a fine pair of smaller Class 1 craters with central peaks. The larger north one is AGRIPPA, 28 miles across and 9800 feet deep. The south one is GODIN, 22 miles across but 700 feet deeper than its larger companion. Both are distorted considerably from the circular, and tonight they show well with more than half their interiors brightly illuminated. Perhaps you can detect a tiny central mountain in each crater if conditions are exceptionally good. This is a bright area. The wall of Godin has the relatively high reflecting power of 16 per cent, and near the full phase each crater may be seen to constitute the center of a bright ray system which has been traced to a distance of 115 miles from Godin and 85 miles from Agrippa. North of Agrippa and west of Menelaus is another sharply defined, irregular, Class 1 crater. That is MANILIUS, 25 miles in diameter and 9300 feet deep. It is about half brilliant and half black tonight. It, too, develops a ray system around full moon which extends outward four times its diameter. Watch it brighten over the next few nights as the sun rises higher in the black sky above it.

About halfway from Hipparchus to the south limb, at the terminator, is the well-formed and conspicuous mountain-walled plain WALTER, 78 by 88 miles across and 12,000 feet deep. It resembles a gaping black hole but the bright crests of its northwest and southwest walls appear perfectly straight and at right angles to each other. A similar relationship of the east walls shows that the enclosure is square rather than circular. The tip of the east-of-center mountain shines brightly against the black interior, but it is difficult to resolve through binoculars until more of it brightens. Walter is one of the very few out of hundreds of craters studied and measured by Baldwin which he placed in his oldest unflooded Class 4.

From a point near the northeast edge of Walter a curious straight string of four tangent craters almost equal in size stretches northwest to the terminator. They are large by general standards but are definitely smaller than Walter. The first one, with a broad bright inner west wall and about $\frac{1}{3}$ of its floor black, is ALIACENSIS, of Class 1, 46 by 55 miles across and 12,900 feet deep. Walter Goodacre (*Splendour of the Heavens,* 1925) has called attention to the considerable departure of its walls from the circular in contrast to those of WERNER, the well-formed crater next in line. Forty-three miles in diameter and 15,000 feet deep, Werner has been described by Wilkins and Moore as "one of the most circular of all lunar rings." It belongs to Class 1, of course, and the crispness of its ⅓ bright and ⅔ black figure indicates that it is considerably younger than any of its notable neighbors. Indeed it contrasts conspicuously with the very old and heavily battered BLANCHINUS, the next in line. The walls of the latter are no more than 5200 feet high and fragmentary, but they still can be traced around the black interior. Blanchinus is a double crater 45 by 51 miles across with a split-level floor, as may be noted tomorrow night when the sun reaches its interior. The eastern ⅓ is higher than the rest of the floor. The fourth crater of the string, with the terminator along its west wall, is the more distinct Class 3 LACAILLE, 40 by 44 miles across and 9100 feet deep. Its smooth floor, all of which is in shadow, ordinarily shows considerable detail but only through large telescopes. As we compare and contrast the members of this four-in-a-row configuration, we realize that in the extremely rough and heavily bombarded southern continent we are privileged to view side by side some of the oldest and some of the newest surface features.

South of Aliacensis a distance equal to the length of the four-in-a-row group may be seen near the terminator a large prominent double crater. Its main part is the mountain-walled plain STOFLER, of Class 2, 68 by 85 miles across and 7500 feet deep (Fig. 51). Although considerably larger than Maurolycus, almost tangent to it on the east, Stofler does not appear to its full advantage since its southeast wall has been transported almost into its center by a large intruding crater. The intruder, which usually may be resolved, is the Class 2 FARADAY, 33 by 44 miles across and 7500 feet deep (Fig. 51). Faraday also has several craters in its walls, the largest of which, on the southwest, has in turn been intruded upon by an almost equally large crater. Those last two, FARADAY K and FARADAY E, are 25 and 20 miles in diameter, and 5600 and 4900 feet deep, respectively. They can be seen and perhaps resolved through binoculars. Here we have an excellent example of four successive epochs of crater formation. In addition, Stofler has several smaller sharper wall craters of Class 1. Thus the Stofler-Faraday complex as it appears today

must have been built up over a long period of time.

Just south of Stofler is a strange, irregular enclosure with the outline of a bent capsule or bean. Its west inner walls are bright, and its predominantly dark interior is divided by a narrow bright longitudinal stripe through the middle. The rounded ends of the mysterious formation are craters. The larger north one is LICETUS, 47 by 50 miles across, 9800 feet deep, and of Class 1. The rest of the enclosure is HERACLITUS, the southwest end of which is a crater 31 miles in diameter and 9500 feet deep (Fig. 51). The walls of the two craters are separated by a distance of 22 miles. While there are numerous small intruding depressions to complicate the picture, the remarkable and unexplained feature of the enclosure is the absence of portions of the walls and the normal surface material between the two craters which has been removed, as if by two scoops of a Gargantuan shovel, leaving a straight ridge between the cuts. Look again at the Licetus-Heraclitus freak later when the sun has illuminated more of the interior.

Tangent to the Licetus-Heraclitus complex on the southeast is another fine Class 1 crater, CUVIER, with a brilliant west wall, a black east wall, and a bright floor (Fig. 51). It is 47 by 53 miles across, and its wall, which rises 12,100 feet above the floor, has been broken away on the north through the combined effect of several small craters of more recent origin.

At the terminator about one Crisium width south of Licetus is the most prominent crater in the limb vicinity tonight. The narrow black enclosure, more triangular than elliptical, with brilliant inner west wall, is CURTIUS, 59 by 63 miles across and 13,000 feet deep. It may appear anywhere along the terminator from a point halfway between Heraclitus and the limb to almost the limb, owing to the effect of the libration in latitude. Curtius is joined on the east by a triangular enclosure which extends another 20 miles and which combines with it in this light to give the appearance of a single nacelle-shaped crater rounded on the west and pointed on the east.

Two nights hence most of the interior of Curtius will be illuminated, and it will show well, but because of its high latitude we never get a good view of the inner surface of the near wall. At best we see it approximately edge on just as we see the normal lunar plains near the limb. Since the longitude of Curtius is only four degrees east of the central meridian we have drawn on the moon, the sun reaches its maximum altitude at the crater around the time of full moon, and its rays graze the inner surface of the near wall at very nearly the same angle as our line of sight when we view that wall. The latitude of Curtius is 67 degrees south, which means that the average maximum altitude the sun can attain at the crater is 90 − 67, or 23 degrees. The actual maximum altitude can vary from 11 to 35 degrees owing to a 5.1-degree angle between the plane of the moon's orbit and the plane of the earth's orbit around the sun and a 6.7-degree angle between the moon's axis and the perpendicular to its orbital plane. Since the inner walls of Curtius that we can observe are terraced, it is possible that the near inner wall has a few narrow strips where the upward slope is greater than 35 degrees. If so, the sun never shines on those strips, and they must remain very cold as well as dark. Ten degrees closer to the pole than Curtius, in latitude 77 degrees either south or north, the noon altitude of the sun is about 13 degrees plus or minus up to 12 degrees. Furthermore, those areas are still too far from the pole to experience the midnight-sun effect. In such locations it is quite likely that there are numerous craters, rills, pits, and cracks from which the warming rays of the sun have been excluded for several billion years. "So what?" you may well ask at this point if not before. On the answer may depend not only the exploration of the moon but the later exploitation of its resources. If internal activity, like that of the earth, long ago brought great quantities of water to the lunar surface—water which eventually evaporated into outer space—those high-latitude dark spots may shelter stores of ancient water preserved as ice.

EIGHT-DAY MOON

(Chart III)

A leading feature of tonight's moon is the finest mountain range which our satellite has produced on the readily observable portion of its surface. There are higher mountains and perhaps more extensive ranges. Beyond Schiller and over the southwest limb, for example, lie the great Doerfel Mountains visible in profile from time to time under favorable libration. On such occasions binoculars reveal at the limb some of the highest peaks rising to a measured 26,000 feet, the greatest lunar elevations recorded. Yet it is generally agreed that the APENNINES con-

stitute the outstanding lunar range (Fig. 7). Tonight they are shown in full, running from the west shore of Mare Serenitatis southwest some 450 miles to the dark side. A magnificent spectacle, they rise through a series of bright foothills to the brilliant crest set off by short black shadows along the plain to the northwest. The Apennines, which I have described previously, continue the gentle curve begun by the Caucasus to the northeast. Indeed it would appear that these are but two parts of one original range separated later by the lava flows which produced the present floors of the maria.

Nearly ½ of MARE IMBRIUM may be seen as a shaded half ellipse enclosed by the great mountain walls of the Apennines, Caucasus, and Alps on the east and the terminator on the west. On its north shore at the terminator stands the mountain-walled plain PLATO, a feature so conspicuous because of its size and the dark shading of its floor that it sometimes is seen with the naked eye (Fig. 9). The stain-darkened floor, which will show more prominently under higher sun, is enclosed between the brilliant inner west wall and a wider black shadow band on the east. Through a small telescope that shadow band, cast by mountains along the east wall, exhibits a fine structure of humps and narrow spires like the skyline of a small town with several churches. Plato is elliptical, 64 by 67 miles across, but owing to its Class 5 status it is only 8000 feet deep. The smooth floor of unique lava, containing 2700 square miles according to Webb, has been studied intensively for three centuries, and numerous unverified changes have been reported there. Plato is a striking exception to the general rule that mountain-walled plains fade away under high sun. It is conspicuous at full moon and at all other times from sunrise to sunset. Easy to find, it is well worth inspection every night.

On the south shore of Mare Imbrium, where the Apennines meet the terminator, look for another prominent crater. That is Class 1 ERATOSTHENES, 37 miles in diameter and 12,300 feet deep (Fig. 4). Tonight it shows a brilliant inner west wall, a gray outer east wall, and a black interior which hides its massive, crater-capped central mountain. A tail-like mountain ridge extends about 50 miles to the southwest. Eratosthenes is a feature of the first rank tonight, but by full moon it will have faded almost completely away. See if you can find a trace of it then. On the Imbrium plain north of the Apennines and west of Autolycus we see a fine ring plain almost as large as Plato. Named ARCHIMEDES, the Class 5 enclosure is 51 miles in diameter but only

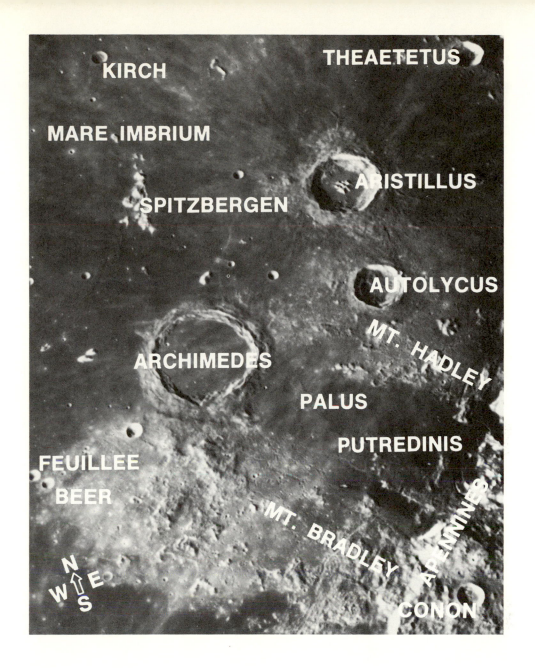

KIRCH

THEAETETUS

MARE IMBRIUM

ARISTILLUS

SPITZBERGEN

AUTOLYCUS

MT. HADLEY

ARCHIMEDES

PALUS

PUTREDINIS

FEUILLEE

BEER

MT. BRADLEY

APENNINES

N
W E
S

CONON

FIGURE 42. Archimedes in the afternoon. On September 13, 1959, Lunik 2 crashed on Palus Putredinis, reportedly at a point midway between the crater Autolycus and the label "Palus." The Soviet space probe is believed to be the first man-made device to reach the moon but possibly not the first terrestrial object to make the trip. Urey has suggested that billions of years ago, when both earth and moon were under heavy bombardment, some of the material blasted out of enormous earth craters may have traveled all the way to the moon. Note the dark stains on Palus Putredinis and the well-defined sinuous Hadley Rill along its southeast border. On July 30, 1971, the Apollo 15 lunar module landed in the area on the right margin where the rill bends northwest away from Mt. Hadley. Astronauts David Scott and James Irwin drove the first Lunar Rover to the edge of the rill. (120-inch Reflector.)

6800 feet deep (Fig. 42). Its lava floor is similar to that of Plato but much lighter and "smooth as a mirror" according to William Beer and John Madler (*Der Mond,* 1837). Appearing as a thin, delicate ring, white on the west and black on the east, it seems to have a large appendage in the form of two ridges of hills that curve off 75 miles to the southwest. North of Archimedes, a distance equal to its width, may be found a short string of white specks, a miniature mountain range 50 miles long named SPITZBERGEN (Fig. 42). Its alignment with a few similar specks to the north suggests that it may be one of the few remnants of what was once a truly enormous crater wall, later melted away by the great lava flows that gave Mare Imbrium its present floor. A peak near the middle of the range rises 4600 feet above the plain. West of Archimedes and north of Eratosthenes the terminator has begun to cross the bright-walled crater TIMOCHARIS, 22 miles in diameter and 9400 feet deep (Fig. 7). A well-formed example of Class 1, its bright east wall looms high tonight, and its black interior soon will be bounded by a narrow bright west crest. Baldwin has traced its splashy ray pattern to a distance of 43 miles from the walls at full moon.

South of Plato, a distance equal to its length, a starlike object shines brilliantly on the dim, flat plain, and just south of it may be seen a fainter one. Those are isolated mountains, and the brighter one is PICO, the best-known object of its exclusive class (Fig. 9). Pico evidently is composed of white rock having relatively high reflecting power for lunar material. Moreover, it stands alone on a rather dark plain. These factors combine to make it appear extremely bright. Tonight it casts a long, tapered shadow toward the terminator which may be resolved under excellent conditions. Examined with a small telescope, it gives the impression of a cathedral-like structure, considerably taller than it is broad. Since it is far removed from the center of the disk, its apparent breadth must be corrected for foreshortening. This makes it an actual 18 miles wide. Since measurements of the height average around 8000 feet, it is clear that Pico is far from cathedral-shaped. Its height is less than $\frac{1}{10}$ its breadth, and its true cross section is probably more like that of a thumbtack head. Pico is a good illustration of the illusion of greatly exaggerated height which we all suffer as we observe lunar features, either visually or by means of photographs, when they are near enough to the terminator to cast long shadows.

Far up along the terminator, almost to the north limb, a crater and a half catch the eye. They are located in the upland beyond the Mare Frigoris shore a distance about equal to the width of the "isthmus" on which Plato stands. The large old one with bright floor and northwest wall and dark southeast wall is GOLDSCHMIDT. It looks like a thin diamond, but it is actually elliptical, 71 by 77 miles across and 7500 feet deep. In its west wall, cut by the terminator, is the much younger and smaller Class 1 crater ANAXAGORAS, 33 miles in diameter and 8900 feet deep. Only its west wall crest is bright, the rest of its interior showing as a thin black ellipse projected upon the white wall of Goldschmidt. The north pole of the moon lies beyond Goldschmidt. In fact this spot is so far north and Anaxagoras is so deep that the sun may never reach the southern part of its inner wall. Strangely enough, the area is exceptionally bright, and you can't miss it if you look for it around full moon. The two craters are then brilliant, with the exception of the shadowed south edge of Anaxagoras, and from them emanate bright rays that can be traced several hundred miles.

Let us now swing southward, stopping first in the newly revealed dark area south of Eratosthenes and the western Apennines. That is SINUS AESTUUM, noted particularly for its exceptionally smooth floor, the east and north parts of which are stained deeply (Fig. 7). Here again Baldwin has identified numerous different lava flows, and he describes the region as "a vast field where old markings are softened and muted." He then goes on to suggest that at one time the area was submerged completely in molten lava which soon flowed back into the moon's interior before it could do more than melt away the outer layers of the old surface features which were there before it came and remained after it had gone.

In the uplands southeast of Sinus Aestuum, between Mare Vaporum and Sinus Medii, may be seen an odd capsule-shaped enclosure, 65 miles long and open at the southeast end. It is a combination of two very old and distorted craters with crumbling walls which appear bright on a light background tonight and provide only a minimum of contrasting shadow to set them off. The larger one to the east is MURCHISON, 35 by 40 miles across and some 3000 feet deep. The somewhat younger one to the west is PALLAS, 30 miles across and 5800 feet deep. Its central mountain and battered east wall may be resolved at times, but actually those features differ little in size and organization from the bright rubble with which the floors of both craters are littered.

About halfway between the Apennines and the south limb is a striking group of three major craters resembling to some extent the Theophilus-Cyrillus-Catharina group except that it is larger. The smallest and youngest of the three at the south end is ARZACHEL, of Class 3, 59 miles in diameter and 13,000 feet deep, with massive terraced walls 10 miles thick. The sunlit floor is framed between a brilliant inner west wall and a black east one, and the bright "central" mountain, which stands five miles southwest of the crater center, should be resolved. North of Arzachel lies the great mountain-walled plain ALPHONSUS. It also shows well in tonight's light, but its much smaller central mountain is difficult to see. Alphonsus is a very old, partially flooded, Class 5 crater 64 by 73 miles across and 10,500 feet deep (6600 feet from analysis of Ranger 9 photographs). Several years before the confirming flight of Ranger 9 Shoemaker found that some of the small craters which occur along rills on the floor of Alphonsus are surrounded by diffuse dark haloes extending from 2½ to 3½ miles beyond the rims (Fig. 28). He observed that they resemble terrestrial volcanoes of the maar variety and that the haloes may be owing to thin layers of volcanic ash. Look again for the little central mountain, and try to make it out. It marks the one-and-only spot on the surface of the moon where an astronomer has observed a change substantiated by photographic proof.

On the night of November 2, 1958, the Russian astronomer Nikolai Kozyrev was observing this mountain with the 50-inch reflecting telescope of the Crimean Observatory. His interest in it had been aroused by the observations of Dinsmore Alter who, on October 26, 1956, had taken a series of photographs with the 60-inch Mount Wilson reflector. On one of them the floor of Alphonsus east of the mountain was obscured as if covered by a patch of haze. Kozyrev was looking for and hoping to find just such a cloud. The process appears to have been somewhat akin to that of the hunter who goes out into the night during the migration season and fires his shotgun up into the sky from time to time on the chance that he might hit a duck flying over. We don't know how many nights the Russian astronomer spent photographing a change-less Alphonsus, but we do know from his account in *Sky and Telescope*, February, 1959, that on this

particular night he had attached the spectrograph to the telescope and he had set the central mountain at the middle of the entrance slit in order to photograph the analyzed light pattern of the radiation coming from the mountain. During such observations one must look constantly through the guiding eyepiece and make minute adjustments of the telescope from time to time in order to compensate for atmospheric irregularities and to keep the object precisely centered on the spectrograph slit. As he carried out those rather monotonous duties Kozyrev was rewarded for his pains. Suddenly the central peak "became strongly washed out and of an unusual reddish hue," according to his report. Later, he obtained a second *spectrogram* (spectrum photograph) while the peak showed "unusual brightness and whiteness." A third spectrogram was taken when "the brightness of the peak had fallen to its normal value." The last spectrogram was entirely normal, but the first two showed unmistakable evidence of a cloud of gas over the mountain composed of various carbon molecules. It now appears that this was not a volcanic eruption but a milder type of disturbance which geologists call "outgassing." Nevertheless, Kozyrev's historic observations, inspired by Alter's earlier work, proved that the moon is not a dead world and that internal changes are in progress there which occasionally result in observable manifestations at the surface.

Alphonsus will be remembered also, of course, as the target selected for the last Ranger photographic mission. On March 24, 1965, Ranger 9 plunged toward the crater floor northeast of the central peak sending back 5814 exceptionally fine pictures of the intriguing region (Fig. 26 to Fig. 34).

The north wall of Alphonsus intrudes upon the enormous, hexagonal, mountain-walled plain PTOLEMY, a Class 5 crater, 93 miles in diameter and 9800 feet deep. Baldwin estimates that the outer walls rise 8200 feet above the surroundings, and he concludes that the interior has been filled with lava to approximately the outside level. Alter, on the other hand, maintains that Ptolemy is a typical mountain-walled plain with inside but no appreciable outside walls. However, he goes on to include both Alphonsus and Arzachel in the same general category. Regardless of how Ptolemy originated, it seems evident that the present floor is a secondary one produced by lava flow. It appears quite smooth through binoculars with the exception of the bright crater PTOLEMY A, six miles in diameter and 3900 feet deep, northeast of center, but the floor abounds in telescopic detail which has been studied extensively because of its favorable location not far from the center of the lunar disk. For example, Baldwin has counted 195 craters there on photographs taken at the Lick Observatory. In addition there are many ridges and shallow depressions which can be observed only at sunrise or sunset. Tangent to Ptolemy on the north is the sharply delineated Class 1 crater HERSCHEL, 22 by 25 miles across and 12,800 feet deep. It shows well tonight, a study in contrast, 1/3 brilliant and 1/2 black with a bright outer east wall and connecting ridge running off to the north. That ridge runs into the west wall of a great groove 5 to 7 miles wide and some 65 miles long. Perhaps you can see it and other such grooves since there are many in the region.

Tangent to Alphonsus on the southwest is a small but sharply defined crater which has brought joy to the hearts of selenographers for generations. By the standards of crater beauty, its figure was considered to be just about perfect before Ranger 9 rudely revealed its defects. Approximately circular ALPETRAGIUS, of Class 1, is 25 miles in diameter and 9800 feet deep (Fig. 26). It has a relatively huge central mountain, almost as massive as that of its big neighbor Arzachel on the southeast, which is located precisely at its center and which appears to be a perfectly symmetrical dome 6200 feet high. Tonight more than half the interior is lost in darkness, but the inner west wall is brilliant. A close examination later may reveal a speck of light at the center when the sun illuminates enough of the peak. If it can't be found, try for the mountain tomorrow night.

About two Arzachel diameters, south of Alpetragius, look for Class 2 THEBIT, 32 by 36 miles across and 10,400 feet deep. Its northwest wall is broken by the younger THEBIT A, 13 miles in diameter and 9700 feet deep, the northwest wall of which likewise is broken by THEBIT L, 6 miles across and 2900 feet deep. Thebit straddles the broad, dilapidated east wall of what once must have been a gigantic crater about 150 miles in diameter. Only its east and northeast wall remnants have survived, the rest of its great ramparts having been melted down ages ago by the lava flows that formed the floor of the mare to the west. When the rays of the sun strike this area at grazing incidence, as they do tonight, an observer with a little more optical power than we are using can trace the rest of the ghost

crater by the rippling ridges that mark the plain where its walls once stood.

About one Arzachel diameter west of Thebit, and through the center of the ghost crater, you may see, with good observing conditions, a straight, black, hairlike line running 70 miles roughly parallel to the terminator. That is the STRAIGHT WALL, a feature so artificial in appearance that in the late nineteenth century it was referred to frequently as "The Railway." It is a fault line, probably similar to those along which are generated terrestrial earthquakes, and extensive subsidence along it has lowered greatly the plain on one side. Since it appears black tonight, we conclude that it is the west side that is lower. An observation around the time of last quarter will reveal it as a bright line. From a series of shadow-length measurements made a few years ago, Joseph Ashbrook found that the Wall rises steadily from the north end until a difference of elevation of 1200 feet is reached. This is maintained for some 20 miles, and beyond the midpoint there is a gradual decline to the south end. The measurements also showed that the "cliff" is by no means a vertical one, but rather a slope rising at an angle of 41 degrees to the horizontal. Even so, that might prove to be one of the steepest slopes to be found anywhere on the moon.

One Crisium width southwest of Thebit the terminator cuts through the remains of what once must have been a major lunar formation. PITATUS, described by Wilkins and Moore as "a magnificent lagoonlike ring," is 60 by 69 miles across from crest to crest although the floor is only 45 miles wide. Lava melting and other forces have reduced most of the ramparts of this mountain-walled plain to less than 2000 feet above the flat interior. Still it shows well under higher sun as a delicate bright ring encircling a dark floor and the bright speck of its tiny mountain, off center to the northwest. Tonight only the east outer wall can be seen. It rises to 2900 feet in some places, but most of it is much lower. Within a few hours the rest of the ring will appear as the sun rises over it, and by tomorrow evening the shadows will have vanished, and Pitatus will be almost washed out.

East of Pitatus lies the truly enormous mountain-walled plain DESLANDRES, the antiquity of which is indicated by the destruction of its east wall through the intrusion of Walter, a Class 4 crater itself. Originally 136 by 152 miles across, the great walls of Galileo's vast enclosure must have been reduced substantially under the equalizing forces of the ages and the destructive effect of literally hundreds of superimposed craters. Yet they still rise to 11,500 feet in places and when near the terminator, the glancing rays of the sun accentuate the wall remnants and reveal clearly the ghostly remains of a once-imposing feature. Notice the unusual brightness of the eastern floor near the outer wall of Walter. That peculiar feature, known as CASSINI's BRIGHT SPOT, will become more intense as the sun rises higher and the vague walls of Deslandres disappear entirely. It was thought at first to be a temporary white cloud by Giovanni Cassini, who noted it in 1671. In the south wall, adjacent to the bright spot, you probably will see the crater LEXELL, which is 36 miles across and 7200 feet deep. Obviously younger than its host, Lexell itself is so old that it is heavily peppered with small craters, and the north quarter of its wall has been leveled completely. Near the west edge of Deslandres' floor rises the sharp, irregular little crater HELL, 22 miles in diameter and 6600 feet deep. If you have ever wondered where in the universe the infamous place of that name is located, you will have to continue to wonder. This spot merely perpetuates the name of an eighteenth-century astronomer and *clergyman*.

The midpoint between Arzachel and Hell lies on the west wall crest of the large mountain-walled plain PURBACH, 62 by 73 miles across and 9800 feet deep. It belongs to Class 5, and the lava flow which gave it a new floor evidently melted away much of its north wall. Even when illuminated by high sun, its floor is darker than the surrounding territory. While Purbach is hexagonal, though not conspicuously so, a comparison of its outline with those of Arzachel and other nearby enclosures indicates that it actually must be nearer elliptical than circular in outline. At the south edge of Purbach and considerably encroached upon by it we find the equally large but less conspicuous mountain-walled plain REGIOMONTANUS. Evidently one of the oldest formations in the area, it has been crowded from the southwest as well as from the north, and now it presents the outline of an onion about 53 by 76 miles across with walls rising 9700 feet above the floor. With excellent conditions you might be able to resolve a bright, stubby mountain ridge running from the north wall to a point just east of the crater center. A small telescope would reveal on the summit of that ridge a crater only three miles in diameter which may be an extinct volcano.

One Crisium length south of Thebit the termi-

nator crosses TYCHO, a Class 1 crater 56 miles in diameter and 13,800 feet deep, with great walls nearly 13 miles thick (Fig. 6). Tonight it is by no means an outstanding object since its interior is completely black, but its bright west wall crest is seen easily, a thin semicircle of light extending out over the dark side. By tomorrow the brilliant interior and central mountain will be conspicuous, and already several of the great rays curve far out across the southern hemisphere. Tycho will continue to grow in prominence until it justifies Webb's extravagant accolade, quoted in a previous section. Webb further claims that it is visible to the naked eye at full moon, a statement which you might test if you have good eyesight. Tycho is undoubtedly one of the youngest of the major lunar craters. Its outstanding ray system strongly supports the impact theory of its origin. Yet it is interesting to note that aside from the effect produced by six smaller craters which almost completely encircle it, Tycho shows very little more in the way of an outside wall than do the mountain-walled plains which are supposed to have none. Its periphery section does not brighten under higher sun nearly so much as do the interior and the ray region beyond. Consequently, Tycho becomes at full a dazzling white oval surrounded by a gray ring that sets it apart from the brilliant ray pattern.

Surveyor 7, last lunar robot of its kind, landed softly on the north outer slope of Tycho January 9, 1968, at sunrise. Its mission was purely scientific since its predecessors had provided the Apollo mission planners with ample data. When the sun set two weeks later, it had transmitted more than 21,000 pictures of its rugged surroundings, determined the physical and chemical properties of the highland soil, and recorded laser beams aimed toward it from two observatories.

A line from the center of Tycho to the center of Walter touches at its midpoint the north edge of the strange kidney-shaped Class 3 crater ORONTIUS (Fig. 6). Once a normal crater 65 by 80 miles across and 9200 feet deep, its unusual shape is owing to the destruction of most of its east wall by the intruding crater HUGGINS, 45 miles across and 8500 feet deep, which, in turn, has suffered the same fate by the encroachment of NASIREDDIN, a Class 1 crater 34 miles across and 8900 feet deep (Fig. 6). The curious top-shaped group, 115 miles wide, shows conspicuously, its bright interior outlined and segmented by the brilliant arcs that mark the west walls and crests of the three craters.

About one Deslandres length south of Orontius,

and an equal distance southeast of Tycho, looms a most prominent feature, the large disappearing crater MAGINUS (Fig. 6). That great Class 3 mountain-walled plain, previously described, must have been truly spectacular in its early years since its old walls are still nearly three miles high in spite of having been broken and scattered by more than 50 smaller craters. The largest of the wall craters, MAGINUS G, is 28 miles in diameter and 7200 feet deep. It can be seen astride the southwest rampart at the terminator. Notice also the weaker crater PROCTOR, 32 by 35 miles across and 4300 feet deep, which intrudes upon the outer north wall of Maginus (Fig. 6). It is likewise ancient and battered but readily seen tonight, ¼ black and ¼ bright. Between Proctor and Orontius lies the Class 2 crater SAUSSURE, 35 by 37 miles across and 6900 feet deep. Like most of its neighbors it has no central peak, but it does have the distinction of almost filling an older crater which is about 60 miles across. Consequently, when near the terminator it appears to have double east walls or double west walls or both (Fig. 6).

MORETUS is another conspicuous crater at the terminator between Maginus and the south limb, and closer to the latter. It is the most prominent crater in its region, and it may be located by running south a line tangent to the west rims of both Tycho and Maginus. Such a line passes through Moretus at a distance beyond Maginus more or less equal to twice the Tycho-Maginus center-to-center separation. Here is a crater much younger than the last few we have seen, as its sharp Class 1 outline indicates. Its 75-mile diameter makes it an easy object, and its massive walls, 13 miles thick at the base, rise 13,800 feet above the floor. They appear to be reinforced by gigantic buttresses on the north which actually are ridges between the walls of four good-sized craters which are tangent to Moretus. Tonight the inner west wall is brilliant, and the interior is a narrow black ellipse against which shines the stellar image of the central peak. If the mountain eludes you, try it again tomorrow night when considerably more of it will be bathed in sunshine. Webb has described that peak as "the loftiest yet measured—6800 feet," and Baldwin agrees that it is one of the highest crater-contained mountains. A recent measure gives it an altitude of 6600 feet in excellent agreement with Webb's century-old survey. Before leaving the south polar region notice the much smaller but equally sharp crater apparently tangent to Moretus on the north. That is Class 1 CYSATUS, 31 miles in diameter and 10,200

feet deep. Between Cysatus and Maginus the terminator detours eastward to form a large rectangular bite out of the disk. That black hole is part of the great crater Clavius which we shall see tomorrow night.

NINE-DAY MOON

(Chart III)

It seems almost superfluous to state that an outstanding feature of the moon tonight is the magnificent crater near the terminator, a little short of halfway from the north limb to the south limb. The graceful arc of the Apennines curves toward it like an inverted index finger. The object is COPERNICUS, a gem of the first magnitude, on which selenographers long ago exhausted their store of superlatives (Figs. 4 and 43). Goodacre calls it "the grandest of all the large crater rings on the moon . . . presenting a superb spectacle under morning or evening illumination." Wilkins and Moore describe it as "the finest ringed plain on the moon's surface." Already half the floor is bright, and the shadow edge passes between the peaks of the brilliant multiple central mountain. The inner west wall is exceptionally bright, and the only difficulty experienced in resolving the central mountain is because of the fact that almost every portion of the crater shines so effectively that little contrast among its components is presented. The curious and largely bifurcated ray system is coming into view, but as yet it is weak. Perhaps you can make out a ray to the east and another to the north across Mare Imbrium halfway to Plato. The rays, which resemble a gigantic splash pattern, will brighten considerably over the next few nights. Copernicus, of Class 1, is not one of the largest craters, being 60 miles in diameter. Its depth of 12,600 feet is exceeded by many others, and its walls, 14 miles thick, are not the broadest of those Baldwin has measured. Its reflecting power is high but not the highest, being 16 per cent on a scale that ranges from 6 per cent for the darkest to 18 per cent for the brightest spots on the face of the moon. Its outstanding appearance is because of a combination of factors. They include its youth, which gives it a sharp outline (actually polygonal), a bright surface, and a conspicuous ray system; its excellent location not far from the center of the disk; and its placement on a relatively smooth, dark plain.

In recent years Copernicus and its environment have commanded the early attention of the Aeronautical Chart and Information Center, U. S. Air Force, and the U. S. Geological Survey. One of the first of the Lunar Aeronautical Charts (1961) and the prototype of the Geologic Maps of the Moon (1961) cover that region in the greatest possible detail. The latter is the work of Shoemaker who has sifted and analyzed an enormous amount of photographic and visual data from the geological point of view. He cites feature after feature in which the great crater resembles terrestrial impact craters, leaving little if any room for doubt that Copernicus is an explosion crater produced by the enormous impact energy of a large meteoritic body in collision with the moon. But much more evidently has happened in the area in addition to the great crash. Among the finer details Shoemaker finds a number of dark halo craters resembling terrestrial maars of the Zuni Salt Lake variety, a number of small domes four to six miles in diameter which look like shield volcanoes, and at least one symmetrical hill about 3½ miles across which resembles a small stratovolcano.

Copernicus is a prominent feature from sunrise to sunset, and its changes from night to night are well worth noting. Much can be seen through binoculars, and much more is revealed with each increase in instrumental aperture right up to the largest telescope in the world. There also are other features of more or less importance visible to the binocular astronomer tonight, and so we must move on. About twice its diameter southwest of Copernicus and close to the terminator is the smaller Class 1 REINHOLD, 28 miles in diameter and 9000 feet deep, with its black interior bordered on the west by a bright wall crest (Fig. 4).

The roughest portion of the Copernican plain lies to the north and northwest of the great crater. There we can follow the CARPATHIAN MOUNTAINS from a point directly north of Copernicus' east wall, 125 miles west to the terminator, and we can trace their bright peaks 25 miles beyond (Fig. 4). The range, which includes peaks 6600 feet high, appears to be a continuation of the Apennines except for a gap of some 60 miles to the west of Eratosthenes. About 50 miles west of the last peaks visible tonight, as Baldwin puts it, "the Carpathians fade away . . . and disappear beneath the lavas." It has been suggested that these mountain remnants were once the peaks of a much more massive range that constituted the original south wall of Mare Imbrium, but Kuiper believes they were formed of

extruded lava in a manner different from that of the Apennines which I shall discuss tomorrow night.

Eratosthenes would appear conspicuous tonight if it were not so close to magnificent Copernicus. Last night the interior was black except for the bright west wall, but tonight only the east wall remains dark, and Eratosthenes shows well in spite of the stiff competition. The weak central mountain is lost against the bright floor unless you have very good eyes and binoculars.

One Crisium width north of Copernicus may be seen the prominent little Class 1 crater PYTHEAS (Fig. 4). Although it is only 15 miles long, 7100 feet deep, and half black inside tonight, its bright half shows well against the smooth, dark background. Through contrast it catches the eye while craters of considerably larger dimensions often go unnoticed. It is definitely not circular, being only 12 miles wide, and the term "rhomboidal" has been used in its description. Continuing north by about ½ the last step we come to the larger but less evident Class 1 LAMBERT, 19 miles in diameter and 7900 feet deep. Lambert is made of darker material than is Pytheas, and it appears as a dark speck bright edged on the west. It stands astride a great ridge which meanders up from Eratosthenes 250 miles to the southeast and continues another 150 miles to the northwest. Tonight part of the ridge appears lighter than the mare background, and, with good conditions, you may be able to make out some of the section between the crater and the terminator. Like its slightly larger but much more prominent neighbor Timocharis to the east, Lambert under higher sun exhibits the hollow stump of a nearly central mountain. Baldwin believes that feature, which is a bit beyond the reach of our optics, resulted from the collapse of a rebound dome formed at the time of the crater's violent birth through meteoritic impact.

One Plato length southwest of Plato and closer to that crater than is Pico, look for several points of light. They are the sunlit faces of the TENERIFFE MOUNTAINS which rise to 6200 feet and show well although they lack some of the luster of Pico (Fig. 9). Close to the terminator and west of the Teneriffe Mountains runs a stubby bright line 55 miles long which fades considerably toward the west end. With higher optical power it resembles a narrow rectangular slice cut from a mountainous region like the Alps to the east and transplanted on the flat mare plain. For obvious reason it is called the STRAIGHT RANGE, and some of its components attain

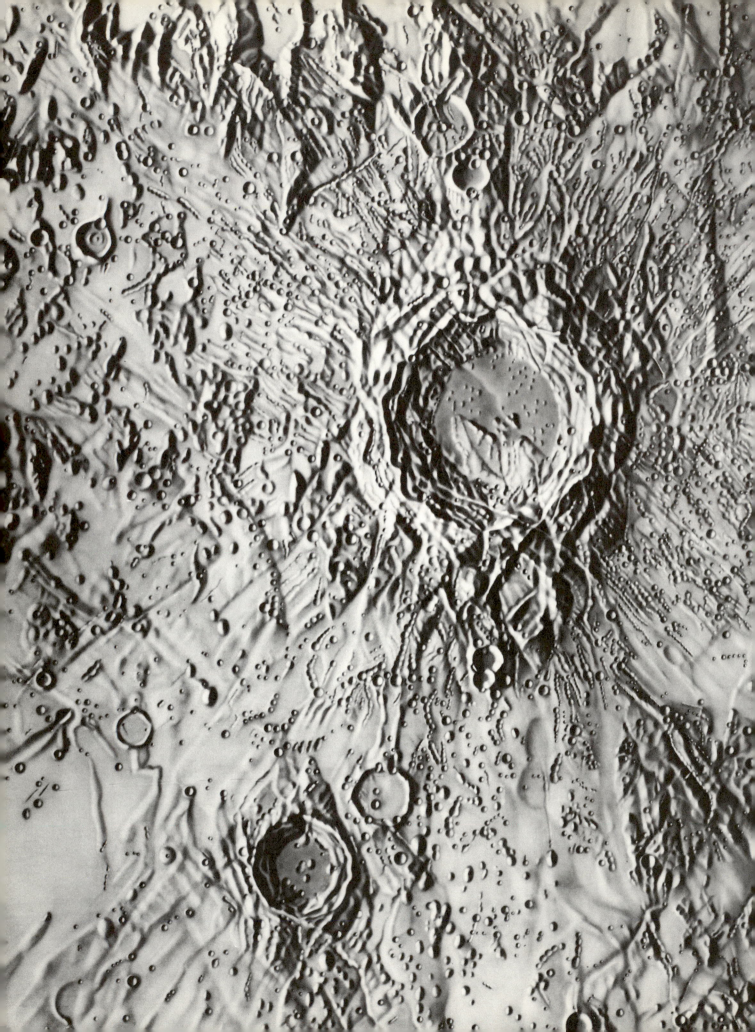

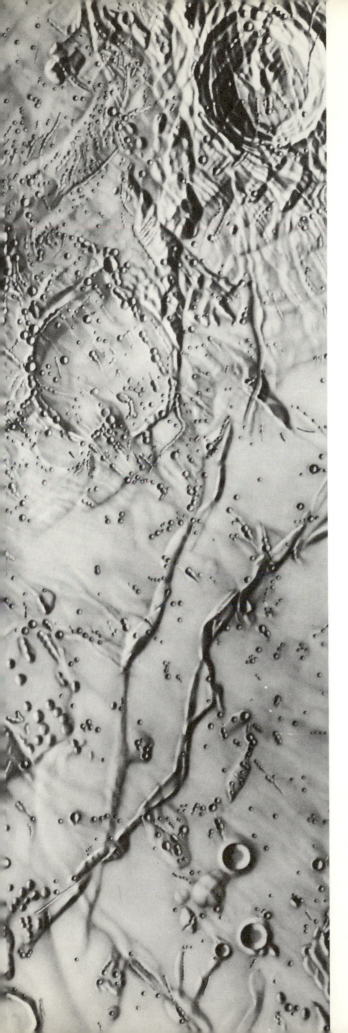

FIGURE 43. The crater Copernicus. Part of the shaded relief drawing for LAC 58 developed by the Aeronautical Chart and Information Center from the best photographic and visual observations available. The Carpathian Mountains are shown near the top, and Reinhold is the prominent crater between Copernicus and the lower left margin. The crater at the upper right corner is Eratosthenes. The equally large but inconspicuous melted-down ghost crater STADIUS is shown southwest of Eratosthenes and east of Copernicus. (Reduced to about two-thirds actual chart size.)

an altitude of 6600 feet (Fig. 9). Although it is not so brilliant as Pico and Piton, it shows well, and like those peaks it is probably another surviving element of the original Mare Imbrium wall.

On a line from Copernicus to Tycho and ⅖ the way is a group of ancient, flattened, fragmentary, Class 5 craters which shows well only when sunlight strikes it almost horizontally. The smallest and brightest one at the north end is PARRY, 28 by 30 miles across and 3800 feet deep. Its south wall is much lower, a 15-mile section having been removed almost entirely. What is left shines conspicuously tonight, but the floor brightness tends to wash out the brilliant walls. It was near Parry (four diameters southwest) on Mare Cognitum that Ranger 7 ended its glorious flight at 8:25 A.M. EST, on July 31, 1964, its difficult mission a magnificent success. Seventy-five miles south of Parry lie the even flatter remains of the larger GUERICKE. It is 36 by 39 miles across and 3100 feet deep where its walls reach their greatest height. It looks like a misshapened horseshoe since the north and northeast portions of its wall have been reduced to a few fragments. You will want to see Guericke, and probably you can resolve it with some effort. It is by no means a strong feature, but it is the crater that appeared in such splendid detail in one of the first Ranger 7 photographs released (frontispiece). If you would care to continue this push into the far-distant past and look for a couple of older and much flatter ghost craters, go back to Parry. Perhaps you can make out the fragmentary, almost imaginary, bright outline of BONPLAND tangent to Parry on the west. It is 34 by 38 miles across and 3300 feet deep on the northeast. Tangent to both Parry and Bonpland on the north is the larger but even more delicately delineated FRA MAURO, 60 miles across. Only traces of its walls remain rising to 2400 feet on the east.

The Parry-Guericke group of lunar antiques (Fig. 22) is located in the northern part of MARE NUBIUM which we see almost fully disclosed tonight. The dark plain reflects only 6 per cent of the light that falls upon it, and it is one of the darkest areas on the face of the moon and a source of highly polarized moonlight according to Audouin Dollfus. That portion of Mare Nubium which stretches west from the old crater group to the Riphaeus Mountains beyond the terminator is the section which was named MARE COGNITUM shortly after the Ranger 7 flight (Fig. 22).

Two Crisium lengths south of Copernicus and a little more than one such unit west of Arzachel stands another of tonight's leading features, the Class 1 crater BULLIALDUS (Fig. 22). Thirty-eight miles in diameter and 11,300 feet deep, this splendid object resembles Copernicus with massive, terraced walls nearly 10 miles thick and seemingly reinforced by numerous radial buttresses extending up to 50 miles across the plain. The interior is ⅓ bright and ⅔ black, and in a few hours the central mountain may be seen with binoculars as a tiny bright speck against the black floor. Bullialdus is of exactly average lunar reflecting power, but it usually stands out conspicuously against its dim background. Concerning it, Webb wrote more than a century ago, "I have seen a large radiation round it, as of ejected matter." Such observations, long discounted by professional astronomers as subjective interpretations, are beginning to take on new significance in the light of much more recent events.

A great change has taken place in Tycho since last night when we saw it first as a black pit at the terminator. It is now a splendid object fully lighted by the sun except for the broad inner east wall. The west wall is exceptionally brilliant, and the bright central mountain casts a large enough shadow to make it visible through binoculars. In a few nights the interior will be shadowless and so brilliant that the peak will be lost in the glare. Can you trace some of the famous rays which already have begun to appear?

On the south shore of Mare Nubium, just east of the midpoint between Tycho and Bullialdus, stands Pitatus which last night was lost in darkness with the exception of its massive east wall. Tonight we see all of it but none of it well. With close attention we may trace almost completely the thin light ring of its crest which varies in elevation from 8900 down to only a few hundred feet above the Nubium plain. Just south of Pitatus are twin mountain-walled plains, the illuminated floors of which are enclosed between bright west walls and dark east walls. The one on the west is Class 5 WURZELBAUER, 52 by 60 miles across and 7200 feet deep. The events of its long history have distorted and demolished it to such an extent that it is noticeable only when near the terminator. The somewhat more prominent east twin is GAURICUS, 46 by 52 miles across and 8900 feet deep. Probably younger than Wurzelbauer, Gauricus likewise exhibits a flooded floor without central mountain, and its walls are pitted with many small craters.

About one Crisium width south of Tycho and near the terminator may be seen the most impressive of all the great mountain-walled plains. CLAVIUS, marked out in broad lines of black and white, is 132 by 152 miles across. It is equaled in immensity by the briefly prominent Deslandres which, by a curious turn of fate, evidently was at its best when Galileo sketched the moon long ago. It is exceeded only by Bailly, a vast excavation which is seldom noticed because of its unfavorable location. Baldwin finds that the inner walls of Class 2 Clavius slope gradually downward for 15 miles, corresponding to a vertical drop of 16,100 feet. He also measures outer walls that rise 5400 feet above the surrounding area to give a total wall thickness of 35 miles at the base! As might be expected in that extremely disturbed region, Clavius is punctured by scores of wall and floor craters. One of the larger intruders, seen on the southeast wall, is RUTHERFURD, 30 by 33 miles across and 7200 feet deep. Its twin on the northeast wall is CLAVIUS B, of Class 1, 30 miles in diameter and 10,200 feet deep. Just west of the midpoint between those two is Class 1 CLAVIUS D, 19 miles in diameter and 9500 feet deep. It also can be seen through binoculars. West of it in a gentle arc across the floor lie successively smaller Class 1 craters 14, 10, and 9 miles in diameter, respectively. The last three are good test objects to try when observing conditions are excellent. That the floor of Clavius actually is convex may be demonstrated when it is on the terminator where the sun illuminates the central portion while both the east and west floor sections are still in darkness. Here, indeed, is an object worthy of frequent observation.

Another prominent feature tonight is wedged between the terminator and the southwest wall of Clavius. A major mountain-walled plain 63 by 66 miles across and 12,500 feet deep, Class 1 BLANCANUS appears dwarfed by its enormous neighbor. Tonight it is conspicuous in contrast, but it shows us very little of itself. The interior is completely lost in shadow, and it takes the shape of a black bean bordered on the west by the brilliant crescent of the illuminated wall crest. By tomorrow night more than half the crater will be bathed in sunlight, and the gray floor will appear quite smooth, its few hills and small depressions unresolved through our binoculars. We then shall see it as a normal ellipse definitely separated from the low shelf between its north wall and the southwest wall of Clavius, which area appears to be part of its interior tonight. About one Clavius length northwest of Clavius at the terminator lies another large mountain-walled plain, equally old Class 2 LONGOMONTANUS, 107 miles across, 12,500 feet deep, and pentagonal in outline (Fig. 6). It shows well with a brilliant west wall,

black east wall, and a convex floor that is black except near the center. Probably you will not be able to see the small, bright, multiple "central" mountain that rises from a location west of the floor center.

Just north of Longomontanus and west of Tycho is a slightly less prominent depression which at first glance appears to be another giant crater nearly as large as Longomontanus. Closer inspection shows that it is a complex of three craters. The newest and most complete of the three is the Class 3 mountain-walled plain WILHELM which forms the north section (Fig. 6). Probably it once was elliptical, 57 by 68 miles across and 9500 feet deep. The south section is the much older and smaller crater MONTANARI (Fig. 6). What is left of it is a distorted elliptical depression of little contrast 46 by 57 miles across, 5200 feet deep, and tangent to both Wilhelm and Longomontanus. The west section is a strange enclosure shaped like a slice of pie. It is all that remains of a crater which is the oldest of the three. All of its walls have been destroyed by the other two with the exception of a small section on the west which is 4300 feet high. The name LAGALLA has been given to the remnant, the longest dimension of which is 50 miles. Not only is the group very old, but it also has suffered the penalty of great age on the moon—heavy bombardment. It has numerous wall and floor craters, four of which are between 10 and 15 miles in diameter. Can you make out one of the three on the southwest wall of Wilhelm?

Nearly one Wilhelm width northeast of Wilhelm you will find the approximately semicircular Class 3 crater HEINSIUS, 43 miles in diameter and 6900 feet deep (Fig. 6). Its southwest wall has been destroyed completely by a pair of tangent craters 15 miles in diameter, and a slightly smaller crater occupies a considerable portion of the floor. The first two intruders, HEINSIUS B and C, may be seen tonight, and the third, HEINSIUS A, may be visible at the edge of the shadow of Heinsius' east wall. Before we leave the great southern continent which bears so many scars of one time violent activity let us note one more crater. It is considerably smaller than most of those we have been observing, but it shows well. If a line from the center of Longomontanus to the center of Wilhelm be extended nearly twice its length, it reaches a little crater near the south shore of Mare Nubium. CICHUS, of Class 1, is 25 miles in diameter, 7900 feet deep, and tonight a contrasty object about half bright and half black. It has a younger crater, CICHUS C, 8 miles across and 4900 feet deep, perched precisely at the crest of its southwest wall. You may be able to resolve it as a bright bead. Webb became interested in Cichus C in 1833 when he noticed it was twice as large as the intruder shown by John Schroter on his map of 1792, published half a century before the first lunar photograph was taken. A young man more eager than experienced, he speculated upon the possibility that it had grown, but a quarter of a century later he wisely noted that "a longer acquaintance with selenography leads to the impression that such evidence of change is of little value."

TEN-DAY MOON

(Chart III)

Again Copernicus tends to dominate the scene, and Tycho runs it a close second, their floors now fully illuminated. Southwest of the point midway between them, Bullialdus also is conspicuous on the western Mare Nubium. But tonight the moon presents a prominent feature which, at first glance, bears little resemblance to a crater.

Some years ago, on a hot summer afternoon (the eleventh day of the moon), I received a telephone call from a gentleman who had discovered something most unusual on the moon and wanted to describe it to "the astronomer." When he arrived, I learned that he worked on the night shift and that it was his custom after work to get out his small telescope and study the moon. He and his "buddy" had devoted many hours to that laudable pursuit, but last night both had been startled by a spectacle never before seen. Either the moon had a natural bridge on it or "they are building a runway up there," he declared. Eventually we got the moon oriented with respect to the horizon as he had seen it, and the approximate location of the shocking feature was established. You may view his discovery tonight at the terminator north of Copernicus. A line from Tycho through Copernicus extended ⅔ its length reaches the "bridge." Quite a bridge it would be—more than 100 miles long and some 30 miles high! It is the western half of the JURA MOUNTAIN range which borders SINUS IRIDUM, curving from LAPLACE PROMONTORY on the east to HERACLIDES PROMONTORY on the west.

Laplace Promontory is the tip of a bright triangular upland peninsula, just west of the Straight Range. It juts out onto the dark Mare Imbrium plain from which it rises abruptly 9900 feet. Heraclides

Promontory rises only 5900 feet above the plain, but in this light its apparent elevation is enhanced greatly. The flat, semielliptical area enclosed between the two is Sinus Iridum, the most beautiful of all lunar "bays" and one of the most interesting and controversial regions. If and when we are able to read its complete geological (or rather lunalogical) history, we shall be close to the true story of the moon. Perhaps tonight you can make out the ridge which begins off Heraclides Promontory and runs generally southward some 185 miles (foreshortened to about 140 miles).

The Jura Mountains, which actually form a semicircle 150 miles in diameter, evidently are the remains of an enormous crater wall. They contain numerous peaks, one of which rises 12,700 feet above the plain. Harold Urey has pointed out that Mare Imbrium, all of which is revealed tonight, is a huge circle running from Laplace Promontory through Plato, the Alps, Caucasus, Apennines, Carpathians, and various isolated mountains and hills back to Heraclides Promontory. He also has noted another significant circular ring beginning at Laplace Promontory and continuing through the Straight Range, the Teneriffe Mountains, Pico, several ridges, Spitsbergen, Timocharis, Lambert, and minor features including the ridge mentioned above, back to Heraclides Promontory. He further notes that there are no mountains on Mare Imbrium within the second circle and that the mountain ridges of the great ranges such as the Apennines and the Alps radiate outward from the inner circle. He concludes that all these features were formed by a single event of enormous magnitude. Writing in *Physics and Astronomy of the Moon*, Zdenek Kopal, editor (1962), he pictures an object resembling a minor planet perhaps 125 miles in diameter hurtling into the region from the northwest at a low angle of approach. It "gouged out Sinus Iridum and bored a deep hole within the region of the [inner] circle, raised a wavelike structure all around the region and sprayed out some of its substance and some of that of the moon in a great fan of radiating mountain ridges. . . . Subsequently the wave settled partly and broke at the shore line, that is at the [outer] circle. . . ."

Baldwin's explanation is similar except that he believes Mare Imbrium and Sinus Iridum were formed by two separate events occurring in that order. He visualizes the Imbrium forming planetary missile approaching in about the same trajectory, ranging in size from 40 miles in diameter for an impact velocity of 10 miles per second up to 118 miles in diameter for two miles per second, and striking the surface near the center of the inner ring. There he has traced the ghostly form of an impact crater 420 miles in diameter in close agreement with Urey's inner ring and with only a scattered series of minor features left "to mark the moon's greatest crater." He regards the outer ring of major mountain ranges as a shock ring resulting from the tremendous disturbance produced by the energy of impact. Sinus Iridum "is an independent crater, very similar to Clavius, though much younger," in his view, "definitely . . . formed after the great [Imbrium] crater and before the lavas came."

More than a century before the researches briefly summarized above were initiated, Beer and Madler in their famous work *Der Mond* (1837) wrote that Sinus Iridum is "perhaps the most magnificent of all lunar landscapes." That claim still meets with general agreement, and to it we now might add that the region is perhaps also the most significant lunar landscape. It both delights the eye and excites the imagination, and much remains to be learned through interpretation of the evidence presented there.

North of Laplace Promontory, a distance equal to the length of the Straight Range, and at the base of the peninsula, is a strange, irregular depression 5600 feet deep and 25 by 29 miles across named MAUPERTUIS. It scarcely can be called a crater, and its walls are incomplete and linear. Tonight it shows well because of the black inner east wall and bright west wall, but as the sun rises higher it will fade into obscurity. A little farther north, on the south shore of Mare Frigoris, may be seen the ancient, low-walled Class 5 CONDAMINE, 25 miles across but only 6600 feet deep. It also shows well when near the terminator, its bright inner west wall contrasting with the black eastern quarter of its interior.

North of Laplace Promontory the terminator cuts through a striking formation which appears as a black nick in the disk. It is the large, irregular, and very old crater JOHN HERSCHEL. Difficult to locate under a higher sun, it shows prominently in part tonight. The bright south wall, which rises only about 1000 feet to separate it from smooth Mare Frigoris, and perhaps a bit of the north wall contrast sharply with the dark interior. The unusually rough convex floor is hidden in shadow, but so great is its diameter (105 miles) that in a few hours the central portion will be highlighted with sunshine, giving it a strange, eerie appearance. A

little more than one Sinus Iridum length northeast of John Herschel at the terminator lies the sharply outlined Class 1 crater PHILOLAUS, 48 by 51 miles across and 10,800 feet deep. It stands out conspicuously from its highly disturbed surroundings, its brilliant wall contrasting with the narrow black ellipse of its interior.

South of Heraclides Promontory, about ⅔ the way to the Carpathian Mountains, a small but easily seen crater rises out of the smooth Imbrium plain. Also it may be located by doubling the length of the line from Autolycus through Timocharis. That is EULER, of Class 1, 17 miles in diameter and 7200 feet deep. Its bright west wall shows well against the dim surroundings, and at full it marks the center of a bright ray system extending outward to a distance of 65 miles.

Southwest of Copernicus, a little less than twice as far as Reinhold, we see the small but conspicuous crater LANSBERG (Fig. 4). Of Class 1, Lansberg is 24 miles in diameter and 10,900 feet deep, but its walls are of massive construction a mile thick. One-third of the interior is lost in black shadow tonight, but note the brightness of the rest of it. The walls of Lansberg reflect 12 per cent of the light that falls upon them, and they represent an average of all lunar surface materials in the property of reflectivity.

Continuing southwest of Copernicus a little farther (2 Crisium widths), we reach a small but bright mountain range, the main part of which stretches 110 miles parallel to the sunrise line. In that favorable position the RIPHAEUS MOUNTAIN area stands out conspicuously, but the bright mountains blend with the light background (Fig. 22). One of the peaks reaches a height of 4100 feet above the plain. Except for two spurs that run off to the north, the whole range appears to combine with various isolated hills and peaks to outline in fragments a circle about 125 miles in diameter, probably the remnants of the original Mare Cognitum crater subsequently melted and filled by lava flows.

Southeast of Lansberg, three Lansberg diameters, and northeast of the Riphaeus Range a smooth region of Procellarum cover is found among the low mountain remnants that dot the plain. Surveyor 3 landed there April 19, 1967, rebounded three times, and settled down on the gradual inner slope of a subtelescopic crater. It was the first lunar probe equipped with a miniature power shovel designed to trench the soil to a depth of 18 inches, and it soon put its tool to work as the television camera watched. The appearance of the subsurface material

CHART IV

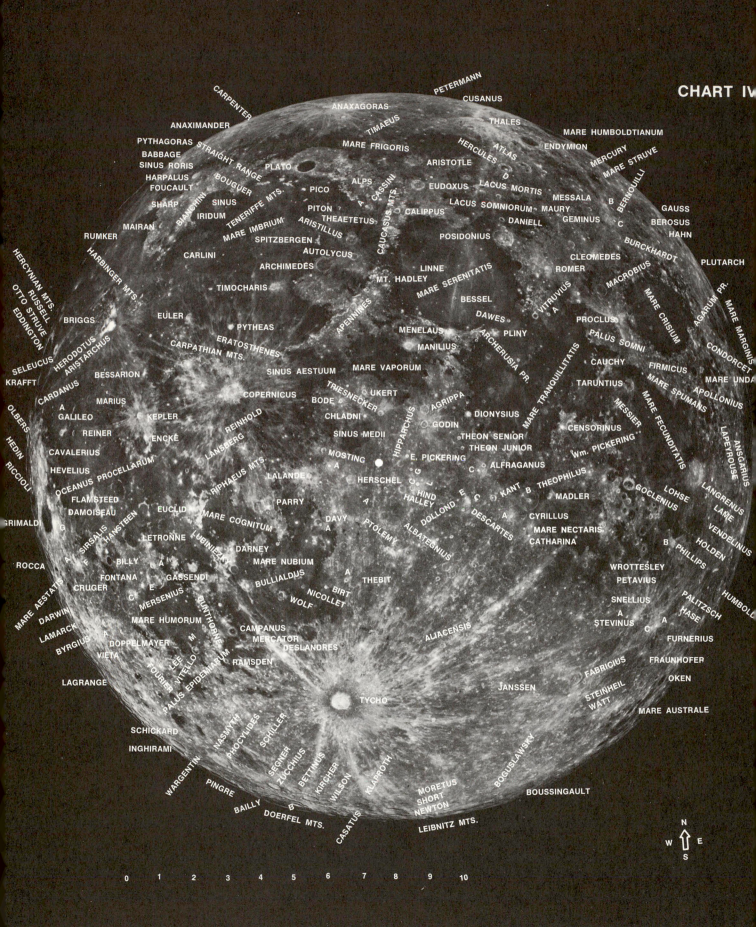

0 1 2 3 4 5 6 7 8 9 10

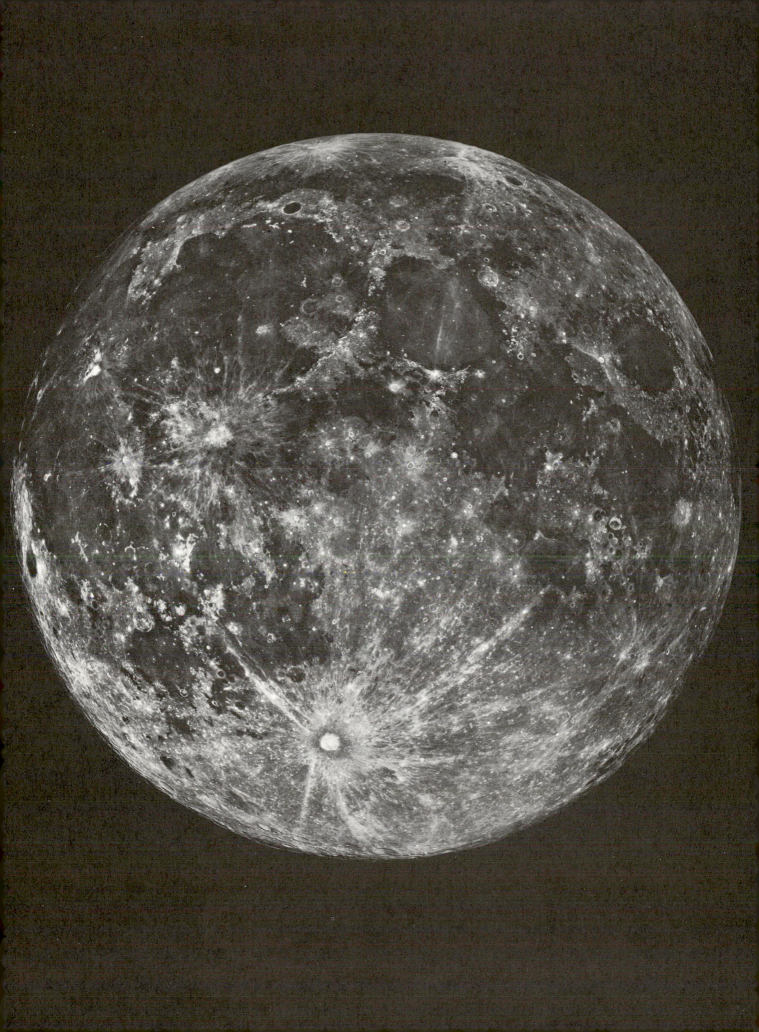

and the clean-cut ditch walls pointed to the conclusion that the lunar soil has the physical properties of fine-grained, damp beach sand. Inspection of dents produced when the shovel was pressed against the surface answered a question of vital importance. The mare *would* support the Apollo space craft and its exploring astronauts.

About one Crisium width southwest of Bullialdus, along the shore of Mare Nubium, are two similarly sized craters connected by a ridge and foreshortened to resemble a pair of bright-rimmed dark eyeglasses. They are the Class 5 craters MERCATOR, on the southeast, and CAMPANUS on the northwest. Their respective diameters are 29 and 31 miles, and their depths are 4300 and 6700 feet. Although not prominent, the two craters may be seen through binoculars, and they constitute one of many examples of pairs of similar craters found on the moon. Here is another clue to the story of our satellite that awaits interpretation. Long ago Webb called attention to such arrangements and noted that often the pair is aligned in the north-south direction, a "rule" which Mercator and Campanus violate. South of the two craters is a triangular dark patch that might be considered part of Mare Nubium except for a few intervening spots of light upland territory. That little plain goes by the fanciful name PALUS EPIDEMIARUM, which constitutes its most striking feature. The largest structure on Palus Epidemiarum, located near its south edge, is Class 5 CAPUANUS, a fig-shaped crater 43 by 47 miles across and 6200 feet deep at the west wall. Its smooth, dark floor has been reported as elevated above the outside level, and at times it does give that impression on the northeast where the wall is extremely low. However, the Army Map Service has found the floor 1000 feet below the palus plain.

About one Clavius length southwest of Capuanus look for a curious bathtub-shaped enclosure with black interior and bright walls. It is a Class 2 multiple crater 44 by 58 miles across and 10,500 feet deep. The object is HAINZEL, and the long-accepted "double" description will suffice for our observations through binoculars. However, visual and photographic studies with good interior lighting and good conditions show that Hainzel actually is a triple crater, the larger southeast portion consisting of two smaller craters side by side. Tangent to Hainzel on the south, at the terminator, is the extremely old enclosure MEE, 85 miles in diameter and 7500 feet deep. Hainzel and numerous smaller craters have broken its walls and encroached upon its floor. Most of its rough interior is illuminated tonight, but it is dim and not easy to see. Under a higher sun its dilapidated walls and pitted floor merge into the general background, and the crater disappears. Neison's excellent map (1876) gives no indication of an enclosure there, but Goodacre shows it on his 1910 chart, unnamed at the time.

Another fine crater deserves our attention. SCHEINER is conspicuous near the terminator west of Clavius, south of Longomontanus, and at the northwest edge of Blancanus. It is a mountain-walled plain of Class 2, 71 miles across and 15,100 feet deep. About ⅓ of its bombarded floor is black, but it is well outlined by bright walls. The inner west wall is considerably broader and brighter, but an appreciable outer east wall is certainly present despite the contention that mountain-walled plains lack such a feature.

ELEVEN-DAY MOON

(Chart IV)

Along the central part of the terminator we notice a large, smooth, darkened plain. It is the southeast section of the largest of all lunar plains, the OCEANUS PROCELLARUM, which stretches from Copernicus south to ancient Parry, from Copernicus northwest to the uplands west of Sinus Iridum, and from those areas westward almost to the crowded, crater-walled limb. Dollfus has found the light from the southeast part of the plain polarized to a greater degree than that from any other part of the moon. Baldwin has measured one of the lowest elevations there, a depth exceeded only by the center of Mare Imbrium. Nikolai Barabashov has detected a greenish hue over most of the Oceanus with a brown tint in the far south.

As the full phase approaches, the maria continue to darken, or rather appear darker as the uplands reflect light more efficiently. The darkest tonight are the great eastern maria—Tranquillitatis, Fecunditatis, and Crisium. The least darkened are Oceanus Procellarum and the nearly linear Mare Frigoris. Far to the east a pair of small, patchy, and poorly defined dark areas may be noted. They are southeast of Mare Crisium and northeast of Mare Fecunditatis. The north one is MARE UNDARUM, and its near neighbor on the south is MARE SPUMANS. The area of each scarcely is larger than that of the bright crater Langrenus south of them on the shore of Mare Fecunditatis, and they seem out of place in the maria class. I did not mention them during the nights of the crescent moon since they were then almost invisible because of the general brightness of the region. Near full, they show with considerable contrast.

Two additional maria sometimes may be seen close to the east limb if the libration is favorable, that is, if the east edge of Mare Crisium is more than ½ Crisium width from the limb. East of Mare Crisium is MARE MARGINIS, appearing as an irregular dark line parallel to the limb and about 150 miles long. South of it, and east of Mare Spumans, is the more obscure MARE SMYTHII, a similar line some 225 miles long. They would show much better if we could view them from directly above, because the east-west extent of Mare Smythii is nearly as great as its north-south dimension, and that of Mare Marginis is half again as great. Over the western edge of Mare Smythii our first Lunar Orbiter took its initial series of photographs on August 18, 1966, from an altitude of 133 miles. The pictures, which it later developed and televised, clearly showed craters less than 100 feet across in a region where the best previous observations had given only vague indications of features several miles in diameter. With favorable libration in latitude, or longitude, or both, MARE HUMBOLDTIANUM may be seen as a weak, narrow, dark ellipse or streak close to the northeast limb. A line from Copernicus through the gap between Mare Serenitatis and Mare Imbrium touches the elliptical plain which is 170 miles across. Do not confuse it with the shorter, wider black ellipse in front of the mare which is the crater Endymion. Under likewise favorable libration, a series of small dark spots may be noted spread over a considerable portion of the southeast limb. They are some of the bays on the edge of MARE AUSTRALE, a strong and really large dark plain, the major portion of which lies beyond the limb.

Tycho is a brilliant spot tonight. Look for the faint, shaded halo which separates it from the bright ray system. The strange halo will seem to darken in the next few nights as the rays grow stronger and longer. Copernicus likewise is brilliant, and its unique ray system is strong and fully developed. The Apennines are nearly washed out against the bright background but can be traced. Eratosthenes, near the west end of the range, has become a test object. Can you find it? Quite a few of the craters we have observed previously appear as bright spots, but more about them at full moon.

West of Copernicus, ⅔ the way to the terminator, is a crater a little larger than Euler with bright

walls and interior, except for the shadow covering its east quarter. It is surrounded by a bright-ray splash pattern that will grow much brighter and, a few nights hence, will extend outward fully 200 miles. You are looking at the important Class 1 KEPLER, 20 miles in diameter and 7500 feet deep. So interesting and varied are the surface features around that modest crater that the U. S. Geological Survey chose it as the first lunar region to be mapped geologically. The superb chart, I-355 (LAC-57), published in 1962, is the work of R. J. Hackman. The area also was one of the earlier sections to be charted for navigation by the Aeronautical Chart and Information Center, U. S. Air Force. The crater shows well tonight, but soon the ray pattern will become so brilliant that Kepler itself will be difficult to locate in the glare. In addition to the striking ray patterns of Kepler and Copernicus, the region is of particular interest because of the large number of domes found there. Ranging up to 18 miles across, they are scattered in the area between the two craters and between Kepler and the Carpathian Mountains. They show only when near the terminator, and none is a good object for binoculars since all are evidently rather smooth and less than 1000 feet high. If you want to look for one, the largest in the region is located about halfway from Kepler to the middle of the Carpathian Range, its distance from Kepler being equal to the Copernicus-Reinhold separation. About one Plato length south of Kepler and not quite so close to the terminator may be seen the much older, hexagonal, flooded crater ENCKE, of Class 5, 18 miles in diameter, but only 2300 feet deep. It is located near the outer edge of the "solid" portion of the Kepler ray system, and so it is not an easy object to resolve.

"Twice on the night of November 1–2, 1963, a large region on the moon near the crater Kepler glowed red, as recorded on photographs made at Pic du Midi Observatory by the authors," wrote Zdenek Kopal and Thomas Rackham in *Sky and Telescope*, March, 1964. They then went on to describe two of numerous photographs in deep red light, taken that night of full moon, on which the surface brightness around Kepler nearly doubled its normal value during brief periods. The two astronomers attribute the striking phenomenon, which affected some 23,000 square miles of lunar surface, to luminescence induced by waves of high-energy particles ejected from the sun during brief outbursts called solar flares.

The HARBINGER MOUNTAINS do not amount to

much if compared with the great ranges that bound Mare Imbrium (Fig. 44). They are thinly scattered over an area some 40 by 100 miles. While they are fairly bright, the group is generally quite inconspicuous. Consequently, the 5800-foot elevation which the U. S. Air Force has found for the highest "peak" seems much more reasonable than the 8000 feet reported by earlier investigators. Tonight the sun has just risen over them, and their individual members shine like stars on the plain one Serenitatis width north of Kepler.

The medium-sized Class 1 crater ARISTARCHUS is seen easily tonight at the terminator north of Kepler (Fig. 44). Kepler, Aristarchus, and the larger Copernicus mark out an isosceles triangle with Kepler at the vertex. The triangle, which actually is delineated by bright rays near full moon, is almost a right triangle, and later you may notice two craters at its northwest corner. Aristarchus, 25 miles in diameter and 11,900 feet deep, is the east one and the more conspicuous by far. In the words of Wilkins and Moore: "This is the highlight of the moon, being the most brilliant object on the lunar surface. At full it is so bright as to dazzle the eye and confuse the details." Webb states that it is "visible to the naked eye on the bright side, and with the telescope on the dark." Tonight we see principally the west inner wall as a bright crescent that is almost a semiellipse. The black interior and the east wall seem to merge into the background. Later the unique feature of the ray system may be seen as a triangular light patch extending westward from the crater across the south rim of Herodotus which is Aristarchus' neighbor.

Not only is Aristarchus the moon's brightest spot, but in recent years it has become one of the more interesting spots because of well-observed outgassing that occurs there from time to time. Kozyrev reporting in *Nature*, June 8, 1963, mentions spectrograms which he had obtained in 1955 that "indicated the existence of a luminescent glow in the Aristarchus crater." He then describes a series of such observations made between November 26 and December 3, 1961, several of which showed the presence of a luminous cloud of gas covering a small area at the center of the crater. After photometric analysis of the spectrograms, Kozyrev concludes that the gas was molecular hydrogen.

On October 29, 1963, James Greenacre, of the Aeronautical Chart and Information Center, U. S. Air Force, was at work on the draft chart of the region. He was observing with Percival Lowell's splendid 24-inch refractor at Flagstaff, Arizona,

adding fine detail to the chart which had been prepared from the world's best photographs. He had been engaged in such work for more than three years and had spent at least 50 hours in the study of Aristarchus and its environment. Rated as a very cautious observer, skeptical of reported lunar changes, Greenacre was astonished to see "a reddish orange color over the domelike structure" just north of Herodotus. Instantly he noticed another such spot a little farther north, and a few minutes later he saw "an elongated streaked pink along the southwest interior rim of Aristarchus." The changes, which he describes in detail in *Sky and Telescope*, December, 1963, were confirmed by his observing assistant Edward Barr, but they could not be seen with the six-inch finder on the Lowell telescope. Within 20 minutes of the first observation the colors had faded away and the area had regained its familiar appearance. A close watch for the next two weeks failed to reveal anything unusual, and the sunset terminator passed over the crater, leaving it in darkness for another two weeks. As soon as sunlight reached Aristarchus again, the observations were resumed, and on November 27 the reddish streak on the rim of Aristarchus reappeared and persisted for one hour and 15 minutes. It was seen by John Hall, Lowell Observatory director, and several other observers, including students using the 72-inch reflector of the Perkins Observatory. Hall suggests that the outgassing is caused by the return of the sun's heating rays to the crater after two weeks of cold darkness. That suggestion fits Greenacre's observations perfectly, but it seems to be somewhat inadequate when applied to those of Kozyrev which were made between 7 and 14 days after the sun had risen over Aristarchus.

Whatever the final interpretation may be, the remarkable events at Aristarchus again dramatically confirm in a general way the long-discounted claims of a host of astronomers, both professional and amateur, who for two centuries have asserted that changes do take place on the moon. However, before we accept with posthumous apologies all the claims of the past, we should weigh these significant facts. First, the photographic observations of Alter (Alphonsus) and Kozyrev were made with reflectors of 60 and 50 inches aperture, respectively. Second, the visual observations of Greenacre and his associates were made with one of the world's larger refractors located in an unexcelled atmospheric environment. Third, Greenacre was unable to see any of the October changes with a professional-quality six-inch telescope, and the November event could not be detected at Flagstaff with a 12-inch telescope! So don't expect to see lunar changes through your binoculars or small telescope. You might do so, but the probability of success is exceedingly low.

HERODOTUS, of Class 5, 20 by 22 miles across and 4700 feet deep, suffers from a most unfortunate location (Fig. 44). At best, it is noted as the dim crater just west of dazzling Aristarchus. Near full moon, when the Aristarchus rays are brightest, Herodotus is overwhelmed by them and virtually disappears as far as the binoculars astronomer is concerned. Tonight it is only a bright speck beyond the terminator, but tomorrow night it should show fairly well. Its smooth floor, fully as dark as the surrounding plain, will contrast sharply with the illuminated portion of Aristarchus' interior, as you will see tonight if you wait a few hours.

Just beyond the north shore of Sinus Iridum, and near the middle of the Jura Range, the Class 1 crater BIANCHINI shows well with about ⅔ of its interior in sunlight. Its 25-mile diameter makes it almost identical in size with Aristarchus, and its distorted walls rise 10,000 feet above the floor. About two Plato lengths southwest of Bianchini and a little farther removed from the sinus than is Bianchini, we find the similar crater SHARP, of Class 1, 24 miles across, and 10,500 feet deep. Continuing in about the same direction for the same distance, we come, near the terminator, to Class 1 MAIRAN, 26 miles in diameter and 11,200 feet deep. To this linear array of three almost identical craters may be added a fourth on the plain northeast of Sharp and northwest of Bianchini. The last is Class 1 HARPALUS, 26 miles in diameter and 9800 feet deep. Of the four inconspicuous craters, Harpalus is the most prominent, partially because it rises from a smooth plain and partially because it has higher reflectivity. If successful with the last four, you might look for two smaller new craters on the north edge of the Jura uplands. FOUCAULT is south of Harpalus, and BOUGUER is east of Harpalus. Their respective dimensions are 15 miles across and 7200 feet deep; 15 miles in diameter and 8200 feet deep. Both are difficult objects. The plain on which Harpalus stands looks like part of Mare Frigoris, but it is known as SINUS RORIS.

Two Crisium lengths south of Kepler the bright mountain-walled plain GASSENDI awaits our inspection (Fig. 20). It may be located also by running a line from Plato to Copernicus and extending it ¾ its length. Gassendi, of Class 5, 70 miles in diameter and 6600 feet deep, is termed by Wilkins and Moore "one of the most beautiful and important of all lunar walled formations." Tonight the terminator is far to the west, and the crater is illuminated fully, appearing as a bright ring. It presents an especially interesting spectacle around sunrise or sunset. At such times the relatively low walls appear high and steep, and the triple central mountain shows well as it does tonight. The north wall is broken by the Class 1 crater GASSENDI A, 20 by 24 miles across and 8500 feet deep, which also may be seen. Note that the young intruder is much deeper than its ancient host. On the south the walls of Gassendi virtually have disappeared over a stretch of some 30 miles, having been melted down by the lavas to a few mere hills. The crater actually is almost exactly circular in shape, but it is foreshortened by its limbward location into an ellipse, the exact shape of which changes noticeably from week to week owing to the effect of the libration. The floor offers a wealth of detail to telescopic observers. It abounds with rills and ridges.

Since Gassendi stands on the north shore of MARE HUMORUM, the cause of its ruined south wall is evident. Baldwin considers Mare Humorum, which is about the size of Arkansas, one of the oldest if not the oldest of the circular maria. He believes it was formed as a huge crater 287 miles in diameter excavated by a planetoid which struck the surface in a nearly perpendicular trajectory. The stupendous impact buckled the surface layers of the moon at a distance twice that of the crater edge and resulted in a circular *anticline* concentric with the mare, 594 miles in diameter, which he has been able to trace in part. Subsequently Gassendi and several other ancient craters along the mare shore were formed similarly by smaller colliding bodies. Then came the lava flows that gave Mare Humorum its present floor. Many years ago the dark floor was described as distinctly green in hue, but more recent (and probably much more accurate) measures have changed the description to reddish. Extensive systems of ridges and rills have been studied on the mare and in the surrounding region, most of them in arcs centered on the supposed point of impact. On the south shore of Mare Humorum may be seen the smaller but equally ancient Class 5 crater VITELLO, 25 by 29 miles across but only 4600 feet deep. One of the more prominent of the ridges mentioned above meanders northward from Vitello across the mare toward Gassendi. Perhaps with excellent conditions

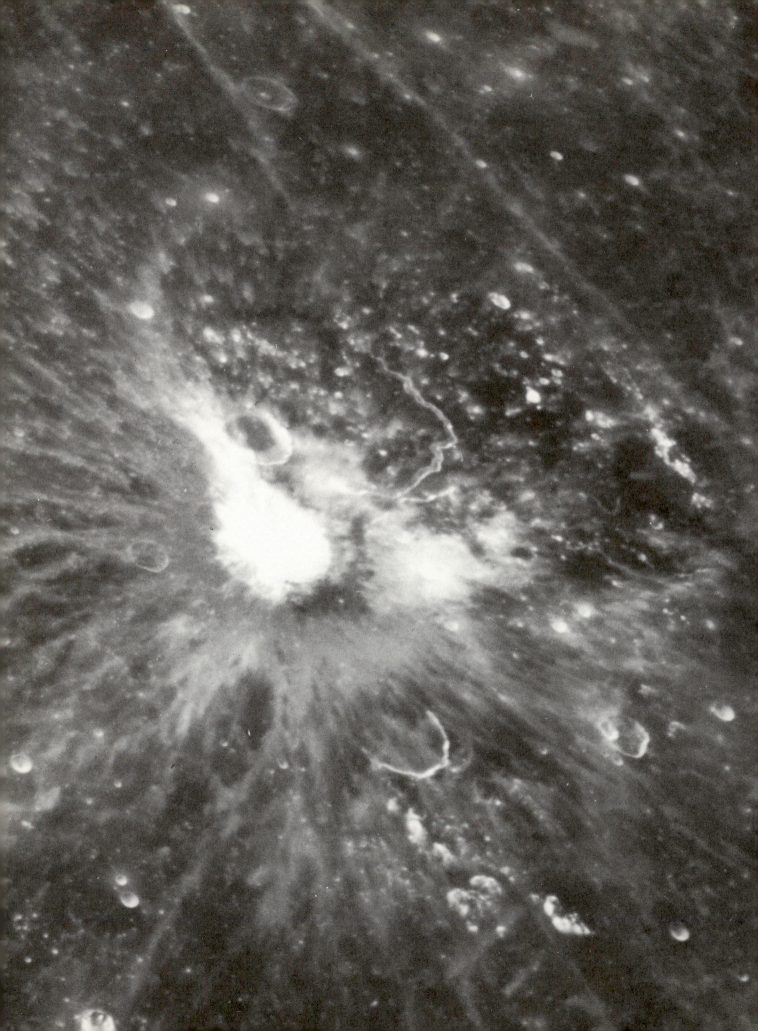

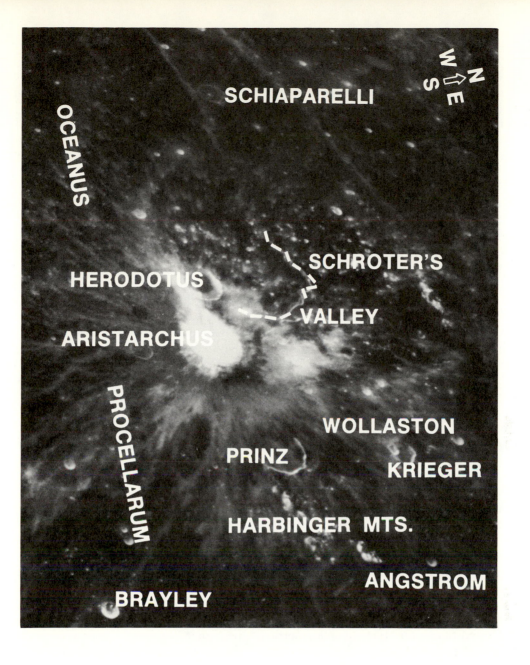

SCHIAPARELLI

N W E S

OCEANUS

SCHROTER'S

HERODOTUS

VALLEY

ARISTARCHUS

PROCELLARUM

WOLLASTON

PRINZ

KRIEGER

HARBINGER MTS.

ANGSTROM

BRAYLEY

FIGURE 44. The brilliant young crater Aristarchus under high sun. Only slightly larger than Herodotus, it appears much larger because of the spreading of its bright image over the plate. In addition to the dazzling plume to the west, it exhibits a fainter splash pattern of rays similar to that of Copernicus. The ancient crater Prinz has been largely destroyed by the Procellarum lavas. Schroter's Valley, a sinuous rill 125 miles long, shows well. Recently Kuiper and associates have suggested that sinuous rills, of which this valley is the largest known example, may be lava drainage channels. They resemble closely such terrestrial volcanic features except that they are 4 to 40 times larger. (120-inch Reflector.)

you might be able to make it out as a fine but crooked black line. It is easier when nearer the terminator.

Two Crisium widths south of Kepler and a little less than one such unit northwest of Gassendi are the remains of a once great crater largely destroyed by the lavas of Oceanus Procellarum. LETRONNE (Fig. 20) and Gassendi stand like ancient fortresses commanding three sides of the broad upland peninsula that juts out into the oceanus and forms the north shore of Mare Humorum. Letronne, of Class 5, is 75 miles across, but the entire north ⅓ of its wall has sunk beneath the surface, leaving only the flattest of ridges and a few light spots to mark its position. The wall remnants now rise only to a maximum of 3000 feet. It is curious that those two large craters, the highest walls of which are separated by less than their diameters, should have suffered the same fate through separate but similar destructive forces which overwhelmed them from opposite directions.

If the Gassendi-Letronne line is extended north by its length, it reaches another equally large crater which has been melted down almost completely. The ghost crater, 70 miles in diameter and marked only by ridges and hills, possibly may be seen tonight since it is visible through binoculars when near the terminator. It has no name, and its highest wall rises only 2100 feet above the plain, but the south edge of its floor is occupied by FLAMSTEED, a Class 1 crater 13 miles in diameter and 5500 feet deep. Flamsteed should be visible as a bright speck since its walls are somewhat brighter than the average. The ghost crater encloses a point of historical importance—the spot 45 miles northeast of Flamsteed on which stands Surveyor 1. As we have seen, Surveyor 1 made the first truly soft lunar landing on June 2, 1966, and sent back more than 11,000 pictures of a surface which resembles that of an essentially flat terrestrial desert pitted with craters and strewn with rocks. The broad floor of the ghost crater was one of the prime touchdown spots tentatively chosen, early in the Apollo program planning, for the first manned lunar landing. However, other Surveyors found equally flat plains located nearer the center of the lunar disk, the best spot for assured contact with Mission Control in Houston, and the Flamsteed area was scratched. Neil Armstrong, the Christopher Columbus of the Space Age, set his spaceship *Eagle* down on Mare Tranquillitatis 65 miles east of Sabine on July 20, 1969. As he stepped down to make the first human

bootprint on that New World, he announced to the waiting millions back home: "That's one small step for man, one giant leap for mankind." Later Apollo landings were made on highland terrain to broaden the scope of scientific investigation, but the first two were confined to relatively smooth mare plains as probably the safest. In 1964, Thomas Gold had taken issue with this view and announced his hypothesis that the maria are covered with a layer of dust at least 300 feet and possibly miles deep! The dust particles were supposed to have been eroded from the mountains and crater walls during the last few billion years and transported to the lowlands in sufficient quantities to provide the deep layers indicated. While most students of the moon agreed that the surface probably was more or less dusty, the extreme views of Gold made but few converts, and they were definitely disproved by the Ranger and Surveyor lunar probes. In fact, as we have seen, the Surveyor probes and Apollo astronauts found a soil-like surface upon which they were unable to detect any trace of dust.

Three craters along the west shore of Mare Humorum deserve our attention. The first one, two Plato lengths southwest of Gassendi, is MERSENIUS, Class 3, 49 by 58 miles across and 7500 feet deep. Compare its shape in Figure 20 and Figure 45 as it varies with the libration. Tonight the terminator bisects it, and all that shows is the light outer east wall and a small black bite out of the disk. In a few hours it will be a magnificent object marking the center of a cruciform structure of mountains and craters, but by tomorrow night it will be illuminated fully, washed out, and difficult to see. The unusually brilliant walls appear quite high as the sun rises over them, and the surrounding area gives the impression of extreme ruggedness. Keep this picture in mind, and you will be surprised tomorrow night at the apparent flatness of the region. South of Gassendi, across the mare and one Gassendi length northwest of Vitello, stands DOPPELMAYER, originally 37 by 44 miles across and described by Goodacre as "a fine specimen of a large ruined ring plain." Scarcely more than a half crater now, it appears as a bright semiellipse centered on a large mountain some 2300 feet high which might be visible tonight through binoculars. Unless conditions are unusually good you will see only the curved bright west wall which rises 2600 feet. Between Doppelmayer and Vitello and intruded upon by both of them are the ruins of a much older ring about 45 miles across with wall remnants up to

1600 feet on the south. About all that can be seen of it in this light is what appears to be a bright mountain spur 20 miles long which runs northwest from Vitello and disappears under the lava. It is known as LEE M since its southwest wall has been destroyed by LEE, a smaller half crater 24 by 27 miles across and 5900 feet deep. There we have a bay upon a bay on the south shore of Mare Humorum.

A line from Copernicus through Bullialdus extended a little less than its length reaches a prominent but strange "gash" in the lunar disk—a wide streak more than 100 miles long with a white border on the southwest and a black border on the northeast. That unusual enclosure is SCHILLER, apparently a triple crater of Class 3, 113 miles long, 48 miles wide, and 12,500 feet deep (Fig. 46). Here we are observing near the limb, and normal circular craters appear foreshortened into ellipses with width to length ratio of about 2 to 3. However, for Schiller the ratio is only 1 to 5, which tells us that it cannot be even roughly circular. Actually it is cucumber-shaped with the long walls nearly linear but not exactly parallel. Wilkins and Moore describe it as "obviously the result of the fusion of two rings," but examination of the best photographs shows no trace of a crater wall across its long, smooth floor, the only feature of which is a longitudinal ridge 30 miles long near the north end. Neison, Goodacre, and Alter all agree that it is a mountain-walled plain, but Baldwin has estimated that its walls rise some 9000 feet above the surroundings, another observation not in harmony with Alter's rule. Schiller is a mystery. It seems likely that it is the result of several unrelated and perhaps dissimilar events.

South of Schiller, at the terminator, you probably will see four conspicuous craters in a line parallel to the terminator and spread over an arc nearly twice the length of Schiller. Their interiors are narrow black ellipses set off by bright walls on the southwest. Generally the most prominent of the four is the west one, ZUCCHIUS, 45 miles across and 10,500 feet deep, but at times the libration in latitude can shift it southward to join the dark side. If that appears to be the case, the group will show better on another night. Just east of Zucchius is BETTINUS, 49 miles in diameter and 12,500 feet deep. Both craters have tiny central peaks which are too small for our binoculars. Southeast of Bettinus is KIRCHER, also 49 miles in diameter but 15,400 feet deep. Tangent to Kircher on the east is the much older WILSON, 41 by 47 miles across and

12,500 feet deep. While its outline is softer it shows almost as well as the other three which are of Class 1.

Just north of Bettinus and tangent to Zucchius may be seen a peculiar, boomerang-shaped enclosure that curves all the way up to the wall of Schiller. In fact, with its light west and dark east walls it appears to be an extension of Schiller. A few hours hence this curious valley may be extended and followed easily northwest from Zucchius along the terminator 125 miles, the whole structure resembling a vast horseshoe with bright Schiller at one end and a dark valley of similar outline at the other end. The curved valley is extremely old since its walls and floor have been broken by many craters ranging in size up to ancient Class 3 SEGNER, 45 miles across and 6600 feet deep. You may be able to resolve through your binoculars the low, rounded contours of Segner which is tangent to Zucchius on the northwest and which separates the "boomerang" from the shaded extension along the later terminator. The horseshoe valley ranks among the oldest surface structures since it antedates craters of Class 3, and its appearance becomes even more striking when examined from directly above as shown in the *Rectified Lunar Atlas*, Chart 23-b. There it appears to be the strongest part of a broad, flat ring formed between two concentric, circular, walled plains respectively 220 and 110 miles in diameter. The northeast wall of Schiller coincides closely with the outer wall which is a scarp of the mountain-walled plain variety and which can be traced in its entirety. Its only departures from the circular are found in two adjacent approximately linear sections, of which one is the Schiller wall and the other is the low, inconspicuous wall connecting the northwest tip of Schiller to the bright wall that runs south to the far end of the horseshoe. The inner wall can be traced over more than ¾ of its course. It runs parallel to and southwest of Schiller; originally it passed near the center of ancient Segner; and it is marked on the northwest by several hills and some low ridges. It is a circular mountain range rising to heights of around 7000 feet above the flat plains both inside and outside of it. On the northwest the inner wall appears to have been destroyed by lava flows, and in the same sector the outer wall exhibits a similar appearance along its weakest portion. The floor of the inner enclosure is stained heavily, suggesting lava flow, as is the northwest portion of the floor of the outer enclosure.

The probability is absurdly infinitesimal that

the double ring marks the walls of two large but independent craters formed at different times by the impacts of two uncommonly large meteoritic bodies, both of which struck the moon in the same spot and both of which came in along essentially vertical trajectories. Consequently, I believe that we are looking at a small circular mare formed in the manner proposed by Baldwin to explain the Maria Imbrium, Nectaris, and others. Extrapolating his data downward, I suggest that the inner enclosure is the true crater wall formed by the impact of an external body from 10 to 30 miles in diameter and that the outer wall is the scarp produced by the ensuing shock wave. If this is indeed a circular mare, it is an extraordinary one, not only because it is small but because it supports a Class 3 crater which, according to Baldwin, was excavated long ages before the impacts that produced the larger circular maria between two and three billion years ago. However, it is not unreasonable to suggest that a heavy impact of the mare-producing type could have taken place shortly after the lunar crust was sufficiently formed to sustain large permanent craters. The walls of the double enclosure have undergone considerable "erosion" or isostatic adjustment and probably are but remnants of their original structure. Were this not so the enclosure would have been observed and described long ago. Is the unexplained Schiller the result of later faulting along a weakened surface structure triggered by subsequent smaller impacts? Another clue is the muted and almost obliterated crater 75 miles in diameter on the floor of the inner enclosure and roughly tangent to its wall on the north. Its floor is on three levels of roughly equal area like broad, shallow steps descending from east to west. It looks far older than any of the Class 4 craters, and it may be the depression through which the lavas issued and later receded hundreds of millions of years after the mare basin was formed. When we have learned enough to read the complete history of the area we should know much more about how the moon got the way it is today.

TWELVE-DAY MOON

(Chart IV)

Tonight the moon appears near full to the naked eye even though it has 2⅓ more days to go. The illuminated disk is close to a perfect circle if the libration in longitude is positive and large (east limb tipped toward us). However, if we compare the east limb with the terminator through binoculars we realize immediately that we have not reached the west limb. The entire south half of the terminator is extremely rough, giving us an exaggerated impression of the rugged territory over which it passes. Tycho with its amazing ray system is the eye-catching feature. Aristarchus and Proclus with their different types of rays are prominent bright spots. Copernicus, of course, is outstanding, and the Kepler rays are growing brighter.

North of Sinus Iridum, at the terminator, peculiar BABBAGE shows well as a huge enclosure 100 miles across but no more than 6200 feet deep. Its four walls are nearly rectilinear, giving it the general shape of a rectangle about 85 by 95 miles on a side. It is very old and is intruded upon by numerous smaller craters including BABBAGE A, 17 miles in diameter and 8900 feet deep, which you might be able to resolve as a black speck on the south sunlit floor of the old formation. Northeast of Babbage, about twice its length, the terminator cuts across another strange enclosure. Its outline, well delineated by bright or black walls, resembles a distorted heart or a billowy cumulus cloud, and it consists of parts of at least five overlapping craters. The group actually is 160 miles long, 70 miles wide, and its meandering walls range in height above the floor from 2900 to 7200 feet. The newest crater, south of the center and 46 miles in diameter, now is known as ANAXIMANDER although the name previously was applied sometimes to the whole complex and sometimes to the southern pair. Obviously severely foreshortened because of its far limbward location, the true shape of the enclosure is not that of a heart but rather a fat, stubby wiener. Near its northeast extremity Anaximander is intruded upon by CARPENTER, one of the younger craters of the region. It is 38 miles in diameter, 10,200 feet deep, and it shows well as a thin black ellipse with a narrow bright border marking its north inner wall.

North of Aristarchus, west of Heraclides Promontory, and near the terminator, lies an object which is not prominent but should be seen without difficulty. It is a broad, low hill running north-south, and it is slightly lighter on its east side than the surrounding plain and slightly darker on its west side. You are looking at RUMKER, the largest of the domes. It is roughly elliptical, 48 miles long, of irregular profile, and quite lumpy. Rumker is a transitory object that cannot be seen with any power unless the terminator is near it. Hence, it is not surprising that it does not appear on the maps of some selenographers while on others it is shown as a crater of sorts. Estimates of its height range from 200 to 2500 feet. Alika Herring discusses it in *Sky and Telescope,* October, 1960, and shows his drawing made with a 12½-inch telescope. In the drawing it bears a remarkable resemblance to the back of a pitcher's mitt.

A line from Copernicus through Kepler extended west its length meets the terminator, and about ¾ of the way from Kepler to the shadow boundary it passes over REINER, a sharply-defined little Class 1 crater, 18 miles in diameter and 8500 feet deep. It stands out on the plain nearly half illuminated, and its location is marked by a much more conspicuous bright spot just west of it. In fact, one of Kepler's stronger rays points directly at the crater and extends almost to it. About ⅓ the way from Reiner to Aristarchus may be seen the larger but less prominent crater MARIUS, an inconspicuous object except when the terminator is in its vicinity. A flooded enclosure of Class 5, 27 miles in diameter and 5500 feet deep, it exhibits a gray interior set off by a bright west wall and a dark east wall. A ridge visible near sunrise meanders some 200 miles south from its west wall and an equal distance north toward Herodotus, but it is a difficult feature. As Webb has pointed out, the area west and north of Marius shows a great many "gray hillocks, the loftiest little exceeding 1000 feet." The ridges and hillocks vanish under higher sun, but probably even now they are beyond the reach of our binoculars.

Near the southwest shore of Oceanus Procellarum, west of Letronne, is a small triangular bright patch readily seen through binoculars. It is a mountain 20 miles long that rises only 3000 feet above the dark plain. Just south and just northwest of the patch are fully-illuminated medium-sized craters of about the same diameter which, through a small telescope, present striking floor contrasts. The south one is BILLY, 30 miles in diameter and 4000 feet deep, with a smooth floor that is one of the darkest stained spots on the face of the moon (Fig. 45). The northwest one is HANSTEEN, 29 miles in diameter and 4100 feet deep, with a rough bright floor (Fig. 45). Through binoculars you probably can make out the floor shade contrast, but Hansteen tends to merge with adjacent bright areas on both north and south while Billy, in spite of its bright border, resembles the mare in shading.

At the terminator west of Billy and Hansteen stands the small but prominent Class 1 SIRSALIS, 28

by 23 miles across and 8900 feet deep (Fig. 45). It appears as an elliptical black spot with bright west border. Within a few hours you will see that on the west it overlaps the similar but less distinct crater SIRSALIS A, 29 by 22 miles across and 6900 feet deep. Its twin makes it easy to identify. Sirsalis is best known for the remarkable Sirsalis Rill system (Fig. 45) that begins northeast of the crater, runs southwest, and curves south toward Byrgius. It is 285 miles long, the longest yet found, and it has several branches. However, for observation it requires not only favorable libration but more optical power than we are using.

North of Sirsalis, a distance equal to one Mare Humorum width, along an extremely ragged terminator, the old multiple-crater DAMOISEAU may be seen on the shore of Oceanus Procellarum. The main crater is 32 miles across and 3900 feet deep, and its west floor is occupied by a crater 25 miles in diameter which shares its west wall. Both have been there a long time, and they occupy the east floor of a much older enclosure some 45 miles across. Tonight all three are combined in one conspicuous black pit that joins the dark side. Tomorrow night you may notice that they stand at the entrance of what appears to be a broad valley running southwest. Southeast of Sirsalis and southwest of Hansteen is a smaller crater so placed that the three mark the vertices of an approximately equilateral triangle. A very old crater with low walls, FONTANA, 19 by 21 miles across and 8000 feet deep, is usually an obscure object, but tonight it catches the eye because of its bright inner southwest wall (Fig. 45).

In the heart of extremely rough territory, near the terminator, and about one Nubium width southeast of Sirsalis, lies the large Class 2 VIETA, 51 by 54 miles across and 17,400 feet deep. It is conspicuous tonight with bright west wall, black east wall, and floor almost fully illuminated. If the sunrise line were closer to the crater and conditions were excellent, you might be able to detect the long, narrow shadow of a mountain that cuts the northeast wall and sometimes reaches clear across the floor. Just southeast of Vieta is the similar but smaller Class 2 FOURIER, 30 by 35 miles across and 12,200 feet deep. Both craters are polygonal in outline, and they exhibit the characteristics of mountain-walled plains.

If we proceed south to the terminator from the south shore of Mare Humorum, we arrive at one of the more conspicuous features of tonight's moon. There we see the enormous mountain-walled plain

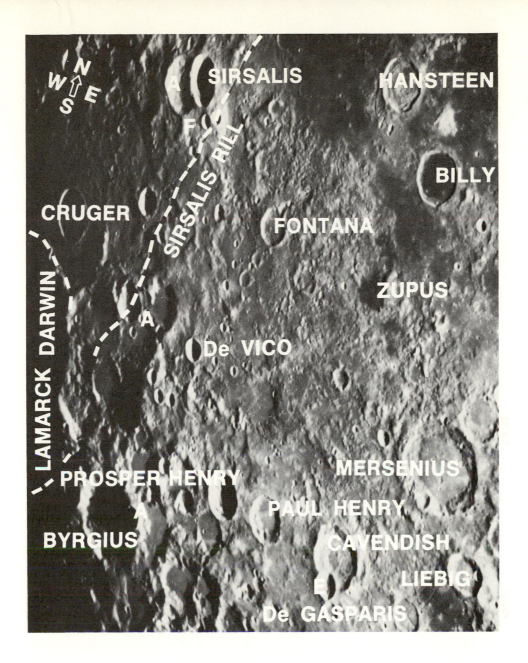

FIGURE 45. Sunrise at the Sirsalis Rill. The dark smooth floor of Billy contrasts with the lighter rough floor of Hansteen. Byrgius appears to be bordered by a group of lava lakes, and Byrgius A gives little indication here of its brilliant system of major rays. The curious enclosure Zupus looks more like a tiny mare than a crater as it cuts into the rugged highland border of Oceanus Procellarum. Only the irregular east wall of the vast multiple crater Darwin-Lamarck is visible in this light. (120-inch Reflector.)

SCHICKARD, of Class 5, apparently 150 miles in diameter and 9500 feet deep (Fig. 46). With a bright white border all around, edged in black along the northeast inner wall, it stands out prominently. A most peculiar feature is the stained floor which will be more evident tomorrow night. The central half has about the same shading as the surroundings, but the ends are much darker as seen through binoculars, and they are 700 feet lower than the central part. It definitely is convex, following the general curvature of the lunar surface, so an observer standing at its center would find that his horizon obscured every portion of the walls. The true shape of the crater can be interpreted equally well as roughly hexagonal or roughly elliptical with axes of 150 and 135 miles.

Just southeast of Schickard, on the terminator, is a similarly outlined crater which is smaller and more irregular. Tonight it resembles closely a print left by a huge shoe in damp earth. The ball part of the footprint is the Class 3 crater PHOCYLIDES, 76 miles long and 6900 feet deep. The heel part, on the side toward Schickard, is the older crater NASMYTH upon which Phocylides has intruded (Fig. 46). It looks much smaller, being 34 miles long, and its 4300-foot walls are considerably shallower than those of its dominant neighbor. Phocylides actually is turnip-shaped, 76 miles high, and 89 miles broad, while Nasmyth is nearly rectangular, 34 by 54 miles on a side. Tangent to Nasmyth on the west and close to Schickard on the south is a dim semiellipse bounded by the terminator, somewhat smaller than Phocylides, but very difficult to see through binoculars because it is almost completely lacking in the highlights and shadows that accentuate lunar features. Under the best conditions it is just visible through binoculars when close to the terminator as it is tonight. It is WARGENTIN, worth hunting because it is one of the moon's most celebrated freaks (Fig. 46). If there is anyone who refuses to believe that molten lava has in bygone ages flowed from the interior of a hot moon, Wargentin appears to confute him. It evidently was once a normal crater 59 miles in diameter. Perhaps it remained so for hundreds of millions of years, or perhaps the impact that produced it triggered the flow. In any case, lava welled up through a fissure and, by strange coincidence, stopped flowing just before it reached the crater rim. By another coincidence, the walls had no significant breaks, and they held the mass until eventually it congealed. Wilkins and Moore have placed the floor level 1400 feet above the surrounding territory and the maximum rim height only 500 feet above the floor. Army Map Service sets the elevation of the lava plateau at 1000 feet above the plain to the northwest. In a small telescope, under favorable lighting, Wargentin looks like an old-fashioned circular cheese, but its thickness is very small in terms of its diameter. Consequently, it has the nickname "The Thin Cheese."

Harold Urey did not accept the above century-old explanation of Wargentin but suggested that the crater may be filled with ice and covered by dust. He argued that if the moon could produce enough molten lava to fill such a large crater, the lunar interior would not be rigid enough to support mountain ranges six miles high. However, igneous lunar rocks gathered by astronauts, observed caldera and lava flow patterns, and the absence of both dust and moisture in all explored areas are strong evidence against the logic of the venerable Nobel laureate in this case.

THIRTEEN-DAY MOON

(Chart IV)

With the exception of the elusive one-day moon, we have proceeded on the assumption that the crescent is bisected by the moon's equator, which passes through Mare Smythii on the east limb, runs north of Delambre, south of Godin, and into small Sinus Medii. Likewise we have expected the terminator of the gibbous phases to move westward along the equator, through Lansberg, so that its midpoint would meet the west point of the limb at full moon. That assumption would require that the moon's axis make an angle of 90 degrees with the plane of its orbit (instead of 83½ degrees). It also would require coincidence between the plane of the moon's orbit and the plane of the earth's orbit around the sun (instead of the existing angle of five degrees between those planes). Since the actual departures from such simplified conditions are not large, it has been true that the terminator has crossed the moon's equator almost at a right angle and that the midpoint of the visible terminator has remained close to the equator. However, such a condition cannot continue through full moon unless we have a total eclipse with the moon passing centrally through the earth's shadow cone.

Ordinarily at the instant of full, the moon is a few degrees either north or south of the shadow cone.

That means that the east and west limbs both are illuminated fully but the sun's rays fall a little short of either the south or the north limb. Thus at full we shall not see the true limb all the way around the disk. There will remain at least the slight semblance of a terminator, and its midpoint will be found at either the north or the south point on the disk. So, during a period of three to four days around full, the terminator leaves the west limb and proceeds to the east limb via either the north or the south limb, and we cannot predict in general even approximately where it will be found on a given night. That will depend upon the details of the lunation.

Tonight you may find the center of the terminator close to the west point on the disk, or you may find it almost anywhere else between the northwest and the southwest points. In extreme cases it could affect the visibility of some of the features I shall list as near the terminator, assuming an average or west centered position for the latter.

A line from the large bright ring of Posidonius, prominent on the northeast shore of Mare Serenitatis, through Copernicus and extended on to the terminator, reaches a conspicuous black ellipse larger than Plato. That is the mountain-walled plain GRIMALDI, of Class 5, 140 by 145 miles across and 10,500 feet deep. It looks as though there is a large crater there with an exceptionally dark floor about 100 miles in diameter and with unusually broad walls. The floor shares honors with several other areas as the darkest spot on the moon, reflecting only 6 per cent of the light striking it. One of the more conspicuous features tonight, Grimaldi is outlined by a bright ring, broken on the north, and broadest on the east where it consists of upland foothills between the crater and Oceanus Procellarum. On the west it takes the form of a bright mountain wall while on the south the ring is bright but edged in black shadow. Goodacre points out that "the walls are very discontinuous," and Dinsmore Alter in his *Pictorial Guide to the Moon* (1963) suggests that "Grimaldi is morphologically truly a connecting link between the maria and the typical walled plains." By tomorrow night the walls will be washed out almost completely, and the black spot will look quite like a small mare. Webb notes "it has sometimes been detected even without a telescope." Can you see it with the naked eye?

Just north of Grimaldi we see Class 5 HEVELIUS, 66 by 69 miles across and 7000 feet deep. It shows well as a bright ellipse, and it is similar to its southern neighbor except that it contains a broad, low mountain or hill, off center to the north, which is difficult to distinguish from the bright floor through binoculars. Tomorrow night the whole crater will be washed out. The north wall of Hevelius is intruded upon by Class 1 CAVALERIUS, 37 miles in diameter and 10,800 feet deep. It readily is seen as a narrow bright ellipse with a thin black border on the east. Sixty miles northeast of Cavalerius, on the edge of Oceanus Procellarum, the first successful landing of a lunar probe took place, as we have seen. There on February 3, 1966, Soviet scientists landed 220-pound Luna 9. Its panoramic views televised to earth revealed an uneven, jagged surface without an appreciable dust cover but characterized by protrusions and depressions down to a fraction of an inch across.

On the terminator one Iridum length north of Cavalerius may be seen Class 2 CARDANUS, 31 miles in diameter and 6600 feet deep. It appears as a narrow black ellipse with a bright border on the west. Just north of Cardanus is its identical twin KRAFFT, the walls of the two craters being separated by the amount of their common diameter. Northwest of Reiner about half the distance to Cardanus, is a smaller Class 1 crater with an even smaller one just north of it. The two may be visible through your binoculars tonight, but they can be seen only when near the terminator, and they are severe tests at best. Their diameters are 10 and 8 miles, and their depths are 6600 and 4600 feet respectively. Why does one pause to look for such tiny bright dots at the limit of visibility? The reason becomes evident when I explain that the larger one is GALILEO and the smaller is GALILEO A. If they don't show tonight look for them next month on the 12-day moon when they will shine like faint stars out of the darkness beyond the terminator.

Here we have dramatic proof that those who named the lunar features either lived too close to some scientists to appreciate their merit or had other purposes in the selection of names. Tycho Brahe, Copernicus, Plato, and even Ptolemy and Kepler were well memorialized, as their achievements entitled them to be, but what was done for Galileo was disgraceful in its niggardliness. The father of experimental science, the astronomer whose observational discoveries are probably unparalleled in importance when evaluated in terms of contemporary concepts, a professor persecuted by the Church for his scientific activities and teachings—all this, and it rated him only one of the smallest of the named craters located almost at the limb! On the other hand, we already have identi-

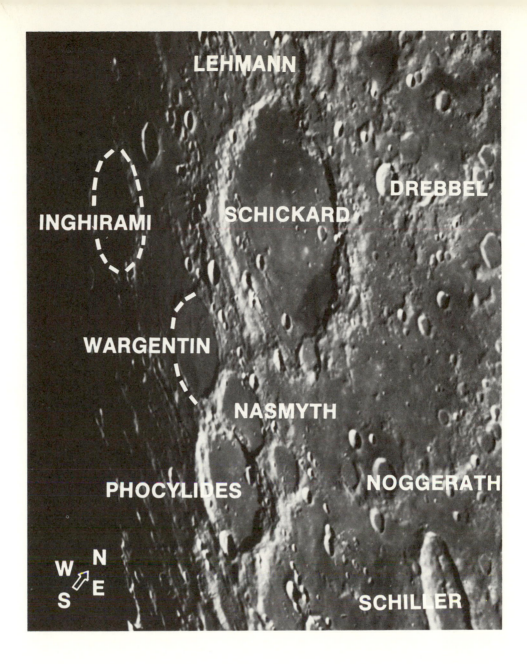

FIGURE 46. Early-morning sun illuminates gigantic Schickard and the double crater Phocylides-Nasmyth which resembles a huge shoe print in mud. (120-inch Reflector.)

fied numerous conspicuous craters with names that are significant only to historians. Such features perpetuate the names of men long forgotten, men whose contributions to knowledge, if any, were either fundamentally revised or rejected generations ago. As the antithesis of Galileo's poor pit we point to the magnificent crater Theophilus, which tonight can be seen as a large, delicate, bright ring nearly halfway from the south shore of Mare Crisium to Tycho. The answer to Theophilus' identity is not found easily. If we go back to *The General Biographical Dictionary,* published 170 years ago, we find that he was the Patriarch of Alexandria around the beginning of the fifth century. His listed accomplishments were the destruction of many non-Christian churches and a mastery of intrigue. So much for the injustice of lunar nomenclature! It never will be corrected for the same reason that a simple and efficient calendar never will be adopted. It is too deeply rooted in the record of civilization.

A line from the north tip of Mare Serenitatis through brilliant Aristarchus passes over a second bright spot just before it reaches the terminator. That spot is Class 1 SELEUCUS, 28 miles in diameter and 7500 feet deep. Seleucus is notable for the long, bright rays that might seem to have some connection with it. The brighter one, which you may be able to resolve, runs along its southeast wall and proceeds to Cardanus on the south and, in the other direction, to a point 125 miles north of Aristarchus. The fainter ray begins west of Aristarchus, converges with the brighter one which it crosses south of Seleucus, and continues to the north wall of Cardanus. About two Plato lengths north of Seleucus you may see a small light spot which is the crater BRIGGS. Goodacre describes it as "a fine regular ring plain." Although it is 24 miles in diameter it is only 3800 feet deep and not so easy to find in the limb area as one might expect.

To the north of Sinus Iridum, near the terminator, the fine Class 1 mountain-walled plain PYTHAGORAS stands out conspicuously as a thin bright ellipse on a bright background with brilliant northwest and black southeast walls. It is 81 by 90 miles across and 16,400 feet deep, and from its center rises a large mountain 6000 feet high which can be resolved through binoculars with good conditions. Webb calls it "the deepest walled plain in the quadrant," that is, the quarter of the moon's disk bounded by imaginary lines from the center to the north and west points on the limb. Pythagoras shows very well from its poor location, and it would

be a spectacular object indeed if we could see it rising from a plain in a location as favorable as that of Copernicus.

North and northeast of Pythagoras may be seen a few cigar-shaped craters that have received names only recently. Their visibility depends, of course, upon favorable libration. You will find most of them marked on the charts that show them and listed in the gazetteer. Southwest of Pythagoras the vast plains of northern Oceanus Procellarum push farther westward, leaving little detail in view for us tonight. Let us, therefore, return to our prominent dark base crater Grimaldi and explore the southern hemisphere. About one Grimaldi length south of its south rim at the terminator we see ROCCA, an old crater 48 by 52 miles across and 7200 feet deep. It is so close to the limb that study of its interior is difficult, but locating it presents no problem since it stands out prominently as a narrow black ellipse with a bright rim. One Grimaldi length southeast of Rocca or about two Grimaldi lengths south of its own south rim you may find a much smaller dark spot which presents greater contrast photographically than visually. That is old, low-walled CRUGER, 26 by 30 miles across but only 3300 feet deep, which Goodacre describes as "remarkable for the darkness of its interior under all angles of illumination" (Fig. 45). It evidently has been flooded with lava, and it resembles Billy to the northeast. With success in the location of Rocca and Cruger, push your skill a little further and see if you can detect a couple of tiny dark markings between the two craters. They are at the limit of visibility, and failure to spot them is no reflection on your selenographic ability. The only reason I mention them is to call attention to the curious fact that they constitute MARE AESTATIS! Here is another example of the absurdity to which the feature namers resorted occasionally. Mare Aestatis doesn't cover enough square miles to make even a medium crater. West of it are two more such "maria" which can be seen only with favorable libration and then merely as thin dark lines parallel to the limb.

Two Crisium widths south of Grimaldi, and close to the terminator west of Mare Humorum, is the bright Class 3 BYRGIUS, 45 by 53 miles across and 15,000 feet deep (Fig. 45). Its black east border may make visible on the northeast wall crest the intruder BYRGIUS A, 10 by 13 miles across, 4000 feet deep, and the center of a ray system which Baldwin has traced 260 miles. It is the latter which has splashed the much older Byrgius with bright ejecta and made the whole area brilliant at full. Byrgius

A is difficult to resolve through binoculars but its bright splash pattern is evident. West of Byrgius, at the terminator, is the south end of a large, irregular enclosure that extends north 165 miles almost to Cruger. It is the ancient, multiple crater long known as DARWIN, which consists principally of three craters of progressively greater size from south to north (Fig. 45). The name Darwin now is restricted to the large northern part of the structure which at one time was an elliptical crater 80 by 89 miles across. The central portion, highly irregular but about 65 by 80 miles across, is called LAMARCK. The south crater, approximately rectangular and 40 by 50 miles on a side, remains unnamed. The heavily-battered meandering structure is conspicuous tonight with its sunlit but stained floor set off by a bright west wall and a dark east one.

Southwest of the great mottled Schickard, near the terminator, is Class 2 INGHIRAMI, 62 miles in diameter and 10,800 feet deep (Fig. 46). It shows well, a thin ellipse with a brilliant southwest wall and a black northeast wall framing a gray floor. Its high wall casts a broad shadow west toward the terminator. Inghirami is almost identical in size and outline with Wargentin, about 1½ lengths to the southeast, but the contrast between its prominence and Wargentin's obscurity is a striking example of our dependence on highlight and shadow in lunar studies.

FOURTEEN-DAY MOON

(Chart IV)

Tonight the moon essentially is full, a brilliant circular disk, enormous, just above the eastern horizon shortly after sunset. A good demonstration of the angular speed with which the earth rotates on its axis is provided by the rising moon. Watch it behind trees 200 to 300 feet away in the winter season when the branches are bare. You can see it creep steadily upward and to the right in the northern hemisphere. I am sure you have noted that the rising (or setting) moon appears a great deal larger than the moon riding high in the sky. Not only is the impression an illusion, but the opposite actually is true. If you had a device for measuring small angles, such as an old heliometer, you might be surprised to find with it that the moon high in the sky is larger in angular diameter than the moon that rose a few hours earlier! The explanation of the paradox goes along with that of the diurnal libration. The rotation of the earth carries us toward the moon in the eastern sky and away from the moon in the western sky. Thus the observer is one earth radius or 4000 miles closer to the moon in the zenith than he is to the rising or setting moon. Since the earth's radius is about 1/60 the average distance to the moon, the moon in the zenith is nearly 2 per cent larger in angular diameter than the moon on the horizon. Our senses give us the opposite impression because of the automatic comparison and judging of size that we go through whenever we see the moon near the horizon together with distant houses, trees, and other terrestrial objects of known size. When we see it high in the sky, we can compare the moon only with the immensity of space that surrounds it. Furthermore, research has shown that when we tip our heads back to look upward, the changing direction of the earth's gravitational pull on our liquid eyes and the muscles that control them causes objects to appear farther away (and smaller) than when we view them at the same distance horizontally.

The average full moon is 14 days, 18 hours, and 22 minutes old. Consequently, our standard terminator for tonight's observation is taken about eight hours before the instant of full phase. However, if new moon for the present lunation occurred shortly after midnight two weeks ago, the moon you see may be already past full. So the terminator, readily located through binoculars, may be centered at *any* position around the limb if we include the possibility that a total lunar eclipse may have taken place a few hours before the time of observation or may occur a few hours later.

Again an outstanding feature is Tycho with its spectacular ray system. The crater is a bright ellipse of such dazzling brilliance that no interior details can be resolved. It is surrounded by a gray ring of uniform width except on the south where it widens considerably. From the ring the bright rays proceed outward in many directions, some of the more conspicuous ones extending to Mare Nectaris, Stevinus, Steinheil, and Bullialdus. A strong ray starts out northeast toward Mare Serenitatis. On the way it joins with those of Hind and Godin, and then with those of Menelaus and Bessel, which stretch across Mare Serenitatis. From there a broader bright streak continues along the east edges of Lacus Mortis and Mare Frigoris to mingle with the rays of Thales, a small crater near the limb. Thus we have the impression of a single ray extending all the way from Tycho to the most remote point of the limb. The most conspicuous ray on the south side of Tycho is narrow but intense. It radiates not from the crater but from the west edge of the gray ring, and it runs toward the nearest point on the limb, almost opposite to the direction of the extended ray just described. If the limb is visible beyond the end of the strong, narrow south ray, and if the libration is favorable, you may see several tiny humps or "pimples" on the terminator. Those are the great peaks of the DOERFEL MOUNTAINS, seldom seen but excellent when viewed in profile. They are among the loftiest known to us since the elevation of one of them has been measured at 26,000 feet. Our own Mount Everest is 3000 feet higher in actual elevation, but when we express the height of a mountain in terms of the radius of the world on which it rises, it is evident that the Doerfel peak relatively is more than three times as high as Mount Everest! If the range is not visible tonight, it is well worth looking for whenever that portion of the limb is illuminated. If you have access to an almanac which gives daily data "for Physical Observations of the Moon," look for the Doerfel Mountains when the libration in latitude (Earth's Selenographic Latitude) is between –3 and –6 degrees.

While at the south limb try for the LEIBNITZ MOUNTAINS, "the most imposing and massive of all lunar mountain ranges," according to Wilkins and Moore. Extend the line from Copernicus to Tycho southward along the bright, broad ray. Where the latter fades out beyond Moretus and Newton, the western peaks of the Leibnitz range may be visible. From there they stretch eastward nearly 600 miles, covering about $\frac{1}{12}$ the circumference of the lunar disk. The mountains generally are less impressive but more frequently in sight than the Doerfels since they do not require strong libration to bring them over the limb. They may be higher than the Doerfels since some of the nineteenth-century observers obtained elevations there of from 30,000 to 33,000 feet.

A conspicuous feature again tonight is the narrow, dark ellipse near the west limb which marks the heavily-stained floor of the large crater Grimaldi. Adjoining Grimaldi on the northwest is the smaller crater RICCIOLI of Class 3 with diameter 97 miles and depth 7500 feet. Its walls are not as high as those of Grimaldi, and already they are washed out against the bright background. Only a small dark spot near the north edge of the old and battered floor rivals Grimaldi's darkness while the rest of the floor resembles the surrounding territory. Consequently, we get the impression that Riccioli is much smaller than Grimaldi when actually their diameter ratio is 2 to 3.

North of Riccioli and west of Kepler (extend Kepler's west ray on beyond Reiner nearly to the limb) lies bright Class 1 OLBERS, 45 miles in diameter and 9800 feet deep. Last night it was in darkness beyond the terminator. Tonight it is illuminated fully and already washed out by ray glare. To see it at all you must observe when it is close to the terminator. Tangent to it on the northwest is the smaller yet much more prominent, narrow, brilliant ellipse of OLBERS A, a young crater 26 miles in diameter and 8200 feet deep, which you may be able to resolve. Olbers A is the center of a major ray system, parts of which are causing the brightness in the area now. Its strongest and second strongest rays are the two we traced last night past Cardanus and Seleucus to the uplands north of Aristarchus. The brighter one extends 650 miles. In another night or two the rays will be so bright that both crater outlines will be lost in the glare, a condition which probably accounts for the common misconception that Olbers rather than Olbers A is the center of the system.

About one Crisium length north of Olbers and west of brilliant Aristarchus a bright feature shaped like a wishbone stands out on the dark plain. That is what remains of an elliptical crater 80 by 90 miles across, much of which was melted away long ago by the Procellarum lavas. Its south wall can be traced under favorable lighting, but it is little more than a broad ridge less than 1000 feet high. The wishbone portion of the wall rises 4300 feet and shows well. Recently it was named EDDINGTON. Its west wall is also the displaced east wall of older OTTO STRUVE, a tomato-shaped crater 97 by 115 miles across and 5600 feet deep. Its south floor appears as a bright spot just west of Eddington, and it looks like a double crater which stretches 165 miles north. There it ends in what resembles a second wishbone slightly smaller than Eddington. The second wishbone is RUSSELL, once a nearly circular crater 62 miles in diameter and now only 3300 feet deep. It evidently is younger than Otto Struve upon which it intrudes, but both are premare, and both have had their south walls almost completely melted away. The west walls of Otto Struve and Russell have their own name, the HERCYNIAN MOUNTAINS. A strange mountain ridge it is that casts no shadow when the sunset terminator is near! Elevations up to 4300 feet above the plains have been measured there, but the upward slope to the west must be slight indeed. At one time the name Hercynian Mountains was applied to an illusory range paralleling the crater wall to the west, a feature that I have sought without success at times of favorable libration.

As we swing back toward the south limb in search of the largest named crater, let us pause for a moment near the southwest limb where we viewed some large craters two or three nights ago. The walls of Schickard have vanished, but the great mountain-walled plain still may be found easily because of the odd two-tone floor, the north ⅓ and the southeast edge dark stained while the central portion is bright. Phocylides, Nasmyth, and Schiller virtually are obliterated, while Wargentin, a difficult binoculars object at best, is invisible tonight even through a telescope. Close to the limb 1½ Crisium lengths southeast of Schickard, and an equal distance southwest of Tycho, lies the vast crater we are seeking. It is about one Crisium width long. Class 5 BAILLY, 200 miles across and 13,800 feet deep, never shows well. Last night the terminator cut through it, and all was black except for the outer northeast wall. Tonight the crater appears illuminated fully and washed out. Perhaps you can distinguish the billiant southwest wall and detect a narrow line of shadow just inside the northeast and east walls—an inconspicuous frame for the bright floor. With favorable libration the huge, roughly circular enclosure opens slightly into a narrow ellipse. Then the intruding BAILLY B on the east floor, about half bright and half black, is easier to spot than its host. It is 44 miles in diameter and 11,500 feet deep, and it is the largest of a great many craters which mar the floor and walls of that ancient excavation. If you found the Doerfel Mountains you may view Bailly under favorable conditions since the mountains lie directly behind it.

FIFTEEN-DAY MOON

(Chart IV)

Tonight, for the first time this month, we see the smooth, sharply defined west limb of the moon bounding the visible disk in place of the less-distinct and often irregular terminator which we have followed for two weeks in its continuous progress. The sunrise terminator now has gone around the west limb, and, hidden from our view, it will continue its tireless advance over craters and mountains which the eye of man never had seen "face to face" but only "through a glass darkly" until early on the morning of December 24, 1968, when Apollo 8 carried astronauts Frank Borman, James Lovell and William Anders into orbit behind the moon on a non-landing but epoch-making reconnaissance mission. While the west limb has been revealed to us, a portion of the east limb has been obscured, another example of the balancing or compensatory action of those great universal forces which our less-sophisticated but more appreciative ancestors referred to as "Nature." The sunset terminator has come into view, and we shall follow it for two weeks as it slowly lowers the curtain over the ever-changing scene of the moon's monthly performance. Again tonight I am unable to specify with any degree of accuracy which portion of the limb the terminator has cut off. It takes a good eye to detect any departure of the moon's outline from the circular, but through binoculars you can locate easily the terminator by its irregularity in contrast to the general smoothness of the limb. Its midpoint may be found near the northeast or the southeast edge of the disk or at any point between. Unless a total eclipse has veiled the moon within the last 24 hours, the midpoint of the terminator will not be found close to the east edge of the disk. However, elementary statistics again remind us that the east edge is the most likely location or, rather, the average over the years of all actual locations for the moon at this age. So I shall proceed on that basis with the reminder that some of the features located near the average terminator may appear somewhat removed from the actual terminator which you see tonight. By tomorrow night the midpoint of the terminator once again will be near the moon's equator along which it will advance westward as before.

Having pointed out the smoothness and regularity of the limb, let us now see if we can detect some slight departures therefrom. Close inspection of the limb west of the conspicuous dark Grimaldi, with favorable libration, should reveal several pimple-like protrusions thinly scattered over a distance of nearly 200 miles. Those are the peaks of the D'ALEMBERT MOUNTAINS, some of which have been measured as 20,000 feet high. They have been described as a range or chain, but they appear as isolated mountains like Pico and Piton. The brightest and most conspicuous of the few D'Alemberts we occasionally see lies west of the south rim of Grimaldi. Only twenty-four people, all American astronauts, have actually seen the far side of the moon, but five Lunar Orbiters did a splendid job of photographing the entire surface in 1966–67, and orbiting Apollo astronauts took hundreds of pictures. Hence, the far side is mapped as precisely as the near side.

Now turn your attention from the western limb, where the early-morning sun is yet low in the sky, to the eastern part of the disk, where the late-afternoon sun is low in the sky and evening is at hand. A line from Copernicus to the gap between Mare Serenitatis and Mare Imbrium extended an equal distance beyond reaches a narrow, dark ellipse with a bright border. It is a little larger than the conspicuous Plato between Mare Imbrium and Mare Frigoris, but narrower owing to foreshortening and not quite so black. That is the lava-flooded crater Endymion which we explored 12 days ago under different light. Beyond Endymion, at the terminator or very close to the limb, may be seen the longer dark patch which is Mare Humboldtianum. It will appear very narrow, resembling a flat isosceles triangle if the D'Alembert Mountains are in view, but if they are not, the mare may show well as an irregular elliptical dark area apparently bounded by mountains except on the east where the terminator cuts across it.

Southeast of Mare Humboldtianum and an equal distance north of Mare Crisium is the large, ancient mountain-walled plain Gauss. In this light it shows splendidly as a narrow, lemon-shaped crater through which the terminator runs, its east wall crest bright and the west one outlined by a thin black arc. Its old walls and floor are broken by numerous small craters. Just south of Gauss lies the smaller but sharply outlined crater HAHN, 50 miles in diameter and 10,200 feet deep, with a large, bright central mountain sometimes visible through binoculars. Northwest of Hahn and southwest of Gauss is the slightly smaller Class 1 BEROSUS, 44 miles in diameter and 10,500 feet deep. The two craters are similar except for the absence of a central mountain in the latter, and both would appear very nearly circular if we could view them from directly above. About ½ Crisium length east of the north tip of Mare Crisium and close to the terminator may be seen Plutarch. It is never one of the more conspicuous objects, but

tonight a strong shadow extending from its east wall to the terminator enhances its appearance.

One of the brilliant spots on the disk catches the eye two Crisium lengths south of Mare Crisium on the east shore of Mare Fecunditatis. Langrenus has been a prominent feature continuously since the sun rose on it 13 days ago. It is larger and brighter than Copernicus, and it would compete strongly with even the splendid Tycho were the latter shorn of its unique rays. It is definitely a young crater of Class 1, and Baldwin has traced its ray system outward nearly 500 miles, but its rays are narrow and weak compared with those of numerous smaller rings. Its eminence is owing almost entirely to the high reflecting power of its floor and walls, and, of course, to its large 85-mile diameter. On the excellent aeronautical chart LAC 80 of the United States Air Force, corrected for the effects of foreshortening, Langrenus appears approximately circular but definitely polygonal in outline. In a bright area at the terminator, east and a little south of Langrenus, you may be able to make out a small elongated black dot. That is the interior of Ansgarius on which the sun is setting. Northwest of Ansgarius and almost tangent to it is the slightly smaller and weaker Lapeyrouse.

At the terminator two Crisium widths south of Langrenus the cigar-shaped outline of the great mountain-walled plain Humboldt stands out under low sun unless the libration in longitude is negative and large. Its 125-mile diameter makes it an easy object even with extreme foreshortening, and its broad central mountain range 35 miles long often is seen through binoculars. Humboldt, you may recall, was the best feature listed for the elusive one-day moon. Tonight it may be found more readily. Southward along the terminator or close to the limb, a little more than one Crisium length beyond Humboldt, you may find a small, narrow black ellipse bordered by bright walls. That is the dark-floored crater OKEN, 45 by 48 miles across and 10,500 feet deep. With favorable libration, quite a few dark streaks parallel to the limb may be seen in the vicinity of Oken, particularly on the south side. Those are the bays and inlets that constitute the visible portion of Mare Australe.

The great ray craters have been discussed as we came upon them in our westward journey of exploration, and most of them have been mentioned again as prominent objects in the last few evenings. But there are a great many more bright objects on the face of the moon around full, as doubtless you already have noted. A complete list of bright rings and spots visible through binoculars at this time would include hundreds of entries. Not only would such a list be excessive for our purpose, but some of the entries might be identified by grid coordinates only since a large number of the smaller bright spots are so inconspicuous during most of the lunation that they have no names. Consequently, I shall not attempt a complete coverage of bright objects but shall point out the more prominent or notable ones. Let us begin at the north limb and work our way back and forth across the disk until we come to the far south where the whole area is so brilliant, and bright little craters are so numerous that even a sampling might become tedious. In order to facilitate identification I shall give the diameter of each crater even though it has been listed earlier.

A line from bright Copernicus across dark Plato extended almost to the limb reaches a brilliant area from which bright rays may be followed several hundred miles in many directions. The center of the system is 33-mile (diameter) Anaxagoras. North of Sinus Iridum and a little more than one Crisium length west of Plato 26-mile Harpalus stands out against the dark background of Sinus Roris. Halfway from Plato to the east edge of Sinus Iridum is the Straight Range. East of the latter and southwest of Plato are the scattered Teneriffe Mountains. About two Plato widths south of that crater is the brilliant mountain Pico.

One Apennine Range length east of Plato, and the same distance southeast of Anaxagoras, we see the bright ring of 55-mile Aristotle, and south of Aristotle, halfway to the shore of Mare Serenitatis, we spot the companion ring of 44-mile Eudoxus at the north end of the Caucasus Range. One Crisium length east of the last pair is the somewhat fainter elliptical ring of 45-mile Hercules which has a bright spot just southeast of its center marking the interior of HERCULES D, 10 miles in diameter and 2600 feet deep. Just east of Hercules and south of dark Endymion is the weak ring of 54-mile Atlas, of which only the east crest shows well. East of Atlas, the same distance as Hercules to the west, is the most brilliant spot in the whole area. It is the work of a tiny unnamed ray crater only five miles in diameter. About one Crisium width north of Hercules, and an equal distance northwest of Endymion, lies the center

of a broad ray pattern, the young crater THALES, 22 miles in diameter and 5900 feet deep. There one night in 1892 the keen-eyed astronomer Edward Barnard at the Lick Observatory saw a localized luminous cloud.

Another ray system that is strong just before full moon seems to be centered on 53-mile Geminus located about halfway from Endymion to Mare Crisium. Actually the rays originate at two 10-mile craters respectively north and east of the main crater. The first, which produces most of the rays, is MESSALA B, and the second is GEMINUS C. Tonight the rays have faded, and Geminus appears as a dark-edged bright oval on a bright background. On the northeast shore of Mare Serenitatis, and separating the latter from small, dark LACUS SOMNIORUM, is the bright ring of 61-mile Posidonius with several brighter spots within it and on it. A little more than one Crisium length west of Posidonius and an equal distance south of the Aristotle-Eudoxus pair look for tiny LINNE on the Serenitatis plain. A line from Copernicus tangent to the arc of the Apennine Range passes right over it at a point about midway between the north and south shores of the mare. The strange history of this apparently evolving feature was summarized in an earlier chapter, but the important fact tonight is that you are observing through your binoculars a lunar crater only one mile in diameter centered on a bright spot seven miles across.

One Crisium length southeast of Plato, measured along the Mare Imbrium shore, is the small, bright, distorted ring of Class 1 THEAETETUS, 16 miles in diameter and 7700 feet deep (Fig. 42). It is seen easily just west of the Caucasus Mountains. Wilkins and Moore report that in 1902 Charboneaux at the 33-inch Meudon refractor saw an "unmistakable white cloud" appear near the crater and that 50 years later Moore at his 12½-inch reflector "saw a hazy line of light crossing the shadow-filled interior."

West of the Caucasus and north of the Apennines a right triangle is marked out by the bright rings of 36-mile Aristillus, 25-mile Autolycus, and 51-mile Archimedes. The first one is strong, but the other two are not distinguished easily from their light backgrounds. North of Archimedes and west of Aristillus, the Spitzbergen Mountains are conspicuous on the dark Imbrium plain. On a line from Copernicus to Plato and almost halfway there a large bright spot may be seen. It is 22-mile Timocharis plus a bright area roughly 30 miles square adjacent to it on the east. Timocharis is another crater over which the appearance of a cloud once was reported. Southwest of Timocharis and an equal distance north of Copernicus shines little 15-mile Pytheas, a brilliant spot in a bright region. It is, in fact, one of the brightest spots on the disk, and, with good seeing, your binoculars will show the ring to be even brighter than the interior.

About one Crisium width west and a little north of Pytheas, and likewise in the midst of the Copernican rays as they thin out over Mare Imbrium, is a larger, irregular bright spot, part of which appears to be an intensification of the Copernicus ray system. The round north end of the patch is 17-mile Euler. West of Euler one Crisium length we couldn't miss the dazzling pendant patch of 25-mile Aristarchus and its triangular train to the southwest over the rim of rayless Herodotus. If we follow two narrow rays that begin at different points north of Aristarchus and curve southwest, we come upon the bright ring of 28-mile Seleucus about one Crisium width west of Aristarchus. The two rays converge and cross over south of Seleucus and continue southward about one Crisium length from that crater to 45-mile Olbers close to the west limb and almost washed out in its bright surroundings.

Southeast of Aristarchus, about one-fifth the distance to Tycho, is the bright splash pattern of Kepler's rays. So bright is the patch that the outline of the 20-mile crater near its center is difficult to discern. An equal distance east of Kepler we recognize, of course, the prominent 60-mile Copernicus with its delicate, spidery, multiple rays spreading in a uniform sunburst pattern over considerable portions of Mare Imbrium and Oceanus Procellarum. There has been speculation concerning the possible significance of what appear to be interlocking ray systems such as the well-defined double rays that mark out a large 45-degree right triangle on Oceanus Procellarum with Copernicus, Kepler, and Aristarchus at its vertices. Are these merely accidental alignments? Is it a coincidence that the Kepler-Aristarchus side of the triangle lies on the arc of the great double ray that runs from Tycho across western Mare Nubium? When such questions can be answered we shall know much more about the mysterious rays. West from Kepler runs a strong double ray which could be a continuation of the strongest ray linking Copernicus and Kepler. If continued, it would reach Olbers and tie the latter into a triangle with Kepler and Aristar-

chus. It stops short of the relatively inconspicuous 18-mile Reiner, just west of which is seen a prominent diamond-shaped bright spot. Telescopic examination of that spot, which has no appreciable relief, discloses an almost complete elliptical dark ring well centered upon it. Here we expect to see elliptical craters such as Reiner because of foreshortening, but the dark ellipse has its long axis rotated about 45 degrees instead of 90 degrees to the direction in which the foreshortening takes place. The marking actually is elliptical and more than twice as long as it is wide.

In the broad southeast fan of the Kepler ray splash, on the way to Tycho, may be seen a brilliant spot that outshines both Kepler and its rays. It is 18-mile Encke. Farther along toward Tycho, a distance from Encke equal to the Aristarchus-Kepler separation, a prominent bright area is noted. The east portion, resembling a white bow tie 100 miles wide, is the small, low Riphaeus Mountain Range which separates Mare Cognitum from Oceanus Procellarum. Just west of the range is an even brighter spot centered on little Class 1 EUCLID, 8 miles in diameter and 4400 feet deep (Fig. 22). The relatively recent impact that formed it covered the surrounding territory with a blanket of highly reflecting ejecta out to distances of 20 to 30 miles from the sharp little crater.

Sinus Medii is the small dark patch nearest the center of the disk. It is shaped like a flattened pentagon, and its length is about twice that of Plato. About one Medii length southwest of its center shines a brilliant circular dot on a bright ray splash background, the brightest of many such dots in the area. That is small MOSTING A, 8 miles in diameter, 7800 feet deep, and ordinarily an inconspicuous feature very difficult to locate through binoculars. Its brilliance, small diameter, precise circular outline, and its location only 115 miles from the center of the disk, for zero libration, have combined to make it a key point in lunar surveys for more than a century. Most of the precision measures of surface features have been made with reference to Mosting A as origin.

Northeast of Mosting A one Crisium length we find a light upland area similar in shape and size to Mare Crisium except that the long axis of its roughly elliptical outline runs east-west. It lies on the intersection of lines from Tycho to the Alps and from Aristarchus to Langrenus, and it is not far from the centers of those two lines. Bounded on the south by small Sinus Medii, on the north by Mare Vaporum, on the northwest by Sinus Aestuum, and on the west by Oceanus Procellarum, it appears to be connected to the main continent only on the southeast. On the upland are four bright spots of varying size and intensity which mark the corners of a crooked keystone. Most prominent is the west one, Class 1 BODE, 11 miles in diameter, 7200 feet deep, and the center of a minor ray pattern. The northeast one is strangely deformed UKERT with bright polygonal walls 14 by 16 miles across and 9500 feet deep. If it reveals any detail at all to us in this light, it looks like a bright "ring" circular on the outside but triangular on the inside. The southeast spot is Class 2 TRIESNECKER, 16 miles in diameter and 9500 feet deep. The small ring has a most unsymmetrical ray system consisting principally of a triangular fan to the northeast. Triesnecker is noted for a splendid pattern of rills that radiate from a tiny crater just east of it, but they require much more optical power than we have in hand tonight. West of Triesnecker and south of Ukert is the smallest but not the faintest of the four spots. It is CHLADNI, 8 miles in diameter and 6900 feet deep, a young crater with brilliant walls.

On a bright area about one Crisium width east of Triesnecker a pair of larger, weaker rings may be seen. A Tycho ray passes just east of them—the streak that appears to connect with the bright single ray which divides Mare Serenitatis in half. The larger north one is 28-mile Agrippa, and the smaller, brighter, south one is 22-mile Godin. The brightening of the area is owing in part to their combined rays. East of the pair, a distance equal to that of Triesnecker to the west, we come upon one of the outstanding spots of the whole disk. Little DIONYSIUS, 12 miles in diameter and 8300 feet deep, stands out like a beacon on the west coast of Mare Tranquillitatis (Fig. 10). It has a ray pattern 86 miles in diameter, according to Baldwin, but essentially all of the light you can see comes from the crater and a fringe a few miles wide around the outer walls.

The brilliant ring of 25-mile Manilius stands on the northeast edge of Mare Vaporum. It may be located by running a line from Kepler through Copernicus and on to the northern shore of Mare Crisium. Manilius is about midway between Kepler and Mare Crisium. A little farther along the same line, where the brightest ray on Mare Serenitatis meets the south shore, smaller but brighter 19-mile Menelaus is a striking feature. Its tiny ring is conspicuous under most angles of illumination and especially bright around full moon, the east half being definitely more brilliant than the west. East

of Menelaus, about one Crisium width, the mottled ellipse of 27-mile Pliny shows well just off ARCHERUSIA PROMONTORY that separates Mare Serenitatis from Mare Tranquillitatis. Halfway from Pliny to the opposite promontory Class 1 DAWES, 11 miles in diameter and 5900 feet deep, may be seen as a bright speck (Fig. 11). Just southeast of the last mentioned promontory is the bright ring of Class 5 VITRUVIUS, 17 miles in diameter and 7600 feet deep. Its floor is exceptionally dark. Just east of it shines VITRUVIUS A, 11 miles in diameter and 8100 feet deep.

About one Crisium width north and a little east of Vitruvius the ring of 25-mile Romer produces a larger, weaker bright spot in the Taurus Mountain area. South and southwest of Romer to the shores of the two maria may be seen numerous bright specks marking small craters and resembling the illuminated circles under suburban street lamps as viewed from a night-flying airliner. One Crisium length southeast of Romer is another of the outstanding features of the disk. The dazzling little ring of 18-mile Proclus in the uplands west of Mare Crisium is so intense that it first appears as a solid oval of light, and only with difficulty do our binoculars resolve it into a thick ring with a tiny dark dot at the center. The bright fan of rays running southeast toward Langrenus is strong, and so is the broader fan in the opposite direction toward Endymion and Posidonius. You probably can make out two weak sheaths extending east and northeast about halfway across Mare Crisium. Fully as conspicuous as the rays is their complete absence to the south, southwest, and west—over an angle of 135 degrees from Langrenus to Plato. Halfway between Proclus and Romer is a pair of bright specks that mark Class 1 craters 11 and 13 miles in diameter. They are MACROBIUS B, 8800 feet deep, and MACROBIUS A, 10,100 feet deep.

South of Proclus, about one Crisium width, where Mare Fecunditatis merges with Mare Tranquillitatis, we see a bright spot produced by the ring and ray center of 36-mile Taruntius. South of Taruntius and west of Langrenus, close to the ragged west shore of Mare Fecunditatis, stands a ring a little larger and brighter than Taruntius, thick on the northeast and thin on the southwest. Note that it appears slightly elongated in the east-west direction in contrast to other rings in the vicinity which appear elliptical but with long axis north-south. That is GOCLENIUS, an egg-shaped double crater 46 miles long, 34 miles wide, and 6100 feet deep.

Just east of the midpoint between Goclenius and Taruntius an elongated bright speck may be seen on the dark Fecunditatis plain with a weak ray running west from it some 70 miles to the edge of the mare and a more conspicuous north-south ray crossing it. With excellent conditions you may be able to resolve this speck into a close pair of dots separated in the direction of the weak ray, and the ray itself may be resolved into two fine lines resembling a comet tail. The east dot is MESSIER, 6 miles wide and 4100 feet deep, and the west one is WILLIAM PICKERING, 7 miles wide and 5700 feet deep. As we have seen, many "changes" in those two little craters have been reported over the years, particularly by the astronomer after whom the west one is named. Evidently they are so formed that the changing angle of solar illumination causes striking variations in outline and general appearance. A study of the plates on which they appear in the *Photographic Lunar Atlas* confirms this rather common but here exaggerated phenomenon. The craters are not circular but greatly elongated on an axis through their centers, their lengths being 10 and 11 miles respectively. Such considerations prompted Harvey Nininger, writing in *Sky and Telescope*, June, 1952, to suggest that the two were formed almost simultaneously by the impact of a single meteoritic body which came in at a low angle from the east, pierced the surface at Messier, and ricocheted out again at William Pickering, scattering a double trail of bright ejecta westward across the plain. Does a tunnel connect the two? We shall look forward to the report of an examination of the curious pair by one of our astronauts someday after space exploration resumes.

A line from Copernicus through Dionysius and continued eastward passes across Messier and William Pickering. On that line one Crisium length east of Dionysius, at the base of a promontory which extends into Mare Tranquillitatis from the southeast, shines a spot of exceptional brilliance. The center of the small but conspicuous area is marked by the tiny crater CENSORINUS, an object difficult to study because of the glare. It is less than two miles in diameter and only 1300 feet deep, but it is sometimes confused with the more evident crater immediately east of it. The latter is CENSORINUS A, four miles in diameter and somewhat deeper. A broad highland valley about 100 miles east of Censorinus is another touchdown site candidate selected early on for exploration by an Apollo team—a site as yet unvisited.

One Crisium length southwest of Censorinus may

be seen the delicate ring and central mountain of 65-mile Theophilus. It is just south of a point ¼ the way from bright Langrenus to dark Grimaldi. The small five-mile crater THEOPHILUS B, in the northwest wall, may be resolved as a brighter spot on the ring. The time-tarnished ring of much older 62-mile Cyrillus, which joins that of Theophilus on the southwest, is too faint to be seen against the lighter background, and the ring of still older 64-mile Catharina, just south of Cyrillus, is difficult to find with a telescope. However, the location of Cyrillus is well marked by the bright little ray crater CYRILLUS A, 10 miles in diameter, 3300 feet deep, and situated near the base of the inner southwest wall (Fig. 5). Just east of Theophilus, near the northwest edge of the Mare Nectaris plain, stands the smaller Class 1 MADLER, 17 miles in diameter and 7900 feet deep (Fig. 5). The brightest of its rays may be traced about 50 miles east.

West of Theophilus, nearly twice the distance of Madler to the east, shines the similar but brighter ring of Class 1 KANT, 19 by 22 miles across and 9400 feet deep (Fig. 5). An equal distance north of Kant (toward Menelaus) is the smaller, brighter ring of ALFRAGANUS, 12 miles in diameter and 9200 feet deep, which brightens its area, particularly to the south, with a ray system (Fig. 50). Just southwest of Alfraganus a bright dot marks ALFRAGANUS C, 7 miles in diameter and 3600 feet deep. Farther southwest (toward Tycho) and west of Kant, a spindle-shaped bright patch some 50 miles long can be seen. Two minute craters, two-mile DESCARTES C and three-mile DOLLOND E, near the southeast and northwest edges of the patch, are the only clues to its origin (Fig. 50). Northwest of Alfraganus (toward Sinus Iridum), and south of Dionysius, may be seen a pair of light dots marking weak young craters. The south one is THEON JUNIOR, 12 miles in diameter and 8900 feet deep; the north one is its father THEON SENIOR, 13 miles in diameter and 10,500 feet deep (Figs. 10, and 24).

About one Crisium width southwest of the Theons, and nearly halfway from Tycho to Posidonius, a brilliant dot on a light patch marks the 11-mile ray crater Hipparchus C. Just northeast is the smaller and much fainter speck of eight-mile Hipparchus L. Just southwest of Hipparchus C you may be able to make out the faint, equally difficult ring of 17-mile Hind, located near the end of one of Tycho's rays. Just west of Hind is the larger, stronger ring of 22-mile Halley. Halley marks the location of the south wall of the invisible

83-by-89 mile Hipparchus, and just southwest of Halley is the almost equally well-camouflaged 81-mile Albategnius. You may be able to see it since its floor is a little darker and its broad walls are a little lighter than the surrounding territory. The 93-mile crater Ptolemy is similarly outlined. It is just west of Halley and Albategnius, on a line from Tycho to the Caucasus Mountains, and it has a faint bright speck on its floor northeast of center which is six-mile Ptolemy A. A little to the north of the Hind-Halley group you may be able to resolve two weak light dots aligned in the same direction as the Theons but slightly farther separated. The north one is EDWARD PICKERING, 10 miles in diameter and 7800 feet deep. The south one is HIPPARCHUS G, 9 miles in diameter and 7700 feet deep, which rises in the east wall of the invisible mountain-walled plain.

Northeast of Tycho, about one Crisium width, on the ray that runs toward Mare Serenitatis, the Cassini Bright Spot is seen easily. If the great double ray running northwest from Tycho is traced across western Mare Nubium, its fainter branch reaches the bright ring and central mountain of 38-mile Bullialdus. East of Bullialdus and north of Tycho, near the edge of the dark Nubium plain, the brightest speck in the neighborhood marks Class 1 BIRT, 11 miles in diameter and 7700 feet deep. West of Birt, about ⅓ the way to Bullialdus, lies the fainter speck of Class 1 NICOLLET, 10 miles in diameter and 4600 feet deep. At an equal distance on the other side of Birt look for 13-mile Thebit A, brighter than Nicollet but fainter than Birt. Southwest of Nicollet, a distance equal to that of Birt on the east, may be seen a bright spot or ring more conspicuous than Nicollet and nearly as large as Bullialdus. That is the strange enclosure WOLF (Fig. 22). It has a flat pear-shaped floor some 15 miles across with walls that resemble clumps of hills rising to 2300 feet on the southwest but declining to a few hundred feet or less at several gaps. A few small craters shine on Mare Nubium north of Birt.

One Crisium length west of Bullialdus, on the north shore of Mare Humorum, we see the large, delicate, elliptical ring of 70-mile Gassendi with the brighter spot of 24-mile Gassendi A like a jewel on its north edge. Just southwest of Gassendi are two small bright spots on a line that passes through the crater. The nearer one is GASSENDI E, 5 miles in diameter and 2600 feet deep; the farther one is MERSENIUS C, 9 miles in diameter and 4900 feet deep (Fig. 20). Several small craters may be

spotted on dark Mare Humorum, ranging in diameter from two to seven miles. Just south of the point midway between Gassendi and Grimaldi the dark elliptical floor of 30-mile Billy contrasts with its brighter surroundings, especially with a triangular patch of bright hills on the plain just north of it.

West of Billy and southeast of Grimaldi a bright spot may be seen in the rugged uplands. Its center lies at the south edge of 28-mile Sirsalis where one of a pair of tangent small craters is responsible for the bright ejecta layer. It is SIRSALIS F, 9 miles in diameter and 5300 feet deep, and it overlaps the Sirsalis Rill (Fig. 45). At the southeast rim of Grimaldi a similar bright area surrounds a cluster of six tiny craters. Those supplying most of the reflectivity are from north to south: GRIMALDI Ga, 7 miles in diameter and 3800 feet deep; GRIMALDI Gb, 5 miles in diameter and 4900 feet deep; GRIMALDI Gc, 8 miles in diameter and 4400 feet deep; and a rill crater, just southeast of these, only 1½ miles in diameter. About one Grimaldi length southwest of the Sirsalis bright area dark 30-mile Cruger stands out in conspicuous contrast to its bright surroundings. A short, crooked dark line running northwest from it is the tiny, obscure Mare Aestatis which appears at its best tonight. West of Mare Humorum and southeast of Grimaldi we find a broader brighter area than any of the last several mentioned. Its rays emanate from 13-mile Byrgius A on the east rim of 53-mile Byrgius.

On the south edge of Mare Humorum the delicate ring and central formations of ancient 29-mile Vitello resemble a small replica of Gassendi on the opposite edge. Close to the midpoint of a line from Gassendi to Tycho are the dark-floored, light-walled, difficult twins, 31-mile Campanus and 29-mile Mercator. Southwest of the pair and east of Vitello a brilliant speck shines out of the bright highland strip between Mare Humorum and Palus Epidemiarum. That is DUNTHORNE, 11 miles in diameter and 4900 feet deep, another young crater of modest dimensions but high reflectivity. South of Dunthorne and near the west edge of the dark Palus Epidemiarum is the small ring of RAMSDEN, 13 by 16 miles across and 5900 feet deep, in the neighborhood of which observers with good telescopes may examine a fine system of rills.

We now have covered all the maria except those which are poorly placed close to the limb, and we have seen that, with a few notable exceptions, their unmelted craters are generally small and brilliant under high sun. This is in harmony with the concept that the lava floors of the maria formed in an era which followed the age of great meteoritic bombardment—an age that probably left the entire lunar surface in a state similar to that which we view today in the south. Subsequent, sporadic impacts of relatively much smaller bodies produced most of the maria craters which show best under high sun. Only the youngest craters still have highly-reflecting interiors, walls, and sometimes ray patterns. We have advanced well into the southern continent, almost all of which is bright, and you should not be surprised to see that it, too, is liberally sprinkled with brilliant specks. Those also are small craters of the most recent formation period. Most of them have no names for ordinarily they appear inconspicuous in contrast to the old giants upon which they have intruded.

We have used Tycho many times tonight as a guidepost to less-conspicuous features, and it seems superfluous to call attention to that magnificent crater, dazzling in its brilliance and fantastic in its ray pattern. In fact, one must resist its compelling demand for attention if he is to see anything else on the face of the full moon. Examine Tycho as long as you wish, for it is well worth the time invested, and then let us conclude tonight's observations with a look at some of its lesser imitators to the east.

On the east edge of Mare Fecunditatis, near its southern extremity, and nearly one Crisium length southwest of broad bright Langrenus, is a small bright splash pattern. It is centered on 23-mile Petavius B., an irregular and evidently double crater. You may recall that one of the prominent features of the two-day moon was 110-mile Petavius, considerably larger than Langrenus, which faded the following evening and has been inconspicuous ever since. Tonight its elliptical outline and central peak are well camouflaged an apparent 75 miles (actually 115) south of Petavius B.

Southwest of Langrenus, about twice the distance to Petavius B, may be seen two brilliant spots each of which is the center of a ray system that can be traced several hundred miles. They lie approximately on a radius of the lunar disk and appear to be separated by about one Langrenus length. A point between them marks one vertex of a large equilateral triangle with Langrenus and Theophilus at the other corners. The nearer (northwest) dot and ray center generally has been identified as 16-mile square STEVINUS A, west of the main crater, but a close study of the area does not confirm the identification. The bright speck and

evident ray center is a much smaller four-mile crater between Stevinus A and Stevinus. (Recently Stevinus A has been redesignated Stevinus R, and the small ray crater now has the label Stevinus A.) Here we observe a strange and puzzling phenomenon—a small crater at the center of a large conspicuous asymetrical ray pattern that extends outward as far as 650 miles! The difficult elliptical ring and central mountain of 43-mile Stevinus stand between the two dots and probably cannot be seen. The farther dot is nine-mile FURNERIUS A, but some of the rays of this system may be due to 12-mile FURNERIUS C, a much fainter speck just west of A. The last two craters are located on or near the north wall of large, ancient, 81-mile Furnerius which also is camouflaged completely tonight with little to mark it even for the telescope user except a large stain on its floor.

Note that one of Tycho's rays points toward Stevinus A and can be traced almost to it. Note also that the Stevinus and Furnerius rays appear to merge on the southwest into a broad ray that extends more than one Crisium length beyond the junction. At its midpoint it crosses another Tycho ray. Near the intersection may be seen the bright speck of a young nine-mile crater on the south floor of enormous, ancient, and tonight invisible 122-mile Janssen. Judging by the extent of the Stevinus-Furnerius rays in other directions, it seems unlikely that they contribute much to the broad east-west ray through Janssen, and the floor crater there appears to be responsible for only a portion of it. There we see a ray as long, as broad, and as bright as some of those of the unique Tycho system, but nowhere along its extent can be found a crater that appears adequate for its production. A satisfactory explanation of this ray might lead to a better understanding of all of them.

SIXTEEN-SEVENTEEN-DAY MOON

(Chart IV)

You may wonder, from the title of this chapter, and with justification, how it would be possible to observe the moon at two different ages on the same night. The underlying idea here is similar to that of a question which my students usually regard as a trick question: "Does the moon rise every day?"

After we have learned a little lunar theory our impulse is to reply confidently, "Yes, of course." However, if we recall that the moon rises, on the average, some ¾ of an hour later each day, we readily can visualize it rising on, say, Monday night at 11:45 P.M. and rising again the following night at about 12:30 A.M. Thus it would rise each night as expected, but the second rising would occur on Wednesday morning, and Tuesday would have come and gone without the event having taken place. The 16- and 17-day moons rise several hours before midnight, but if we wish to observe under the best possible conditions, we have learned that we should select a time when our satellite is near the celestial meridian, or at least several hours' journey above the horizon. So it is likely to be around 11 P.M. when we find the 16-day moon well up in the sky, and, if our studies continue for more than one hour, that same moon will be 17 days old before we finish. So we can observe the moon at two different ages during the same night without any hocus-pocus at all, and we are likely to do so at about this phase.

Tonight we see Mare Crisium fully revealed for the last time as the sunset terminator reaches its eastern shore. Its surface is quite dark, and it appears to be bordered by a light ring, some 50 miles wide, of very rugged mountains except on the east where most of the ring is lost in darkness. These are mountains indeed since they rise to heights as great as 13,500 feet above the plain on the south and 14,400 feet on the north. Thus the mare resembles a huge crater when the sun is low, and tonight it presents to us strong evidence of its origin. Stretched out in a line running north from Mare Crisium toward the terminator Cleomedes, Geminus, and Messala are conspicuous. Their interiors still are bright, but broad shadows have grown along their inner west walls to give them contrast again, and their east crests are brilliant. Between Cleomedes and Geminus, and closer to the former, the smaller BURCKHARDT shows well with bright and dark walls. A Class 1 irregular crater, 27 by 34 miles across and 15,900 feet deep, it appears to have been formed on top of two slightly smaller tangent craters. It has removed about half their walls, and it gives the impression of a face with huge ears when examined under excellent conditions. We have noted many instances of smaller craters in the floors and walls of older ones, but intruders larger than their hosts are relatively rare. Just east of Geminus, the small central mountain of which might be visible tonight, is the smaller Class 1 BERNOUILLI, 31 miles in diameter and 11,500 feet deep. Its floor is far from level, being 1600 feet higher on the south than on the north.

North of Messala, east of weak Atlas, and close to the terminator, we see briefly conspicuous Class 2 MERCURY, 40 miles in diameter and 9800 feet deep. Its east wall has been damaged severely, but that is a detail a bit beyond our reach. We find only a black interior with a bright east rim. Northwest of Mercury, north of Atlas, and about one width inside the terminator, we could scarcely miss the ever-prominent, dark ellipse of Endymion which tonight has acquired a bright rim. Note the distance from Endymion back to the Geminus-Bernouilli pair, and lay it off in the opposite direction from Endymion northwest to the terminator. There you may see a pair of very narrow and rather irregular dark ellipses with bright north walls. The larger one on the west is PETERMANN, 48 miles in diameter and 12,100 feet deep. The east one is CUSANUS, 40 miles in diameter and 11,800 feet deep. A line down the backbone of the Caucasus Mountains through Aristotle passes over the pair which is close to the limb even under favorable libration.

The southeast shore line of Mare Crisium is broken by the bright AGARUM PROMONTORY which juts conspicuously northward for 40 miles over the dark plain and continues another 40 miles before disappearing beneath the lava floor. The ridge rises to heights of 12,000 feet above the plain. Wilkins and Moore write: "On several occasions, a mist-like appearance has been witnessed near Agarum, especially when the Mare is bisected by the terminator under sunrise illumination." At the terminator east of the point where the Agarum Promontory meets the normal shore line you may find the bright-rimmed dark ellipse of temporarily prominent Class 5 CONDORCET, 49 miles in diameter and 10,100 feet deep. Its hidden floor is said to be convex. If the present lunation began early, Condorcet will be lost completely in darkness tonight.

Proceeding southward to the east shore of Mare Fecunditatis, we view for the last time another feature which has been an outstanding landmark every night since it first appeared on the thin crescent of the two-day moon. Langrenus is magnificent with the terminator in the rough highlands just east of it. Its glare has faded, but highlights and shadows have returned to bring out its details and show it at its best. The double central mountain is seen, and some of the structure of the brilliant

CHART V

ANAXIMENES
PHILOLAUS
ANAXAGORAS
GOLDSCHMIDT
CHALLIS MAIN
SCORESBY
DeSITTER
EUCTEMON
BAILLAUD
BARROW A
EPIGENES
Wm. BOND
NEISON
MAYER
ARNOLD
FONTENELLE
SINUS RORIS
MARE FRIGORIS TIMAEUS
HARPALUS
ARCHYTAS
METON
SHEEPSHANKS
KANE
DEMOCRITUS
GARTNER
De la RUE
MAIRAN
PLATO
PROTAGORAS
MARE FRIGORIS
GALLE
MITCHELL
BAILY
LACUS MORTIS
ENDYMION
SINUS IRIDUM
PICO
ARISTOTLE
EUDOXUS
A
HERCULES
STRAIGHT RANGE
ALPS
CALIPPUS
BURG
MASON
ATLAS
TENERIFFE MTS.
AGASSIZ PR.
PITON
PLANA
GROVE
CEPHEUS
MARE IMBRIUM
CASSINI
THEAETETUS
ALEXANDER
MAURY
SHUCKBURGH
LICHTENBERG
SPITZBERGEN
ARISTILLUS
CAUCASUS MTS.
HOOKE
RUSSELL
HERCYNIAN MTS.
BRIGGS
HARBINGER MTS.
ARCHIMEDES
DANIELL
HALL
FRANKLIN
OTTO STRUVE
HERODOTUS
EULER
TIMOCHARIS
AUTOLYCUS
POSIDONIUS
G. BOND
EDDINGTON
ARISTARCHUS
PALUS PUTREDINIS
LINNE
CHACORNAC
NEWCOMB
SELEUCUS
PYTHEAS
MARE SERENITATIS
SERPENTINE RIDGE
DEBES
KRAFFT
HAEMUS MTS.
BESSEL
LEMONNIER
TAURUS MTS.
ROMER
CARDANUS
APENNINES
MENELAUS
ARCHERUSIA PR.
ARGAEUS
TRALLES
CLEOMEDE
ERATOSTHENES
MARE VAPORUM
MANILIUS
PLINY
VITRUVIUS
DAWES
MARALDI
TISSERAND
LBERS
OCEANUS PROCELLARUM
KEPLER
SINUS AESTUUM
BODE
TRIESNECKER
BOSCOVICH
ROSS
JULIUS CAESAR
PROCLUS
PALUS SOMNI
MACROBIUS
RICCIOLI
COPERNICUS
ARIADAEUS RILL
CAYLEY
ARAGO
MARE TRANQUILLITATIS
LAVINIUM PR.
OLIVIUM PR.
PEIRCE
PICARD
GRIMALDI
SINUS MEDII
AGRIPPA
GODIN
MANNERS
CAUCHY
Da VINCI
MARE CRISIUM
FLAMSTEED
HIPPARCHUS
D'ARREST
DIONYSIUS
MASKELYNE
WICHMANN
HORROCKS
THEON SR.
RITTER
SABINE
TARUNTIUS
MARE AESTATIS
EUCLID
LALANDE
THEON JR.
DELAMBRE
SECCHI
HANSTEEN
LETRONNE
RIPHAEUS MTS.
HALLEY
TAYLOR
HYPATIA
TORICELLI
MESSIER
CRUGER
MARE COGNITUM
HIND
ALFRAGANUS
ISIDORUS
Wm. PICKERING
MARE FECUNDITATIS
ZUPUS
BILLY
PTOLEMY
ALBATEGNIUS
CAPELLA
GASSENDI
ALPHONSUS
KLEIN
KANT
GAUDIBERT
GUTENBERG
BYRGIUS
MARE NUBIUM
ARZACHEL
ABULFEDA
THEOPHILUS
MADLER
GOCLENIUS
MARE HUMORUM
BULLIALDUS
WOLF
ALMANON
CYRILLUS
AVENZELE
MAGELLAN
VITELLO
CAMPANUS
GEBER
CATHARINA
BEAUMONT
COLUMBUS
CROZIER
MERCATOR
ABENEZRA
MARE NECTARIS
ROSSE
BOHNENBERGER
COOK
PITATUS
PLAYFAIR
FRACASTOR
SANTBECH
MONGE
WERNER
ARZACHEL
BORDA
ALIACENSIS
SACROBOSCO
PONTANUS
FERMAT
SNELLIUS
SCHICKARD
NONIUS
FONS
GODDARD
REICHENBACH
GEMMA FRISIUS
RABBI LEVI
POLYBIUS
STEVINUS
PALUS EPIDEMIARUM
FERNELIUS
RICCIUS
ROTHMANN
MAUROLYCUS
PICCOLOMINI
ANDEL
TYCHO
STOFLER
NICOLAI
RHEITA
FARADAY
SYBORIUS
RHEITA VALLEY
MAGINUS
LICETUS
K
BAROCIUS
METIUS
YOUNG
HERACLITUS
BREISLAK
NICLAS
FABRICIUS
ZUCCHIUS
LILIUS
A
BACON
PITISCUS
VLACO
STEINHEIL WATT
CLAVIUS
CUVIER
CLAIRAUT
G
HOMMEL
BIELA
RUTHERFURD
JACOBI
MUTUS
ROSENBERGER
DOERFEL MTS.
ZACH
PENTLAND
HAGECIUS
MORETUS
CURTIUS
HOMMEL
MANZINUS
SHORT
SIMPELIUS
NEARCH
BOUSSINGAULT
SCHOMBERGER
DEMONAX
ZUCCHIUS
BOGUSLAWSKY

N
W   E
S

0 1 2 3 4 5 6 7 8 9 10

walls almost can be resolved through binoculars. Just beyond the outer wall on the plain to the northwest look for a cluster of three smaller craters like a rabbit's tracks in the snow. Beginning with the largest and moving clockwise, they are LANGRENUS F, LANGRENUS B, and LANGRENUS K (Fig. 47). Their respective diameters are 30, 21, and 19 miles, and their depths are 8400, 8500, and 8200 feet.

Just south of strong young Langrenus, equally large but very old and dilapidated Vendelinus has returned to prominence for one night. Where its northeast wall should be we note the large, dark-floored intruder Lame which also is quite old and battered. West of the latter and astride the north wall of Vendelinus is the young crater LOHSE, 26 miles across and 9700 feet deep, which has a relatively large off-center peak (Fig. 47). Tangent to the south wall is HOLDEN, 28 by 33 miles across and 13,500 feet deep, which has no central peak. Holden is about ⅔ black while Lohse is only ½ black.

Continuing south along the terminator about the same distance as the last jump we come to conspicuous Petavius which rivals Langrenus in splendor on tonight's final appearance. The rough terraced walls and large central mountain readily are seen, and a small telescope would show a broad bright rill running from the mountain southwest to the wall and resembling a highway across the floor. Tangent to the northwest wall is Class 1 WROTTESLEY, 37 by 43 miles across and 11,300 feet deep. Just beyond the east wall of Petavius look for a bright straight line about 90 miles long which runs approximately north-south. Darkness separates it from the crater wall. That is the east wall of the curious funnel-shaped valley PALITZSCH, 11,200 feet deep, concerning which Wilkins and Moore write: "It is generally described as an irregular, gorge-like formation 60 miles long and 20 wide, and it has often been suggested that it was formed by a meteor ploughing its way through the still-plastic lunar surface. On 4 October 1952 Moore, using the great 25-inch Newall refractor at the Cambridge University Observatory, had a superb view of Palitzsch, and its true nature was at once evident. It is not a gorge at all, but a vast crater chain." Running south from Petavius is a more conspicuous bright line than the wall of Palitzsch, about the same length and wavy. At first it appears to be part of the same sunlit crest, but it is offset slightly to the west. That is the east wall of the double crater Hase, most of the floor of which is now dark.

Again we go south about the same unit distance to the last of the four vast craters that lie approximately equally spaced along the same lunar meridian. Furnerius shows well, more prominent than Vendelinus but inferior to Petavius and Langrenus. Close to Furnerius on the southwest is Class 3 FRAUNHOFER, 33 miles in diameter and 7900 feet deep. North of Furnerius and southwest of Petavius we see the smaller closer pair Snellius and Stevinus. Farther south along the terminator we encounter an increasing profusion of craters, a few of which are large. Abundance and similarity, extreme foreshortening, and the libration in latitude complicate identification. We can enjoy the view perhaps even more if we do not feel called upon to name everything we see, a point that should be kept in mind as we explore from night to night.

EIGHTEEN-DAY MOON

(Chart V)

Now that we are observing the waning moon you doubtless have noted that we have returned to our *modus operandi* for the waxing moon. Concentrate your attention on the features in the strip along the terminator where most objects show to the greatest advantage because of contrasting highlights and shadows. As before, there are numerous strong objects elsewhere on the disk, but all have been pointed out previously. For example, tonight's prominent features include Tycho, Copernicus, Aristarchus, the Apennines, the Caucasus, Aristotle, Eudoxus, Plato, Grimaldi, Byrgius, Theophilus, Anaxagoras, Manilius, Kepler, and several others. These more obvious objects are already familiar, and you are encouraged to check their development from night to night. Of course, a good many of the features along the terminator also have been presented earlier, but when first seen it was early in the morning. Now it is late afternoon or early evening, and the rays of the setting sun often give them a different quality. But we are doing more than merely renewing old acquaintances. I am introducing a good many new features as well. Some of them could have been seen easily at sunrise, but they are much more impressive at sunset and in the waning phases. Others, generally craters, are definitely smaller than most of those taken up previously.

That familiar landmark and useful measuring stick Mare Crisium is almost gone. Night has overspread most of it, and the dimly-lighted western remnant looks like an unnatural extension of the terminator or a bite out of the disk. A couple of larger but similar smooth areas are seen to the south where the terminator passes through the two sections of Mare Fecunditatis. At the western extremity of Mare Crisium stands the bright gate formed by two mountain ridges about 20 miles long that run north-south and almost meet like a stalactite and its stalagmite. The broader north ridge is the Olivium Promontory, and the narrow south one is the Lavinium Promontory. The two ridges show well, but they appear to run together so that the gate is not easy to resolve through binoculars. East of Lavinium Promontory and close to the terminator is small Picard, its outer west wall appearing as a tiny light dot. Even though it seems to be most favorably placed for observation and large enough to be seen through binoculars, you will be doing exceptionally well if you see it tonight. In the chaotic surroundings of the uplands, small craters show best at the terminator because their black interiors contrast sharply with the bright sunlight scattered by the myriad surfaces of the debris piled all around them. On the other hand, the relatively smooth surfaces of the maria near the terminator generally reflect very little light in our direction, and the area around a small crater is almost as dark as the black interior. Picard tonight is a severe test.

Cleomedes is seen as a great black hole at the terminator just north of Mare Crisium. All is in darkness except the narrow crest of the east wall and the dimly-lit outer west wall, and we have the illusion of gazing into a vast, bottomless pit instead of the relatively shallow old ring we have known there. Tangent to Cleomedes on the west may be seen a tiny black dot which is the dark interior of irregular Class 2 TRALLES, 27 by 28 miles across and 11,000 feet deep. If you find Tralles look for a weaker dark dot just northwest of it. That is older DEBES, 20 miles in diameter and 7100 feet deep, which is the north half of a double crater, the south half being about the same size and tangent to Tralles on the west. Southwest of Cleomedes and northwest of the two Crisium promontories Macrobius shows well with bright east wall and black west wall. Tangent to it on the east you might be able to resolve smaller Class 2 TISSERAND, 22 miles in diameter and 9500 feet deep, the black interior of which contrasts with its bright

east wall and that of Macrobius. Nearly halfway from Cleomedes to Posidonius stands Class 1 NEWCOMB, 23 by 26 miles across and 9100 feet deep. While not conspicuous, it is the largest and brightest crater in the eastern Taurus Mountains.

A line from Theophilus to Posidonius, extended beyond the latter about 1½ Sinus Iridum lengths, reaches a conspicuous pair of almost equally large craters—Atlas and Hercules. About equally distant from Posidonius to the northeast is the smaller pair —Franklin and Cepheus. Northeast of these two, at the terminator, you may be able to detect what appear to be their ghosts or weak copies. They are both irregular old craters with low walls. The slightly smaller one on the southeast is HOOKE, 21 by 25 miles across and 3400 feet deep. Its more irregular neighbor is SHUCKBURGH, 24 by 33 miles across and 4300 feet deep. The latter is described by Goodacre as "a pear-shaped ring plain," by Wilkins and Moore as "a somewhat square enclosure," while the U.S. Army Map Service gives it an outline resembling a heart. Such variation well illustrates the difficulty of determining the true shape of surface features located far from the center of the lunar disk.

Farther north, between the terminator and the east section of Mare Frigoris, Class 1 DEMOCRITUS, 20 by 25 miles across and 8900 feet deep, shows well. A line from Sinus Iridum through Plato, if continued, crosses it. Democritus long has been recognized as one of the deepest craters in its region. Tonight it appears as a tiny brilliant ellipse with a black southwest border. Look also for weaker but similar HERCULES A, 20 miles in diameter and 4900 feet deep, on a line from Atlas to Democritus. It marks out with Hercules and Atlas an almost equilateral triangle.

At the junction of Mare Fecunditatis and Mare Tranquillitatis, two Iridum lengths south of Macrobius, stands Taruntius, a bright narrow ring with a dark interior. An equal distance south and a little west from Taruntius, and near the west edge of Mare Fecunditatis, is Goclenius, the double crater which looks almost circular while its neighbors appear elliptical. A shore-line section of the mare about halfway from Goclenius back to Taruntius was selected as another of the tentative touchdown spots for the Apollo mission. Two Plato lengths south of Goclenius may be seen battered and distorted Class 5 COLUMBUS, 56 miles across and 7300 feet deep. Actually this crater is shaped like a square with rounded corners. Tangent to it on the northwest is the more normal Class 1 COLUMBUS A,

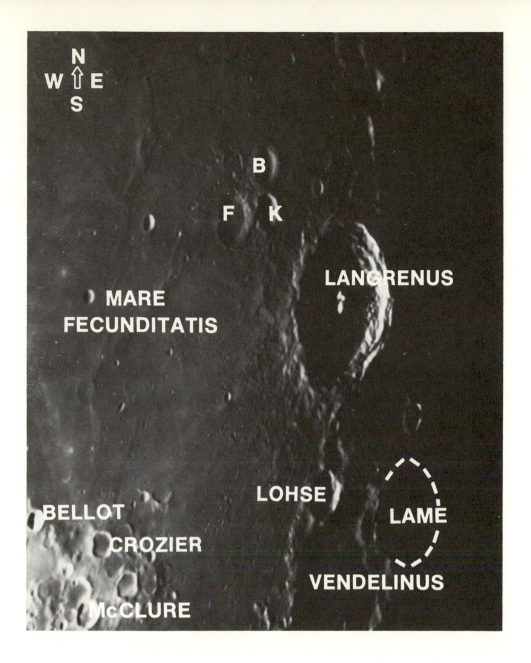

FIGURE 47.   Sunset at Langrenus. Note the chains of craters and linear depressions radiating from the great crater across Mare Fecunditatis, particularly to the southwest. Such radial features, some of which are 50 to 65 miles long, have been regarded generally as scars or grooves cut by large flying blocks blasted out of the crater in low trajectories. However, Kuiper believes they are due to collapse of the surface over subterranean crevasses torn open by the explosive energy of impact which produced the crater. Similar features surround Aristotle (Fig. 8) and Aristillus (Fig. 42). (120-inch Reflector.)

28 miles in diameter and 6000 feet deep. If the latter can be seen without strain, look for a figure "8" formed by two similar tangent craters between Columbus A and Goclenius. The larger north one of the pair is irregular Class 5 MAGELLAN, 23 by 25 miles across and 6400 feet deep. The other one is weaker, older MAGELLAN A, 21 miles in diameter and 5700 feet deep.

One Plato length southeast of Columbus is the old flooded crater COOK, 26 by 30 miles across and 3800 feet deep. Its low walls reflect very little light from the setting sun, but you may be able to make them out as a thin, light ring. Just southwest of Cook and south of Columbus is slightly smaller, younger MONGE, 23 miles across and 8500 feet deep. Monge is much easier to see than Cook because the broad gradual slope of its inner east wall reflects much more light at this phase. Continuing southwest from Monge about 1½ times the last distance brings us to the larger and more conspicuous Santbech which shows well tonight. A line from Macrobius to Santbech runs along the east face of the Pyrenees Mountains, and if it is extended beyond Santbech a distance almost equal to the length of that range it reaches Class 1 REICHENBACH, 45 miles across and 11,200 feet deep. It is the largest crater in the area, but it is narrow, distorted, and well camouflaged in the heavily-blasted neighborhood. Easily seen after it has been located, Reichenbach is of interest chiefly because of the broad, shallow valley that runs south from it to the terminator and beyond. The unnamed valley, prominent tonight, is fully as broad as the crater is long, and near the terminator it divides into two narrower valleys. Note that it appears to radiate from Mare Nectaris.

Southwest of Reichenbach about ¾ the Reichenbach-Santbech separation may be seen Class 1 RHEITA, 44 miles across and 14,500 feet deep, which tonight is more than half illuminated. Running south from its west rim is a conspicuous black streak with a bright border on the east. That is the Rheita Valley which extends across the terminator and which combines with several small craters to resemble a dagger with a curved blade. The valley must have been there a very long time one concludes after close examination with more optical power. Not only does the crater Rheita encroach upon it, but so do at least a half-dozen others, one of which has acquired some floor craters of its own. Note that the Rheita Valley also radiates from Mare Nectaris, the second such valley we have seen tonight. Baldwin has found several others. He believes that Mare Nectaris was once an enormous explosion crater produced by the impact energy of a colliding minor planet and that the valleys were formed in connection with that spectacular event.

Southwest of the crater Rheita, and across the Rheita Valley, we find the larger and more conspicuous Metius-Fabricius pair which shows well tonight with broad bright and black borders. The outline of huge, ancient Janssen, which includes Fabricius, readily is seen under the low sun, but it will be more conspicuous tomorrow night when the terminator passes through it. Just south of Janssen's bright south wall we see the prominent twin ovals of the side-tangent pair Steinheil and Watt.

The Metius-Fabricius axis extended through Janssen about one Janssen length beyond the wall of that enclosure reaches the conspicuous young Vlacq. Tangent to Vlacq on the south is the larger, dimmer, and less well-defined Class 3 ROSENBERGER, 55 by 58 miles across and 7200 feet deep. Just south of Rosenberger is HAGECIUS, 50 miles across and 7500 feet deep, which in this light combines with the crater upon which it intrudes to the northeast to produce a thin black gash about 75 miles long with a bright south edge. Close to the terminator northeast of Hagecius and south of Janssen is a stronger, narrow, black ellipse with light southern border. That is Biela, always seen as a thin ellipse and notable only near the times of sunrise and sunset. Southwest of Hagecius the terminator approaches the limb in the vicinity of the south pole. There, depending upon the libration in latitude, you may or may not see several larger craters that resemble narrow valleys or black-and-white-bordered grooves gouged into the rough surface. Seen almost edge on, they merge one into another, and individual identification becomes rather tedious.

NINETEEN-DAY MOON

(Chart V)

The terminator now has advanced across the eastern edge of Mare Tranquillitatis, and near the north boundary of that plain it slices through PALUS SOMNI, an upland area of considerable roughness which is exaggerated by the nearly horizontal rays of the setting sun. It is characterized by elevations of from 3000 to 9000 feet above the Tranquillitatis plain, and sometimes just beyond the terminator one may glimpse some star-like points of light where hilltops and low peaks rise out of the darkness around them. One Plato length southwest of the southwest extremity of Palus Somni your binoculars may pick up a bright speck on the dark mare plain. A line from Aristarchus across almost equally brilliant Manilius passes over it if extended half its length. The speck is the brilliant inner east wall of the small young crater CAUCHY, 8 miles in diameter and 5900 feet deep. Even with most of its interior in shadow it still reflects enough light to be seen. While your binoculars will probably not reveal it, a small telescope tonight would show a hairlike white line resembling a minute scratch on the dark plain just south and west of Cauchy running southeast-northwest. The brightest portion is about 60 miles long (one Plato length), but this major rill has been traced some 200 miles to a point beyond the terminator. It is bright in this light because the mare plain along the southwest edge of the crack has dropped, exposing the northeast wall to direct sunlight. Baldwin attributes the feature to cracking and subsidence of the main mare floor as the lava layers cooled and hardened, and he finds this to be the only prominent rill with a significant elevation difference between its two edges. A second rill, parallel to the first, and in detail a mirror image of the first, crosses the plain just north of Cauchy. It is a normal rill, appearing dark under most angles of illumination, and more optical power is required to reveal it.

The north shore of Mare Tranquillitatis consists in part of a broad, rough peninsula which separates it from Mare Serenitatis. Near the south tip of the peninsula are two difficult craters separated one Plato width in an east-west line. The larger west one is ancient Vitruvius, and the smaller but brighter east one is young Vitruvius A. As previously mentioned, Vitruvius is noted for its dark floor which is actually no darker than the mare at its outer walls. An even darker floor, enhanced by somewhat lighter surroundings, is exhibited by extremely old and dilapidated MARALDI. Maraldi, 24 miles across and 4300 feet deep, is seen easily as a small dark spot northeast of Vitruvius and north of Vitruvius A. The bright area north of Maraldi, between Mare Serenitatis and the terminator, is the upland we recognize as the Taurus Mountains. Its brightest spot, just south of center, is the ray crater Romer, half bright and half black.

At the terminator north of Posidonius a black ellipse with brilliant east border catches the eye.

It is the fine crater Hercules, the interior of which is almost completely dark. It looks strange there by itself, but the sun already has set on Atlas, its larger companion to the east. Perhaps you can see a short bright streak on the dark side marking the outer west wall of Atlas.

Near the terminator, northwest of Hercules and northeast of Aristotle, the sharp little crater Democritus shows well again tonight with black interior and brilliant east wall. Just southeast of it you may be able to detect in this light the bright northeast and dark northwest walls of the ancient half crater GARTNER, 66 miles in diameter and 4300 feet deep. The entire southern half of the structure is missing, evidently melted away by the lavas that formed Mare Frigoris. With larger optical equipment one can trace the missing south wall of the hexagonal crater as a perceptible ridge on the mare plain. In the opposite direction from Democritus and a little farther removed, you may find the similar but stronger old crater ARNOLD. That ring, a flattened hexagon 50 by 64 miles across and 6600 feet deep, did not fare so badly as Gartner, but portions of its south wall have been melted down. Its identification may be checked by noting that it lies on a line down the Caucasus Mountains and across Aristotle.

Moving onto the southern hemisphere, we find at the terminator, southeast of Mare Tranquillitatis, the ragged coast line which is all that remains of Mare Fecunditatis. There the strange, distorted crater Gutenberg forms a conspicuous black bay with a bright outline resembling a test-tube clamp. The most prominent part of the crater, its bright western wall, continues south about twice its length in a straight line. That is the western slope of the Pyrenees Mountains. Crowded between the mountain scarp and the edge of Mare Nectaris is dark BOHNENBERGER, an old crater 21 miles in diameter and 7700 feet deep. Tonight it merges with a broad, shallow, shaded valley along the east shore of the mare, but when well illuminated it appears to have been penetrated from the north by a mountain ridge parallel to the Pyrenees and 45 miles long. Two Plato lengths west of Gutenberg and three such units northeast of Theophilus are the more conspicuous twin craters Isidorus and Capella.

About one Mare Nectaris length south of Gutenberg the terminator passes through Santbech. Its interior is completely black, and the crater that was a conspicuous study in contrast only a few hours ago now reveals itself merely as one of the smaller nicks along the terminator. West of Santbech, on the south shore of Mare Nectaris, we find the bay formed by large, old, horseshoe-shaped Fracastorius, its gray floor brightly outlined tonight. A line from Capella through Fracastorius continued south half its length reaches the splendid crater Piccolomini. Between its black-and-white walls the central peak shows well. East of Piccolomini and near the terminator you will see the light-edged black ellipse of smaller Class 2 NEANDER, 27 by 32 miles across and 9900 feet deep. Just north of it a broad shadow cuts in from the terminator running northwest.

South of Piccolomini nearly as far as Neander to the east, and marking out with those two a right triangle, Stiborius shows well—a smaller copy of Piccolomini, with a tiny central mountain. South of Neander, Metius and Fabricius are conspicuous at the terminator, their black interiors separated from the dark side by bright east-wall crests which are narrow and broad respectively. Janssen is seen best in this light which brings out maximum contrast between its huge but low and dilapidated walls.

Near the terminator, two Janssen lengths southwest of Metius, Vlacq is a prominent feature, but its tiny central peak is a severe test. Tangent to Vlacq on the west is huge distorted Hommel with its two large intruders, Hommel A on the east and Hommel C on the west. Its actual shape is elliptical. Just north of Hommel is well-defined and more normal Pitiscus, and a little farther removed to the south of Hommel is Nearch, which shows well near the terminator. Southwest of Hommel the black-and-white bordered narrow ellipses of Mutus and Manzinus also show well. Several other large craters probably can be made out in the vicinity of the south limb. However, look about over the disk for some of the splendid familiar features already mentioned. For example, Theophilus is an excellent object, and its central peak is easy to resolve.

TWENTY-DAY MOON

(Chart V)

"Some are born great, some achieve greatness, and some have greatness thrust upon 'em," read Malvolio from the mysterious letter in the second act of *Twelfth Night*. If we substitute "prominence" for "greatness" we could apply that famous declaration to the lunar surface features. In the first group are craters and mountains which are outstanding objects from sunrise to sunset two weeks later. Most of the objects we have been observing, however, belong to the second class. Their "finest hour" lasts from a few hours near the terminator to several days, depending upon the size and construction of the feature. The last category consists of a steadily growing number of X's on lunar navigation charts— X's that mark areas of no particular interest until space probes crashed or landed gently upon them and spots selected as touchdown targets for Apollo landings.

Craters such as Aristarchus, Copernicus, Grimaldi, Plato, and Tycho are outstanding objects night after night, and they should be noted and enjoyed by the student of the moon even when not specifically called to his attention. In the case of Tycho, you probably have noted that the great ray system has been fading the last few nights. Soon it will be gone.

Due east of Tycho, a little more than halfway to the terminator, the great crater Maurolycus has risen again to prominence, its brilliant east wall and black west wall giving it an arresting appearance tonight. Tangent to Maurolycus on the southeast, Barocius also shows well. In the northern hemisphere the vast arc of the Apennines and Caucasus leads to the conspicuous crater pair Eudoxus and Aristotle. Close to the terminator, between the remnants of two maria, the magnificent trio Theophilus, Cyrillus, and Catharina contends strongly for our recognition. These are the outstanding features of the 20-day moon, but they are by no means all that should claim one's attention tonight.

In the dim zone between Theophilus and the terminator you may be able to spot smaller Madler as a tiny light circle before darkness overtakes its crest. Two Theophilus widths north of Madler, pear-shaped TORRICELLI makes a little better showing. That odd enclosure, 6900 feet deep, is the union of two craters, the east one being 12 miles in diameter and the west one about half as large. East of Torricelli the terminator coincides with one edge of the great square peninsula (200 miles on a side) that is bounded by Maria Fecunditatis, Tranquillitatis, and Nectaris. The sun has set on all of it except for a few peaks to the north which shine like stars out of the blackness of the dark side.

The main body of Mare Tranquillitatis is bisected by the terminator, and the low-angle illumination brings out some of its smaller markings. At the terminator and north of Torricelli, a distance equal to that of Theophilus to the south, you might be able to resolve irregular little MASKELYNE, which measures 15 miles across and 8200 feet deep. Near the west shore line, midway between Theophilus and bright Menelaus in the Haemus Mountains, may be seen the tangent rings of SABINE and RITTER (Fig. 10). Both are Class 5, irregular, and about 18 miles across. Sabine on the southeast is 4600 feet deep while Ritter is 4300 feet deep. They are fully lighted by the sun, and their lava-stained floors are framed by delicate rings—light on the east and dark on the west. Far brighter than either is much smaller Dionysius northwest of Ritter about twice the separation of the latter from Sabine. While in the neighborhood don't miss prominent Delambre about one Theophilus length southwest of Ritter and south of Dionysius.

On the dark plain northeast of Ritter, a distance equal to that of Delambre in the opposite direction, you may find the bright inner east wall of smaller but brighter Class 1 ARAGO, 16 miles across and 5900 feet deep. Between Arago and Ritter you might be able to see smaller Class 1 MANNERS, 10 miles across, 5900 feet deep, and a difficult test (Fig. 10). North of Arago, a distance equal to that of Sabine in the opposite direction, is similar Class 1 ROSS, 17 miles across and 5900 feet deep (Fig. 10). Have a try at much fainter, smaller, shallower, polygonal MACLEAR, 13 miles in diameter and 2300 feet deep, just southwest of Ross (Fig. 10). Northeast of Ross one Theophilus length, and just off the Archerusia Promontory, larger, brighter Pliny stands out prominently, about ⅓ bright and ⅔ black.

On western Mare Tranquillitatis are three of those unmarked but important spots that have had "greatness thrust upon 'em." Midway between Pliny and Dionysius unresponsive Ranger 6 crashed on February 2, 1964—a splendid shot up until the time it was supposed to send back pictures. One year and 18 days later perfect Ranger 8, its mission accomplished, struck the surface south of Pliny and east of Dionysius at a point which marks an equilateral triangle with Arago and Sabine. The plain east of Sabine, scouted by both Ranger 8 and Surveyor 5, was the touchdown area for Apollo 11 (3½ crater diameters from Sabine), as we have seen.

The terminator coincides with the east shore of Mare Serenitatis, and tonight is an excellent time to look for the Serpentine Ridge that wriggles across it from Pliny north to Posidonius. There may be just enough light on its west slope and shadow to the east to bring it out for us. Look for a faint, light, wavy, discontinuous line, stronger toward the north end. Posidonius at the terminator is conspicuous as a bright ring with a mottled gray floor. Just south of it, on the rugged shore, a bright arc on the east edge

of a black spot marks the location of Class 5 LEMONNIER, 36 miles in diameter and 7300 feet deep on the east (Fig. 11). Its west wall has been reduced to a ridge only a few hundred feet high. The floor is hidden in shadow tonight, but when illuminated it is found to be unusually smooth and free of detail. From the west end of Lemonnier's wall the brightest portion of another ridge extends south some 80 miles parallel to the shore line and to the Serpentine Ridge.

Near the terminator east of Eudoxus is a small, irregular bright spot which is the sunlit massive east wall of PLANA, 26 by 31 miles across and 6700 feet deep. The mountain mass that forms the east wall is higher than the rest of the wall, and it separates Plana from old Class 5 MASON on the east. The latter, 20 by 25 miles across and 6300 feet deep, is on the terminator and a very dim object if visible at all. Just north of Plana on smooth Lacus Mortis we find the much deeper but no larger Burg, all black except for a bright east crest and a light outer west wall.

North of Aristotle, about halfway to the terminator, the young crater GALLE, 15 miles in diameter and 6600 feet deep, may be seen as a bright speck on the Mare Frigoris plain (Fig. 8). At the terminator north of Galle you may be able to resolve the narrow elliptical ring of ancient KANE, 34 miles in diameter and 4300 feet deep, the light, broken walls of which appear continuous tonight but show large southern gaps under higher sun. One Plato length northwest of Kane look at the terminator for similar NEISON, 31 by 35 miles across and 7900 feet deep. Nearly tangent to Neison on the northwest is huge, shamrock-shaped METON, 110 miles across and 8200 feet deep. It is the union of three or four major craters, but the vast floor shows little if any trace of overlapping walls. Recently the name has been restricted to the northern crater of the group. The only strong markings on the floor are several light wall-to-wall streaks that cross from east to west. They are visible through small telescopes around full moon and are part of the splash pattern of the ray crater Anaxagoras 150 miles west. Even if the present floor resulted from a lava flow, as it probably did, its uniformity is most unusual. In a line west of Meton lie the three craters Barrow, Goldschmidt, and smaller Anaxagoras. The first two are washed out bright with only traces of wall shadows except for the strong mountain ridge between them, but the younger Anaxagoras shows well.

Let us now return to the southern hemisphere. Two Theophilus widths west of that crater much

FIGURE 48. The Altai Scarp cuts through rugged
territory bordering Mare Nectaris, the southwest edge
of which is shown beyond Beaumont and Fracastorius.
The prominent irregular cliff, which rises 2 miles high
in places and curves northward some 600 miles, is
part of a vast ring concentric with circular Mare Nectaris. Baldwin has traced it completely except where it
is buried under the lavas of other maria. He believes
that the shock wave resulting from the impact which
excavated the Nectaris basin caused the surface to
buckle, break and drop away along the scarp. See Fig.
5 for more of the cliff. (120-inch Reflector.)

smaller Kant stands out in its mountainous surroundings. It is still ⅓ bright, but its shadow and those of its neighbors are growing long. Southwest of Cyrillus, northwest of Catharina, and marking out an almost equilateral triangle with the two, is weaker, irregular, Class 1 TACITUS, 28 by 30 miles across and 9300 feet deep (Fig. 5). Two Theophilus widths south of Tacitus is similar but older and even weaker Class 2 FERMAT, 24 miles across and 7400 feet deep (Fig. 48). Just east of Fermat a conspicuous, black, somewhat irregular line catches the eye. It comes south from a point near Tacitus, turns sharply to the southeast near Fermat, and continues on to the terminator. That is the Altai Scarp which was prominent as a bright line on the five-day moon.

South of Catharina and halfway to the Altai Scarp may be seen the small black ellipse, bright bordered on the east, of irregular POLYBIUS, 25 miles across and 7700 feet deep (Fig. 48). Two Theophilus lengths south of Polybius, beyond the Altai Scarp, the similar but younger Class 1 ROTHMANN, 26 by 28 miles across and 9900 feet deep, stands near the terminator (Fig. 48). More than half its interior is black, but what is illuminated shines brightly. One Theophilus width southwest of Rothmann a bright crescent stands out. It is the inner east wall of larger Lindenau. The rest of the crater is lost in black shadow, partly its own and partly that of its larger, higher neighbor tangent on the west. That neighbor is Zagut, which shows well tonight and resembles the head of a bird with a stubby beak. The beak is the intruder on the east, Zagut E, which gives the crater a pointed outline on that side. The eye is a difficult young floor crater eight miles in diameter not far from the center of Zagut.

Tangent to Zagut on the south is a similar enclosure in both size and shape. It, too, is pointed on the east where the "beak" section is marked by a confusion of at least 10 small craters. Instead of one eye it has five. That crater, Rabbi Levi, wasn't just hit; it was sprayed. Immediately southeast of the wreckage of Rabbi Levi you may notice a black crescent. That is the inner west wall and just about all that is left intact of ancient, irregular RICCIUS, 45 miles across, 5900 feet deep, and heavily damaged. Two Theophilus widths south of Rabbi Levi is the smaller and much younger crater NICOLAI, 24 by 28 miles across and 7000 feet deep (Fig. 49). With dark interior and bright east wall, it is the strongest of the numerous pits that are scattered throughout the area. An equal distance south of Nicolai, at the terminator, look for Pitiscus, a large black oval with bright borders on both east and west. Its greater neighbor on the south, Hommel, is enveloped in darkness with the exception of its bright outer west wall and a few high spots along other parts of the wall.

West of Pitiscus and south of Barocius our armless snow man shows well again tonight, but upside down without an inverting telescope. His torso is the largest crater, Bacon. His head is Bacon A, and his legs are Breislak on the east and Bacon B on the west.

Continuing south along the terminator, we may find three more large craters before we reach the limb near the south pole. Southwest of Pitiscus, about twice the combined widths of Pitiscus and Hommel, is the light-outlined black ellipse of Mutus. Just beyond Mutus is larger and more conspicuous Manzinus. Its broad inner east wall is illuminated completely and shining brightly, and its outer west wall also is strong. Beyond Manzinus, after an interval equal to its length, we come to somewhat smaller Class 1 SCHOMBERGER, 53 by 56 miles across and 11,200 feet deep. If you don't find the last two or three craters don't be concerned about it. Under average conditions, the last three craters are each approximately tangent to the terminator, but the libration in latitude can introduce conspicuous variations in their positions not only with respect to the limb but with respect to the terminator as well at this phase.

TWENTY-ONE-DAY MOON

(Chart V)

What is the most conspicuous feature of the moon tonight? Aristarchus is the brightest, but through a small glass it is a dazzling spot with a tail and little more of interest. The Kepler splash pattern is brilliant, but the crater itself virtually is lost in the glare. Copernicus and Tycho still are conspicuous, but their luster is fading. Some of the large craters of the south near the terminator stand out prominently because of contrast between their black interiors and brilliant walls and crests. Plato and Grimaldi attract the eye because their floors, while well illuminated, are made of dark lava and their surroundings are bright. Picking the *most* conspicuous object on the face of the moon is not just difficult; it is impossible. On what basis do we select "the most" unless it is measured in tons, bushels, dollars, or some other definite unit of quantity? As soon as we apply the term to judgments in questions of appearance or relative evaluations in such vast, intangible realms as human achievement, leadership, contribution, and service, our measuring instruments become subjective and our data inadequate.

Having shown that it is impossible to pick *the* outstanding lunar feature, I now shall call attention to *one* of the outstanding features. The great arc of the Apennines continued, after interruption, by the Caucasus Mountains is a magnificent sight tonight. The foothills are not so bright as they were a few nights ago, but the broad faces of the higher mountains, rising to the crest of the range on the Mare Imbrium border, are in full sunlight, and once again they shine with a sort of super brilliance. Tomorrow night, at about the time the moon rises, the standard terminator will begin its 40-hour journey across the Apennines, and by observing time the following night it will have crossed most of the mountains and reached the narrow tail between the main body of the range and Eratosthenes. If the current lunation happens to be out of step with my average or most probable lunation by as much as 6 to 12 hours, you may be able to observe when the terminator is not far from the center of the range. At such time the mountains, because of their great height, easily are seen extending at least 100 miles beyond the terminator. Concerning the range, Webb writes: "Its projection into the dark side, which may be seen without telescopic aid, probably gave rise to the early idea mentioned by Plutarch [first century, A.D.], that the moon was mountainous." The standard terminator for the seven-day moon bisected the Apennines and revealed the same interesting phenomenon, but I did not mention it at that time, preferring to introduce the Apennines the following evening when the whole range came into view. Remember to look for the range on the terminator, and see if you agree with Webb that it is visible to the naked eye insofar as its bright peaks extend onto the dark side.

Tonight Mare Tranquillitatis is gone, and only half of Mare Serenitatis remains. As we trace the terminator across the latter, we may be able to see, about ¼ the way from the south to the north shore, Class 1 BESSEL, 10 miles in diameter and 5200 feet deep (Fig. 11). Its interior, of course, is black, but the east and west walls are bright. Perhaps also we can make out a faint, light ridge running north from the little crater. It swings slightly to the east for 45 miles, disappears, continues north after a break of

some 25 miles, and ends 120 miles from Bessel. If the terminator still is several Bessel widths east of the crater, look for a second ridge curving gently to the southeast of Bessel about 90 miles and reaching almost to the shore.

Just west of the terminator, on the south shore of Mare Serenitatis, Menelaus remains conspicuous even though most of its interior is black. On either side of it the Haemus Mountains show well as they delineate sharply the south border of the mare. Across the mountains, two Plato lengths west of Menelaus, larger Manilius shines brilliantly. Tonight less than ½ its interior is dark, but by moonrise tomorrow the sunset line will have crossed it and advanced on Mare Vaporum to the west. As we look again at one of the maria, note that they do not appear as dark as they did around full moon. Each night their contrast with the bright uplands will diminish, and when we reach the crescent stage very little intensity difference will remain between plain and highland.

West of the Caucasus, Aristillus and Autolycus are prominent, their fully-illuminated floors framed by bright east and black west walls. Northeast of the Caucasus, the terminator has reached the foot of Eudoxus' outer east wall and the foot of Aristotle's inner east wall. Both appear as large black pits with brilliant east walls. Near the middle of the Caucasus, southwest of Eudoxus a little farther than the distance of Aristotle to the north, distorted, pentagonal, Class 1 CALIPPUS appears (Fig. 8). With a black interior outlined in white, it is 19 by 21 miles across, 9800 feet deep, and in this light it bears close resemblance to a square. It is partly covered by the shadow of a mountain near its west wall which, according to Goodacre, rises 18,000 feet, a measurement made about a century ago but in good agreement with Army Map Service's elevation of 17,400 feet above the Imbrium plain.

On Mare Frigoris, about two Plato lengths northwest of Aristotle, the small, young crater PROTAGORAS, 12 by 14 miles across and 6900 feet deep, may be seen as a black dot (Fig. 8). One Eudoxus length northwest of Protagoras is the white dot of large Class 1 ARCHYTAS, 20 by 22 miles across and 7500 feet deep (Fig. 9). Continuing in the same direction one Plato length from the last we come to similar Class 2 TIMAEUS, 20 miles in diameter and 7400 feet deep (Fig. 9). There should be enough contrast to bring out huge WILLIAM BOND, tangent to Timaeus on the northeast and fully illuminated (Fig. 9). It is bounded by four nearly straight walls laid out in the apparent shape of a diamond 100

FIGURE 49. Late afternoon in the heavily cratered region near the southeast limb. Bacon and Bacon A exhibit approximately hexagonal outlines, the most prominent linear sections of which run northeast—southwest. A similar alignment is shown by many other craters here, particularly by Mutus and its smaller neighbors to the north and northwest. In the lower right section of the picture, linear crater walls, chains of small craters, and linear depressions are so numerous that they seem to trace out a series of parallel northeast—southwest lines across the surface. Such lines are called lineaments, and they mark the direction of breaks or faults in the moon's crust. Other lineaments, less conspicuous here, can be traced north—south and northwest—southeast. The lineaments constitute the grid system or global cracking pattern of the moon. Similar patterns are found on earth in the few places where erosion has not obliterated them, and they are expected on other planetary bodies as well. (120-inch Reflector.)

miles across and rising at one point to a height of 9800 feet. The wall on which Timaeus stands is low, broken, and inconspicuous, but the other three should be seen. The southeast and northwest walls, bright with dark edges, are double over most of their lengths, and through binoculars they resemble parallel bulldozer cuts. Actually, the enclosure is a pentagon distorted into a close approximation to a rectangle 90 by 107 miles on the sides. Just north of William Bond is the thin ellipse of bright-floored Barrow with the young intruder Barrow A in the southwest wall. Barrow A, dark except for its east crest, probably is merged with the strong shadow that covers the west quarter of Barrow's floor. North of Barrow a distance equal to approximately its separation from William Bond, or half its separation from Archytas, Class 1 SCORESBY, 35 miles in diameter and 11,800 feet deep, may be seen. Although it appears as a tiny black ellipse tonight, its outline is enhanced by a bright northeast crest. Tangent to Scoresby on the northwest is what appears to be another weak dark crater of about the same length but definitely wider and broader at the ends. It is an overlapping pair of craters older than Scoresby. The south one is CHALLIS and the north one is MAIN. Their respective diameters are 37 and 32 miles, and both have walls that rise 9200 feet above the floor. At the center of Main we still are nine degrees from the north pole, but let's not push our luck any further tonight. If there is a large negative libration in latitude we are not likely to get this far north before the craters become flat streaks and lose their identities. So let's turn around and head for the southern hemisphere.

One Iridum length south of Menelaus, and an equal distance southeast of Manilius, the horseshoe outline of ancient Julius Caesar shows well, and at least one of the giant parallel "claw marks" in the area can be traced along its northeast wall. Two Plato lengths southwest of Julius Caesar, Agrippa stands out in the midst of rugged territory, and so does the smaller Godin just south of Agrippa. Their interiors are half black, but their inner east walls are brilliant. Midway between Julius Caesar and Agrippa, the ARIADAEUS RILL may be traced in an approximately straight east-west line for 145 miles with a small telescope (Fig. 10). You might spot it through your binoculars, but it is an extremely difficult test. West of it lies the equally difficult HYGINUS RILL. About one Humorum length south of Julius Caesar, at the terminator, Delambre is seen for the last time as a light-bordered black ellipse. Between Delambre and Godin is an unused tentative touch-down site, but in the highlands 130 miles south of Delambre, Apollo 16 landed on April 21, 1972. One Plato length southwest of Delambre are two similar but smaller craters. The southeast and stronger one is Class 1 TAYLOR, 21 by 31 miles across and 8100 feet deep. The northwest one is older, deformed TAYLOR A, 20 by 25 miles across and 10,000 feet deep (Fig. 50). Taylor, near the terminator, is a small black pit ringed in light, but Taylor A doesn't look like a crater tonight. It appears more like a valley bounded by bright mountain walls.

Southwest of Delambre, approximately the Copernicus-Kepler distance, lies Albategnius, the strongest of the several huge craters in that area. Its central peak may be seen, and the crater can be identified by its intruder Klein which occupies most of the floor between the peak and the southwest wall. Just north of Albategnius is the larger but less-prominent Hipparchus. Its intruder, Horrocks, on the northeast floor shows well. The contrast between the walls and floor of Hipparchus will increase considerably in the next few hours as the sun goes down. On the south wall of Hipparchus, Halley and its slightly smaller neighbor Hind also show well. Again you might check your resolving power by looking for smaller Hipparchus C just northeast of Hind and even smaller Hipparchus L northeast of Hipparchus C. The four are about equally spaced, but the last two are difficult.

Note the distance from the south rim of Albategnius to the north rim of Hipparchus (or to Horrocks). That distance east of the south rim of Albategnius we should find Class 1 ABULFEDA, 40 miles across and 10,500 feet deep (Fig. 50). Its flat floor, sharply bordered by a broad, bright east wall, is only half in shadow, but the sunlit half appears dark also. In the telescope it resembles a piepan. Just south of its east wall is similar but smaller and older Class 2 ALMANON, 31 miles in diameter and 6600 feet deep (Fig. 50). Southwest of Almanon, a distance equal to its separation from Abulfeda, we see Class 1 GEBER, 27 miles in diameter and 9600 feet deep (Fig. 50). Because of its greater depth and bowl shape, most of its interior is black even though it is a little farther from the terminator than Almanon, more than half of which is still illuminated. Continuing southwest of Geber, not quite so far as the last step, we find a strong pair of tangent craters. The south one is irregular, Class 1 AZOPHI, 29 by 34 miles across and 11,200 feet deep. Its companion on the north is similar Class 1 ABENEZRA, 25 by 28 miles across and 10,500 feet deep. While both belong to the youngest class, the former is older since its wall is overlapped by that of the latter. Southeast of the last three craters and south of Almanon may be seen the much larger and very much older Sacrobosco, most of which is now in darkness, with its bright east wall close to the terminator. A few hours earlier the three large floor craters still were visible, and they marked the eyes and mouth on the face of a startled monkey, inverted as seen through binoculars.

Southwest of Azophi and Abenezra, about one Humorum length, is the much larger and more conspicuous pair Aliacensis and Werner, fully illuminated except for their west walls. Through binoculars the last pair closely resembles a doubly-enlarged image of the first pair since the relative sizes and orientation are about the same for both pairs. Midway between the two Class 1 pairs we see a more widely-separated north-south pair of Class 2 craters apparently connected by a bright ridge. The ridge is the west wall of a much older, inconspicuous crater wedged between the two, and it is also the east wall of a much larger ancient crater upon which both have intruded. The larger, southern crater of the pair is APIANUS, 38 by 44 miles across and 10,100 feet deep. The north one is PLAYFAIR, 27 by 30 miles across and 7800 feet deep. Both have smooth, flat-appearing floors which remain fully illuminated tonight. East of Apianus and south of Geber, ancient, distorted PONTANUS may be distinguished from its jumbled and broken surroundings. It is 35 by 38 miles across and 6900 feet deep, but its outline is more nearly square than circular (or elliptical in projection). One Albategnius width south of Pontanus is slightly smaller but more prominent Goodacre. The larger and more conspicuous crater there, the north wall of which has been partially destroyed by Goodacre, is Gemma Frisius. Its broad east inner wall shines brilliantly near the terminator.

One of the more conspicuous features is the great crater Maurolycus near the terminator and one Iridum length south of Gemma Frisius. Its broad bright inner east wall is the chief highlight of the whole southern continent tonight, outshining by far the celebrated Tycho. The rough arc of the west wall is outlined well by the shadow which covers most of the floor and forms a backdrop for the bright central mountain. Perhaps you can resolve the peak through your binoculars. Tangent to Maurolycus on the southeast is a large, bright-bordered black ellipse that merges with the shadows to the north. That is Barocius. A few hours earlier we would have been able to resolve its northeast wall intruder Barocius B, since its complete crest was then in sunlight, the west half a delicate, bright semiellipse separating the black interiors of the two craters.

Nearly tangent to Barocius on the southwest is the similar but less prominent Class 2 CLAIRAUT, 37 by 59 miles across and 9900 feet deep (Fig. 51). Like Barocius, its true outline is elliptical, or, rather, it was before several subsequent impacts broke and distorted its walls and floor. Tonight it resembles a black bean near the terminator, and you may be able to make out the intruder on its south wall— CLAIRAUT A, 22 by 24 miles across and 4900 feet deep. The Barocius-Clairaut line, extended southwest by its length, reaches more normal and more prominent Cuvier which shows well, about ¼ bright and more than ½ black. Tangent to Cuvier on the west is that curious Heraclitus-Licetus enclosure that appears to consist of two craters connected by a broad double trench. It shows well with broad walls bright on the east and black on the west. The bright ridge down the center of the trench may be visible although it casts a very narrow shadow as yet. North of Licetus and west of Maurolycus may be seen large Stofler with its southeast wall destroyed and replaced by Faraday which in turn displays two overlapping craters on its southwest wall. The group is not prominent tonight. The black west wall of Stofler is narrow, but the Faraday west wall shadow is much stronger and it merges with the half-black intruders on the southwest wall. It is around this phase that the whole combination resembles a sketch of a baby hamster when viewed through a telescope. Tangent to Stofler on the north is the smaller Class 3 FERNELIUS, 41 miles across and 6200 feet deep (Fig. 51). It is distorted and elongated by FERNELIUS A in its west wall, a crater 3000 feet deep and 18 miles across. North of Fernelius, and about halfway from it to Aliacensis, is another strange depression shaped like a flattened pentagon 39 by 47 miles across and 5200 feet deep. NONIUS may be the remnant of an extremely old crater since it appears to be intruded upon by the walls of large, ancient, Class 4 Walter on the northwest. The east wall of Walter is part of a conspicuous, bright, north-south scarp line about 100 miles long tangent to Nonius on the west.

South of the Heraclitus-Licetus complex, a distance equal to its length, we find Class 1 LILIUS, 38 by 42 miles across and 7900 feet deep. The bright crescent of the inner east wall shows well, and perhaps you can make out the bright central peak rising from the edge of the shadow that covers the west half of the crater. Tangent on the southeast is the

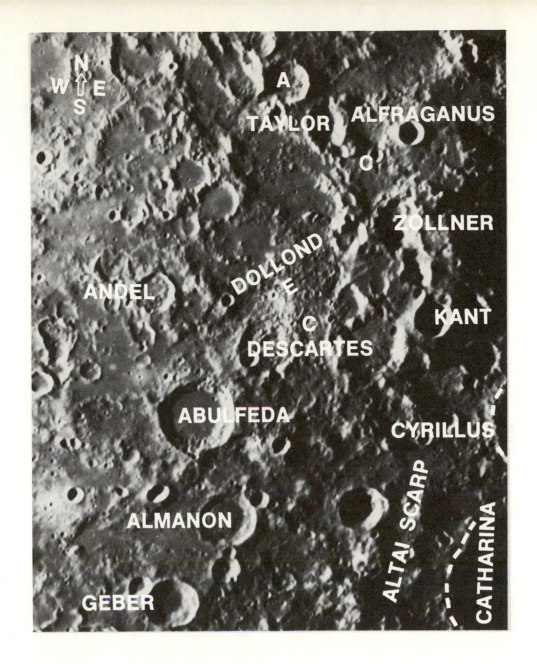

FIGURE 50. Numerous lava lakes and craters of many sizes, shapes, and ages crowd the upland west of Theophilus in this afternoon view. A tiny crater close to the left border west of Almanon has blanketed the area around it with bright ejecta. Note the chain of small craters running from the south wall of Abulfeda toward Catharina. It can be followed more than 100 miles, and it probably is due to volcanic action that produced a series of vents along a fault or crevasse. A shorter chain containing larger craters parallels it northwestward from a point under the label "Cyrillus." These and other linear features here mark the northwest-southeast lineaments of the grid system. (Compare with Figs. 49 and 51.) On April 21, 1972, Apollo 16 astronauts John Young and Charles Duke landed at a point slightly above the second L in Dollond. (120-inch Reflector.)

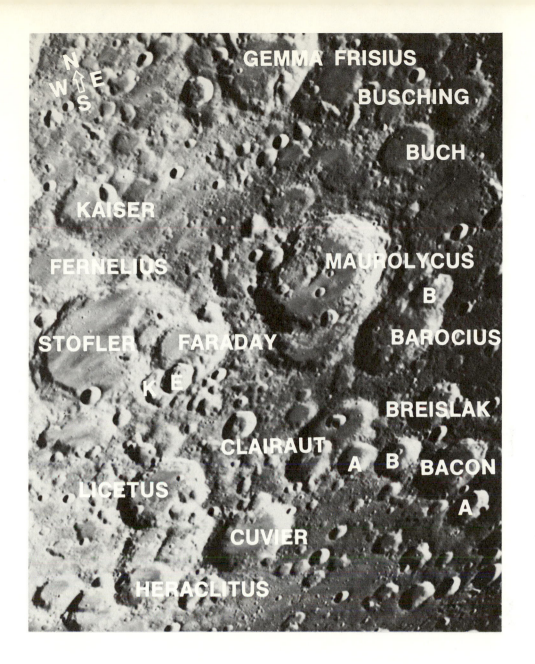

FIGURE 51. Two battered old giant craters, Maurolycus and Stofler, stand out among their smaller neighbors in the rough country east of Tycho. Many ages
are represented. Faraday E, for example, overlaps
Faraday K, which overlaps Faraday, which overlaps
the half crater beyond Faraday's northwest wall, which
overlaps Stofler. The strange Licetus-Heraclitus complex, with its shallow connecting double valley, may be
explained by a combination of collapse and impact
events. Here again the northeast—southwest lineament
pattern is prominent. (Compare with Figs. 49 and 50.)
(120-inch Reflector.)

younger, sharper Class 1 LILIUS A, 23 by 26 miles across and 6200 feet deep. It appears as a dark speck since it is about half bright and half black. Continuing this line of tangent craters in the same direction is weaker Class 2 JACOBI, 43 miles across and 12,500 feet deep. In walls and outline it resembles Lilius, but it has no central peak. Instead, it exhibits through a telescope six floor craters from three to five miles across, all of which are hidden tonight by the shadow that covers most of the enclosure. South of Lilius, a distance about equal to that between it and Jacobi, stands the half-illuminated Class 2 mountain-walled plain ZACH, 44 miles in diameter and 11,800 feet deep. According to Wilkins and Moore: "Zach in its general appearance bears some resemblance to the terrestrial crater-lakes of Coatepeque and Ilopango in Salvador, or to Haleakala in E. Mauri, and may have been formed by similar forces; it is evidently of volcanic origin."

South of Zach a giant step, nearly equal to the last one taken, brings us to larger Curtius, easily seen as a narrow ellipse half bright and half black. If the libration in latitude is positive and large, Curtius will be foreshortened almost to a line, and craters farther south will be difficult to identify. Just southwest of Curtius is larger Moretus, its fully-illuminated interior marked off by sharp arcs of brilliance on the east and blackness on the west. Possibly you may resolve the central mountain which is brighter than the floor but which, as yet, casts only a short shadow. By tomorrow night it will attain greater contrast. Northeast of Curtius, as far as Moretus in the opposite direction, is smaller Class 1 PENTLAND, 36 miles across and 10,200 feet deep. It is more prominent than its size would indicate because it forms a double crater with PENTLAND A on the south. The latter is 25 miles across and 10,800 feet deep, and tonight both craters are about half bright and half black. South of Pentland, and forming an equilateral triangle with it and Curtius, we may see near the terminator a small irregular black oval trimmed in light. It is Class 1 SIMPELIUS, 43 by 54 miles across and 10,800 feet deep. It appears to be overlapped on the northwest by smaller but sharper and more regular SIMPELIUS E, 30 by 33 miles across and 12,800 feet deep. We now have reached latitude 73 degrees south, and the craters beyond are both narrow and closely packed. The moon's south pole, of course, lies south of everything in view, "lunigraphically" speaking, but for our purpose of location it is due south of Moretus and Curtius. It is very close to if not beyond the limb.

TWENTY-TWO-DAY MOON

(Chart V or VI)

Tonight the moon has almost reached the last quarter phase, and it is time for the midterm test of the second semester. To pass it, one must get up for the observation period between 3 and 5 A.M. each morning during the next week, weather permitting. Although the moon will rise around midnight at the beginning of the day on which it becomes 22 days old, it cannot be observed to advantage until a good three hours later. In fact, from now on it would be best to set your alarm clock to awaken you early each morning so you can get your period of observation in before daylight. You then will view the moon under the most favorable conditions.

Tonight the conspicuous Apennine Range meets the terminator at a sharp angle and makes a splendid appearance before bowing out for the next two weeks. The Caucasus are in darkness except for the peaks of their highest mountains. Those may be found on the dark side by extending the Apennine arc to the northeast. At the west end of the Apennines, Eratosthenes shows well with just the right amount of highlight and shadow. You may be able to resolve its central mountain. Beyond Eratosthenes, Copernicus still shines prominently, its multiple central peak lost against the bright floor background, but its once-brilliant ray pattern has faded considerably. The splash area around Kepler remains prominent, and the crater itself still is difficult to resolve because of the glare. Aristarchus continues to resemble a dazzling teardrop pendant. Tycho in the south likewise stands out among its neighbors, but only short remnants of its once-stupendous ray system can be found. The large maria, with the exception of Mare Nubium and Mare Humorum, have lost much of their contrasting pigmentation, but the dark-floored mountain-walled plains, Grimaldi and Plato, are seen easily. Grimaldi is a very strong feature and will continue so for several nights.

North of the central Apennines, about ⅓ the way across Mare Imbrium to Plato, a group of three conspicuous craters marks out a right triangle. The largest and least prominent of the three is Archimedes on the west. The two near the terminator are Aristillus on the north and Autolycus on the south. Note that the interiors of the last two are black except for the bright walls while the floor of Archimedes, only about 90 miles farther from the terminator, is illuminated completely with the exception of a narrow border in the shadow of the west wall.

Since the altitude of the sun at Archimedes is only a few degrees higher than its altitude as viewed from Autolycus or Aristillus, we see here an illustration of the extreme shallowness of some of the great ring plains. In the case of Archimedes, the diameter is 40 times the depth and 50 times the height above the surrounding plain. The mountain arms that curve southwest from the crater wall look like a misplaced section of the Apennine foothills. North of the crater is the much smaller group of Spitzbergen Mountains. Southeast of the crater, against the Apennines, lies the small dark plain of Palus Putredinis. West of the mountain arms and north of Eratosthenes, Timocharis shows well as a bright spot.

The individual mountain Piton shines brightly about two Plato lengths north of Aristillus, and its long, black shadow stretches out toward the terminator. Just beyond the end of that shadow, at the terminator, you may be able to resolve the thin elliptical rings that mark the weakly illuminated crests of the double crater Cassini, most of which is in darkness. Between Cassini and Plato curve the Alps, the sunlit faces of their mountains shining conspicuously and the tips of their extreme eastern peaks gleaming like stars out of the blackness of the dark side. Between Plato and Aristillus, and between Plato and Archimedes, look for the faint curved ridges that may indicate the edge of the great Imbrium impact crater. The strongest of the ridges includes the Spitzbergen Mountains, swings some 60 miles southwest of the group, and extends about 100 miles north of it, ending at a bright speck of a mountain. About halfway between that speck and the south wall of Plato is the prominent mountain Pico. Southwest of Plato and northwest of Pico lie the scattered members of the Teneriffe Mountains.

In the highlands north of narrow Mare Frigoris two of the smaller craters may be distinguished from their jumbled surroundings near the terminator. At the northwest outer wall of the huge, partly darkened enclosure William Bond stands the Class 5 EPIGENES, 35 miles across and 8900 feet deep (Fig. 9). North of Epigenes is the younger, more prominent Anaxagoras, which is about the same size. It rises out of the west wall of the large excavation Goldschmidt, which is black except for a dimly-illuminated strip of floor along the southeast wall. The interior of Epigenes is ⅔ darkened, and that of Anaxagoras is completely lost in shadow. Their visibility is owing primarily to their brilliant inner east walls. About two Plato lengths west of Anaxagoras

the larger Philolaus stands out, its hilly floor framed between a black west wall and a bright east one. Perhaps a few more craters might be identified in the far north, but most of those in view tonight are extremely old, and their low walls present very little contrast beyond the impression of general roughness.

Near the midpoint of the terminator we find an isolated, square section of upland about one Sinus Iridum length on a side. It is bounded by Sinus Aestuum, Sinus Medii, and Mare Vaporum. Southwest of its center is the capsule-shaped double crater Murchison-Pallas outlined in white and black. Can you detect small Bode close to the northwest rim of the enclosure and slightly larger but weaker Ukert about three times as far from the northeast rim? Both show bright-bordered black interiors. On the southeast edge of the section look for slightly larger Triesnecker seen as a small bright ring. A little more than one Murchison-Pallas length south of Triesnecker and right at the terminator is the light-bordered black old RHAETICUS, 32 miles long (north-south), 25 miles wide, and 5200 feet deep. That strangely distorted enclosure, shaped like a peach pit, evidently has been flooded with lava since its floor stands 3000 feet above the level of nearby Sinus Medii. As Webb long ago pointed out, it "marks exactly the moon's equator, and is one of the few spots to which the sun and earth may both be vertical." The reason why the earth can appear in the zenith as viewed from Rhaeticus is that the latter is located only five degrees (1½ Copernicus lengths) east of the average center of the lunar disk —within the range of places which the libration in longitude can bring momentarily to the exact center.

South of the Murchison-Pallas enclosure, a distance nearly equal to that between Archimedes and Eratosthenes, the great hexagonal mountain-walled plain Ptolemy shows well, its smooth floor fully illuminated and bounded by six linear wall sections of equal length that are black, gray, or white. Tangent on the north is much younger and smaller Herschel, about ⅔ black and ⅓ brilliant. Tangent to Herschel on the north is ancient, dilapidated, flooded SPORER, 15 by 17 miles across and 2300 feet deep, the low, broken walls of which can be seen only when the terminator is nearby. We probably would not see it now were it not for a black valley at the foot of its outer east wall and the high, broad wall remnant of a former larger crater about 10 miles west of the west wall. Those markings give the general impression of a crater about 30 miles in

CHART VI

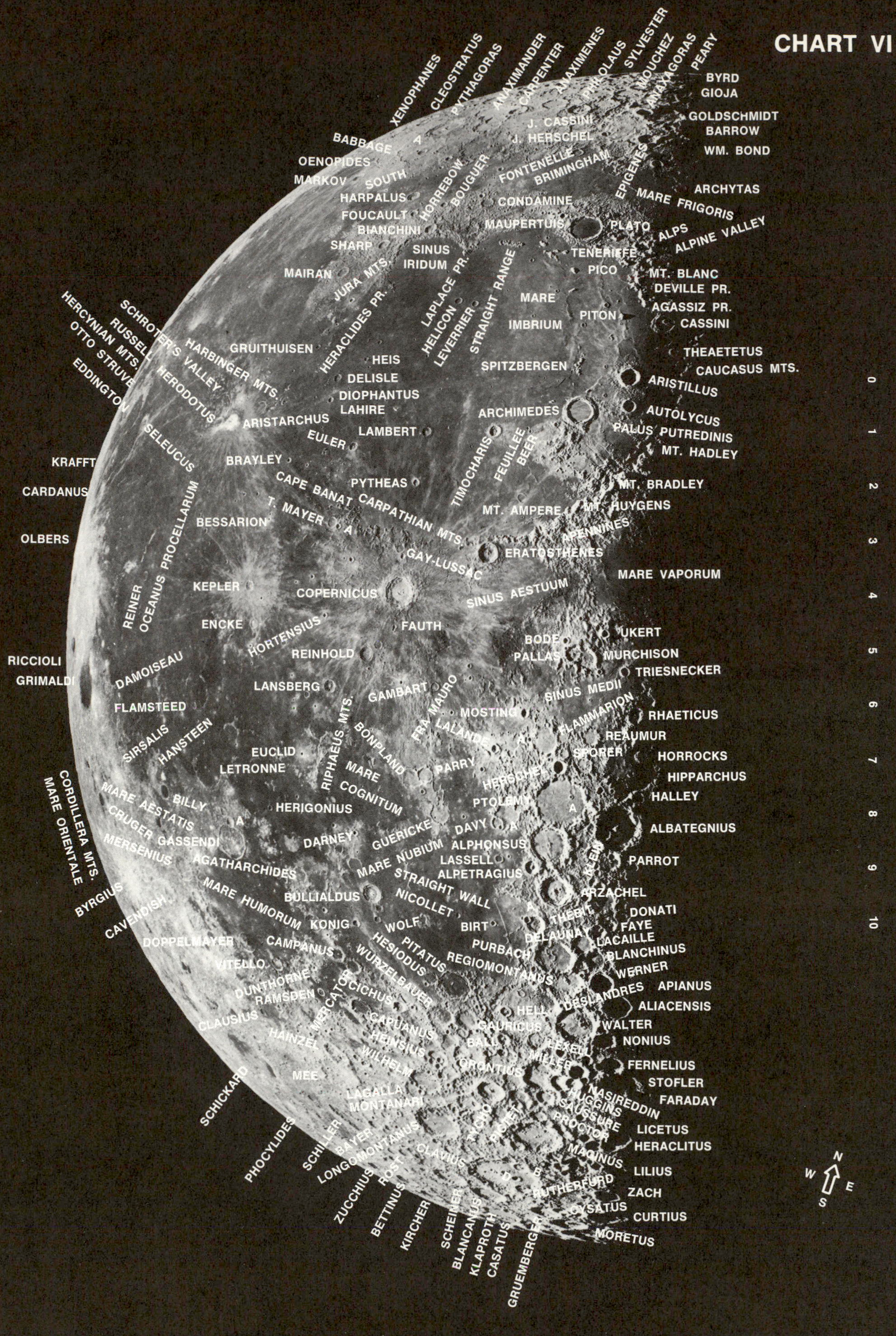

XENOPHANES
CLEOSTRATUS
PYTHAGORAS
ANAXIMANDER
CARPENTER
ANAXIMENES
PHILOLAUS
SYLVESTER
MOUCHEZ
ANAXAGORAS
PEARY
BYRD
GIOJA
BABBAGE
J. CASSINI
J. HERSCHEL
GOLDSCHMIDT
BARROW
WM. BOND
OENOPIDES
MARKOV SOUTH
HARPALUS
FOUCAULT
BIANCHINI
SHARP
HORREBOW BOUGUER
FONTENELLE BRIMINGHAM
CONDAMINE
MAUPERTUIS
EPIGENES
MARE FRIGORIS
ARCHYTAS
PLATO
TENERIFFE
PICO
ALPS
ALPINE VALLEY
MT. BLANC
DEVILLE PR.
AGASSIZ PR.
CASSINI
MAIRAN
SINUS
IRIDUM
JURA MTS.
HEIS
DELISLE
DIOPHANTUS
LAHIRE
LAPLACE PR.
HELICON
LEVERRIER
STRAIGHT RANGE
MARE
IMBRIUM
SPITZBERGEN
PITON
THEAETETUS
CAUCASUS MTS.
ARISTILLUS
AUTOLYCUS
PALUS PUTREDINIS
MT. HADLEY
SCHROTER'S VALLEY
HERCYNIAN MTS.
RUSSELL
OTTO STRUVE
EDDINGTON
HARBINGER MTS.
HERODOTUS
MERODOTUS
SELEUCUS
GRUITHUISEN
ARISTARCHUS
BRAYLEY
HERACLIDES PR.
EULER
LAMBERT
ARCHIMEDES
FEUILLEE
BEER
TIMOCHARIS
PYTHEAS
MT. BRADLEY
MT. HUYGENS
MT. AMPERE
APENNINES
KRAFFT
CARDANUS
OLBERS
CAPE BANAT
T. MAYER
BESSARION
CARPATHIAN MTS.
A
GAY-LUSSAC
ERATOSTHENES
MARE VAPORUM
REINER
OCEANUS PROCELLARUM
KEPLER
ENCKE
HORTENSIUS
REINHOLD
COPERNICUS
FAUTH
SINUS AESTUUM
BODE
PALLAS
UKERT
MURCHISON
TRIESNECKER
RICCIOLI
GRIMALDI
DAMOISEAU
FLAMSTEED
SIRSALIS
HANSTEEN
LANSBERG
GAMBART
FRA MAURO
MOSTING
SINUS MEDII
FLAMMARION
RHAETICUS
REAUMUR
SPORER
HORROCKS
HIPPARCHUS
HALLEY
EUCLID
LETRONNE
RIPHAEUS MTS.
BONPLAND
LALANDE
MARE
COGNITUM
PARRY
HERSCHEL
PTOLEMY
A
ALBATEGNIUS
CORDILLERA MTS.
MARE ORIENTALE
MARE AESTATIS
CRUGER GASSENDI
MERSENIUS
BILLY
HERIGONIUS
DARNEY
GUERICKE
DAVY
ALPHONSUS
AGATHARCHIDES
MARE NUBIUM
LASSELL
ALPETRAGIUS
PARROT
BYRGIUS
CAVENDISH
MARE HUMORUM
KONIG
BULLIALDUS
NICOLLET
STRAIGHT WALL
WOLF
BIRT
A
THEBIT
DELAUNAY
ARZACHEL
DONATI
FAYE
LACAILLE
BLANCHINUS
WERNER
DOPPELMAYER
VITELLO
CAMPANUS
DUNTHORNE
RAMSDEN
MERCATOR
CICHUS
HESIODUS
PITATUS
PURBACH
REGIOMONTANUS
HELL
DESLANDRES
APIANUS
ALIACENSIS
WALTER
NONIUS
CLAUSIUS
HAINZEL
WURZELBAUER
CAPUANUS
HEINSIUS
WILHELM
BALL
GAURICUS
KIESS
MILOS
ORONTIUS
FERNELIUS
STOFLER
FARADAY
MEE
LAGALLA
MONTANARI
HIGGINS
SAUSSURE
PROCTOR
NASIREDDIN
LICETUS
HERACLITUS
MAGINUS
LILIUS
ZACH
SCHICKARD
PHOCYLIDES
SCHILLER
BAYER
LONGOMONTANUS
CLAVIUS
ROST
RUTHERFURD
CYSATUS
CURTIUS
MORETUS
ZUCCHIUS
BETTINUS
KIRCHER
SCHEINER
BLANCANUS
KLAPROTH
CASATUS
GRUEMBERGER

N
W E
S

0 1 2 3 4 5 6 7 8 9 10

diameter, and the valley on the east is but one of hundreds of scars found throughout the vast upland area bounded by the Maria Nubium, Imbrium, Serenitatis, and Tranquillitatis. Most of the scars appear as valleys or cuts approximately radial to Mare Imbrium. Baldwin and Urey have expressed the belief that they were plowed out by debris hurled from the enormous Imbrium crater following the impact and explosion which produced it. Alter believes they "were opened along faults at the time of the sinkings that produced both the Maria Imbrium and Tranquillitatis" (*Pictorial Guide to the Moon*, 1963).

A little more than one Ptolemy width northwest of Herschel may be seen similar but smaller Class 1 MOSTING, 16 miles across and 9800 feet deep. Between the two look for the low walls of ancient, lava-flooded FLAMMARION, 44 by 50 miles across, 8200 feet deep, and a dim object. There was another possible landing site for Apollo astronauts. East of Herschel, at the terminator, a black bay with several bright specks in it joins the dark side. That is Hipparchus, which soon will disappear completely. About one Ptolemy length south of Hipparchus is a slightly smaller but more sharply defined black enclosure with a bright east wall and central peak. You will recognize it as Albategnius, and if you can discern a narrow bright dash between its central peak and southwest wall you are about to view sunset on the east crest of the intruder Klein.

Tangent to Ptolemy on the south is smaller but better-delineated Alphonsus, the enclosure explored by Ranger 9 in March, 1965. Its broader walls stand out clearly, and you might be able to resolve the relatively tiny central peak which rises from the east prong of a bifurcated ridge or spine running from the south wall ¾ of the way across the floor. It will show better in a few hours after it has acquired a longer shadow and the floor has darkened. Just south of Alphonsus is smaller and even better defined Arzachel with a central peak that should be visible through binoculars tonight. Nestled between the southwest wall of Alphonsus and the shaded plain of Mare Nubium, smaller Alpetragius easily is seen, half bright and half black with its oversize central mountain at the dividing line. Thebit and its black wall intruder stand on the Nubium shore one Alphonsus width southwest of Arzachel. An equal distance west of Thebit we may find the tiny bright-edged black speck that is Birt, and between the two we might perceive the Straight Wall. Again it appears as a hair-thin line, pointed toward Copernicus and about as long as the distance from

Thebit to Birt, but this time it is white instead of black. It is a difficult test tonight, but tomorrow night it will be much easier to see.

Southeast of Thebit, about the distance of Arzachel to the northeast, large, ancient, dark-floored Purbach is seen readily, and its hexagonal shape is evident. At its outer northeast wall begins the line of four tangent craters running southeast to the terminator. They are Lacaille, Blanchinus, Werner, and Aliacensis. The interiors of the last two are completely black, but the narrow bright arcs of their east wall crests separate them from the dark side. A broad "V"-shaped shadow covers nearly ⅔ of Blanchinus' floor while that of Lacaille is more generally illuminated. Tangent to Werner on the west and intruded upon by the south wall of Purbach is old, distorted Regiomontanus, which appears to be more of a space between craters than a crater itself. It shows well with its light and dark bordered sunlit floor. Can you make out the central mountain which appears bright against the gray floor? Just south of Regiomontanus is the prominent Walter which intrudes upon the enormous enclosure Deslandres stretching off to the west and well outlined tonight. Its largest intruder, Lexell, and second largest, Hell, are seen easily on the floor at the south wall and near the west wall, respectively. Tangent to Deslandres on the southwest, south of Hell and west of Lexell, is Class 2 BALL, 25 miles across and 5900 feet deep (Fig. 6). Wilkins and Moore describe Ball as "a very remarkable object" with walls "which are magnificently terraced on the inner slopes" and with "a deep and broad groove which descends the inner south slope from the crest to a crater on the floor south of the central mountain." Ball is half bright and half black tonight, and its mountain is difficult.

South of Walter, a distance equal to its length, is a shadow-filled crater of ½ Walter's dimensions. Class 1 MILLER, 36 by 40 miles across and 7500 feet deep, appears as a circular black hole with broad, bright walls (Fig. 6). On the northeast its wall joins that of MILLER A, another Class 1 crater, 21 by 25 miles across and 4900 feet deep. On the south Miller is tangent to Nasireddin, which destroyed the east wall of larger Huggins, which previously had destroyed the east wall of still larger Orontius. Saussure, with its double east wall, is tangent to Huggins and Orontius on their south rims. If identification is uncertain in this exceedingly disrupted region, note that Saussure is due east of prominent Tycho and separated from it by twice the latter's length. South of Saussure, a little less than the Tycho-Saussure separation, large Maginus has lifted its veil of invisibility to become once again a conspicuous feature. A little farther south, giant Clavius shows well, its floor fully illuminated except for the interiors of the smaller craters that have broken it. One of its lengths southeast of Clavius, and near both the terminator and the limb, the cigar-shaped outline of Moretus stands out among the many bright arcs and angular shadows. Half its interior is black and may present enough contrast to bring out the bright dot of the central peak.

Again I have pointed out only the more prominent features of tonight's moon. Those to the west of the terminator belt generally will be enhanced by the increasing contrast of approaching sunset, and I shall include many of them in our observation lists for the next few mornings as we pursue the waning crescent eastward to dawn's horizon.

TWENTY-THREE-DAY MOON

(Chart VI)

It often seems that the waning crescent is a dim moon compared with one's recollection of the waxing crescent at the corresponding *phase angle* (angle between incident and scattered rays; roughly the terminator longitude measured from its full-moon position). The five-day moon, for example, would have a phase angle corresponding to that of the $29\frac{1}{2} - 5 = 24\frac{1}{2}$ day moon. Indeed, it could be argued that the waning crescent should be weaker than the waxing crescent at the corresponding phase angle, since the dark maria cover a large portion of the western hemisphere and a smaller portion of the eastern hemisphere. Yet precise measures of moonlight intensity as a function of phase angle reveal no such differences. The brightness curve after full is almost an exact mirror image of the curve before full. The very slight departures from symmetry in the curves indicate that the waning crescent actually sends us a little *more* light than the waxing crescent. Why, then, the contrary illusion? The apparent brightness of the moon is sensitive to changes in atmospheric transparency, and such changes grow large at low altitudes where the thin crescent generally is seen. Moreover, our ruling habits are such that we probably tend to observe the new crescent in the early evening at a higher altitude than the old crescent near the end of the night. The most important factor, however, is

atmospheric humidity. During the clear, windless nights that are best for observing, the humidity in many locations increases steadily, an effect which often results in fog at dawn if not before. As humidity goes up, transparency goes down, and hence the evening crescent moon at a given altitude is likely to appear brighter than its morning counterpart at the same altitude. Almost every kind of astronomical measurement is subject to similar variable error introduced as light passes through our atmosphere, and much of the astronomer's work in the past three centuries has been devoted to efforts to find the law of error and then to determine how much correction should be applied to each individual observation. We shall look for improvement in these matters in the future from the astronomer of the Space Age who already has begun remote observation via television through electronically controlled telescopes orbiting the earth above our atmosphere.

If we happen to be observing at the right lunar time (terminator near the east wall of Archimedes and across the floor of Maginus), our attention will be drawn to Deslandres, a large enclosure at the terminator on the southeast edge of Mare Nubium. The west portion of its vast floor still is illuminated and is bounded by the black main wall, but the east floor is in shadow and is outlined by the brilliant irregular east crest that separates it from the dark side. Review the first few pages of the section on Selenography, and compare what you see with Galileo's drawing of 1610. Look at the left upper drawing of Fig. 16, and turn the book until the bright half of the moon is oriented parallel to the one you see through binoculars. The huge crater is likely to be found a little closer to the south limb, and it will appear somewhat smaller than shown by Galileo. Notice how the Apennines and Alps extend over the dark side 100 miles or so as bright pincers closing on darkened Mare Imbrium just as he sketched them more than $3\frac{1}{2}$ centuries ago. A bright patch on the dark side between the Alps and the north limb, shown on both left-hand drawings, is seen on the moon. It is the broad east crest of the large crater William Bond. A smaller but conspicuous bright streak, likewise shown, is seen beyond the terminator between Deslandres and the Apennines. It is the east crest of Alphonsus. The bright outlined but centrally shaded protrusion along the terminator, best shown in the left upper drawing, halfway from Deslandres to the south limb, is seen to be the large crater Maginus. The next protrusion shown south of it is the east rim of Clavius. Deslandres is not the strongest enclosure visible by any means, but it shows well, and its size and shape evidently made a deep impression on Galileo. Of course you see many more features than the few he drew, and his pattern of shadows or maria on the western hemisphere is far more symbolic than realistic. But taking Galileo's drawing for what he must have intended it to be—a record of several impressive features along the terminator—what is your verdict? Is it merely a collection of doodles, or did he represent the moon fairly well as far as he attempted to do so?

The contrast in brightness between the smooth maria and the rough uplands continues to diminish, but tonight the plains are well defined and appear darkest along the south and west shores of Mare Humorum and the long west coast of Oceanus Procellarum near the limb. Dark-floored Grimaldi remains conspicuous in the midst of its bright surroundings, and Plato, with the standard terminator at its east wall, is one of the stronger features in this its final hour.

Copernicus also presents a splendid spectacle on the last night before sunset overtakes it. Its floor is illuminated fully, its east inner wall is brilliant, its west inner wall is about ⅔ darkened, and there is enough shading along the outer east wall to restore the third dimension of depth which the great crater has lacked for more than a week. The extensive ray pattern, conspicuous only a few nights ago, virtually has vanished, but you will be able to find traces of it here and there. The Kepler ray splash, on the other hand, still is very bright, and the crater itself has little contrast. However, shadow and highlight are growing along its small walls, and tonight you probably can see it again. Northeast of Copernicus the smaller Eratosthenes has risen to prominence with bright walls and ⅔ of its interior black. That shadow forms an excellent backdrop for the central peak which still may be visible as a bright speck. It reaches maximum visibility when ½ the interior of the crater is in shadow, a few hours before standard terminator time tonight. That mountain appears to have an oblong summit crater four by six miles across which presents an interesting clue to its origin. The mathematics of probability offers us little hope of ever finding an impact crater right at the top of a mountain where its peak should be. Consequently, since there are quite a few such central mountains on the moon, it seems likely that they once were active volcanoes. About 100 miles of mountains connect Eratosthenes with the terminator. That narrow bright band you will recognize as the western foothills of the Apennines now lost in the intense blackness of the lunar night.

North of Eratosthenes nearly two Iridum lengths,

Timocharis may be seen, its broad bright walls contrasting with the growing interior shadow and the dim plain on which it rises. West of Timocharis and north of Copernicus, the smaller, isolated Lambert should be visible as a black-edged light dot since more than half its interior is illuminated. One-third the way from Lambert back to Copernicus, small Pytheas may be detected as a tiny bright ring with a black center. Close to the terminator and about halfway from Timocharis to the north limb, a brilliant spot catches the eye. That is the bright west face of Pico looming out of the dim background. With a small telescope we could resolve its long shadow stretching eastward onto the dark side. Northwest of Pico the Teneriffe Mountains appear weak by comparison, and west of them you may see the long, narrow rectangle of the weaker Straight Range. A little farther west the graceful semi-ellipse of the Jura Mountains enfolds Sinus Iridum.

A line from Timocharis through the Teneriffe Mountains extended about half its length reaches FONTENELLE on the north shore of Mare Frigoris (Fig. 9). An old Class 5 ring plain, 25 miles in diameter and 9200 feet deep, Fontenelle is visible as a black speck contrasting with the bright mountain mass at the west wall. Wilkins and Moore report that the former "has seen the walls almost as brilliant as Aristarchus" through a large refractor. This remark reminds us to note the conspicuous Aristarchus as we return to the southern hemisphere. Not only does the brilliant crater show well tonight, but the delicate light outline of its normally overwhelmed neighbor Herodotus also can be made out.

If the atmosphere is very clear and steady, you probably can see a part of SCHROTER'S VALLEY (Fig. 44), a difficult test for binoculars. Telescopically it is perhaps the easiest of the rills, having been discovered by Christian Huygens in 1686 and rediscovered a century later by John Schroter. It will be easier to resolve three mornings hence when close to the terminator. To describe its location and course let me take the foreshortened width of Aristarchus (17 miles) as unit distance. It lies on the north side of the two craters, and it resembles a snake, beginning at the "Cobra Head" about one unit from the wall of each crater. It runs north one unit, northwest one unit, west with ripples one unit, and southwest one unit. The bright east and northeast walls of the first two legs are the more conspicuous parts tonight, and along them the valley ranges from two to five miles wide. At the junction of the first two legs the Aeronautical Chart and Information Center has measured a depth of 4500 feet. Along the second leg their chart (LAC 38) shows depths of 1500 and 1700 feet. That is pretty fair confirmation of Schmidt's measure of 1600 feet in that area made visually with a six-inch telescope more than 100 years ago.

Far south of Copernicus, on western Mare Nubium, Bullialdus shows conspicuously, its broad, terraced inner walls black on the west and bright on the east. Its multiple central mountain is bright against the light floor and may be visible through binoculars tonight. Northeast of Bullialdus and ⅓ the way to Eratosthenes, the delicate, bright rings of ancient, flooded Parry and Guericke may be traced since there is just enough shadow along their gentle eastern slopes to bring them out. Tonight the moon is about 20 hours older than it was on the morning of July 31, 1964, when Ranger 7 began televising the area as it headed for impact on Mare Cognitum about halfway from Guericke to the Riphaeus Mountains. Apollo 14 landed 150 miles north on February 5, 1971, at a point in the Fra Mauro uplands, a point which you can locate by running an imaginary line from the center of Guericke to the center of Parry and extending it by its length.

At the terminator northeast of Bullialdus, a distance equal to that between Copernicus and Kepler, you probably can find weak Class 5 DAVY, 20 miles in diameter and 4900 feet deep (frontispiece). The loss of its south wall and the mountain mass that joins it on the north combine to give it the appearance of a short-handled tongs like those used by the iceman of a bygone age. The jaws of the tongs close on DAVY A, a sharp young crater 9 miles in diameter and 6000 feet deep, which doubtless is too weak for our binoculars to reveal. If you found Davy you may want to look south of it one Copernicus length along the terminator for smaller, flooded, polygonal LASSELL, a curious enclosure 14 miles in diameter with gaps in its 3100-foot wall on both the north and south sides (frontispiece). Near the terminator east of Bullialdus (south of Davy and Lassell) the Straight Wall is at its best as sunset draws near. The plain is dark, and the wall with its 41-degree slope is reflecting sunlight in our direction at just about peak efficiency. I am sure you can see the narrow bright line pointing toward Copernicus and equal in length to that crater. A line of sunlit peaks beyond the terminator to the southeast resembles an extension and makes the wall appear twice as long as it really is.

The south coast harbor on Mare Nubium formed by the bright, broken ring of ancient Pitatus shows well, and its tiny remnant of a central peak, off center to the northwest, may be seen as a bright speck. Just south of Pitatus are tangent twin craters, both of which depart considerably from circular outline. Light and dark-bordered Wurzelbauer on the west has a mottled floor which is entirely illuminated but much darker on the east than on the west. Gauricus shows a bright east wall, a broad black west wall, and a smooth gray floor. West of Wurzelbauer, on the east edge of Palus Epidemiarum, the smaller, younger Cichus may be seen, its broad, bright walls contrasting with its half-black, half-gray interior. Cichus C, a bright speck on the dark southwest wall, may be too small to resolve. Farther west, larger Capuanus forms a bay near the middle of the Palus Epidemiarum south shore. Its heavy bright west wall resembles a peninsula that juts out into the middle of the plain, and its low east wall is washed out. North of Capuanus, at the north edge of the palus, the bright-rimmed dark pair of craters Mercator and Campanus can be seen. The latter shows well, but the former, which is just as prominent, is difficult to resolve because of the brilliant triangular mountain arm on its south edge which draws our attention.

Near the terminator and southeast of Pitatus, about one Humorum length, a black ellipse marked by a brilliant crescent on the east is seen easily. That is the recently spectacular Tycho which has shed most of its rays and is about to retire into complete darkness for the next two weeks. One Humorum width west of Tycho is the equally large but usually inferior Wilhelm, the floor of which is illuminated fully. Perhaps you can see two of the intruders on the southwest wall. They are WILHELM A, 12 miles in diameter and 3900 feet deep, and WILHELM B, 10 by 12 miles across and 3600 feet deep (Fig. 6). Both appear about half bright and half black. At an equal distance south of Wilhelm larger Longomontanus shows well. West of Wilhelm, as far as Eratosthenes is removed from Copernicus, boat-shaped Hainzel stands out from its mottled and jumbled surroundings.

Between Tycho and the southern limb the sun is setting on gigantic Clavius. The broad black west wall and the brilliant east wall, extending over the dark side, frame the floor which has taken on a strange appearance. We have seen many crater floors near the sunset terminator, and we are accustomed to find them sunlit to the east and shadow covered to the west. But Clavius is a nonconformist. Like that of equally large Deslandres, its western floor is illuminated, but its eastern floor is black! There is the proof of a statement I made earlier that the floor of the great crater is convex. It follows the general curvature of the moon's surface, and while the sun still is above the horizon as seen from the west floor, it already has set for an observer on the east floor. Clavius B may be seen as a bright ring on the northeast wall, and the brilliant east crest of Rutherfurd on the southeast wall shines out of the darkness. Between them a thin white broken ellipse marks the crest of Clavius D.

Apparently tangent to Clavius on the southwest is Blancanus, a really large crater (diameter 66 miles), which always looks small in contrast to its huge neighbor. The interior is half black, and the remaining floor is a dark gray, but the broad east inner wall shines brightly. Just northwest of Blancanus, equally large but less prominent, Scheiner shows well with its inner wall bright on the east and black on the west. South of Blancanus, with an interval between walls of one Blancanus width, are two similar but overlapping craters. Their line of centers runs approximately north-south, and their floors, which are more than half illuminated, reflect sunlight only feebly in our direction. They are not easy to resolve, their most conspicuous feature being bright east inner walls. The nearer one is KLAPROTH, and the south one is CASATUS. The west walls of both appear blasted and distorted by intruding craters, which, combined with the foreshortening of their near-limb location, makes accurate measures of their dimensions difficult. However, both are approximately 65 by 77 miles across, and their depths are 10,200 and 12,800 feet, respectively. Klaproth has been classified as a mountain-walled plain while the younger Casatus, which is the overlapper, is called a ring plain. Casatus definitely is deeper, and it does appear to have a slightly higher outside wall than Klaproth, but the distinction between a mountain-walled plain (no outside walls, or almost none) and a ring plain (outside walls) becomes hazy when we compare closely various examples under excellent observing conditions. The situation becomes even more confused if we accept the interpretation that mountain-walled plains are subsidence craters and ring plains are explosion craters.

TWENTY-FOUR-DAY MOON

(Chart VI)

Tonight the terminator, near its midpoint, cuts across the east wall of black Copernicus which is separated from the vast dark side by a bright east inner wall. The outer west wall is seen to be a little lighter than the Procellarum plain. A telescope would show a forest of hills stretching 150 miles west of Copernicus over territory which recently looked quite flat except for the pits of hundreds of tiny craters. Although most of the Copernicus rays have disappeared, the Kepler splash pattern still is brilliant with the crater itself now visible as a black-edged bright dot. We are once again near enough to the date of new moon (full earth) to notice that the dark side is illuminated faintly by earthshine. It is definitely brighter than the sky behind it, and you may be able to discern the vague outlines of some of the eastern maria. The strength of this twice-reflected sunlight will increase considerably each night during the remainder of the lunation.

Southwest of Copernicus and about half as far away as Kepler, Reinhold stands out in strong contrast. Most of the interior is black, but both sections of the wall that decline to the west are bright. A little farther southwest is Lansberg, about half black and half bright. Just north of a point ⅓ the way from Reinhold to Kepler you may be able to see a black speck, the interior of Class 1 HORTENSIUS, 9 miles in diameter and 7800 feet deep (Fig. 4). Its walls have an albedo of 15 per cent, making it one of the brighter spots when well illuminated. North of Copernicus the Carpathian Mountains can be seen stretching west from the terminator about 200 miles. Their northern extremity is a bright spot known as CAPE BANAT (Fig. 4). Thirty miles south of Cape Banat, the highest mountain of the range rises 6600 feet above the Imbrium plain, its position marked by the base of the strongest, longest shadow. One Copernicus length southwest of the cape, near the west end of the Carpathians, look for the bright east wall of TOBIAS MAYER, a shallow, distorted crater 20 by 22 miles across and 5200 feet deep (Fig. 4). In spite of its size, it is not an easy object, and under higher sun it fades away. Generally it is more difficult to spot than the newer TOBIAS MAYER A, tangent on the southeast and only 11 miles in diameter but 8200 feet deep. The latter, which Webb calls "a fine specimen of subsequent eruption" (!), refuses to show itself tonight, being blacked out except for its small inner east wall. North of Tobias Mayer and east of Aristarchus stands almost equally large Euler, about half bright and half black. More conspicuous than the crater is the adjacent bright area two to three times its size that trails off to the southeast. That is part of the Copernicus ray pattern.

Northeast of Euler, and near the terminator, Laplace Promontory shines brightly where Mare Imbrium merges with eastern Sinus Iridum. One Grimaldi length south of the promontory you may resolve two weak light specks on an east-west line and separated by about one Grimaldi width. Those are the bright walls of two dark Class 1 craters isolated on the open plain. The east one is LEVERRIER, 13 miles in diameter and 8000 feet deep. The west one is HELICON, 16 miles in diameter and 7400 feet deep. Just south of the two and forming an equilateral triangle with them is an impossible, tiny crater only four miles across, which, according to Baldwin, stands on the spot where the great interplanetary projectile crashed and formed Mare Imbrium, "the moon's greatest crater." Near the middle of the graceful arch that marks the Sinus Iridum shore line, the Jura Mountain Range is broken at its crest by the irregular crater Bianchini, which tonight appears half bright and half black.

North of Laplace Promontory, near the south shore of Mare Frigoris, look for ancient, low-walled Condamine, about ⅓ shaded. It is never a strong feature, but tonight its contrasting walls present it at about its best. In the north horn of the crescent, and roughly midway between the terminator and the limb, Philolaus, youngest and sharpest of the large craters in the polar region, shows well.

Let us now go south. On southern Oceanus Procellarum, one Iridum length south of Lansberg, the bright Riphaeus Mountains show well against their shadows to the east. Just west of the main group a difficult black dot centered on an easy bright patch marks the young crater Euclid. Webb calls it "the best specimen of an infrequent variety, the light-encompassed crater . . . closely surrounded by a luminous cloud not resembling the white streaks in character." Such a "cloud" now is known by its less-familiar Latin equivalent *nimbus,* since the term cloud carries a gaseous and temporary connotation, and "streaks" is a synonym for "rays." A nimbus, or halo, is simply a bright patch of indefinite extent that surrounds a relatively recent crater. It is interpreted as a surface cover composed of ejecta from the crater. The high albedo (0.149) of Euclid's walls also fits the interpretation of unsullied youth. From the Riphaeus Mountains southeast to the terminator Mare Cognitum shows darkly. Like other mare plains, it has only a few craters broad enough to be seen through large telescopes (no more than two dozen of diameter one to four miles), but Ranger 7's television frames showed it heavily pocked with smaller craters down to ten feet across. Numerous winding ridges wrinkle its surface.

A line from Kepler across the Riphaeus Mountains extended almost its length reaches Bullialdus near the terminator. There is another long-conspicuous feature now standing in the dim sunset zone blacked out except for its bright east-wall crest and its barely-perceptible outer west wall. Halfway from Bullialdus to Gassendi, on the broad peninsula that separates Mare Humorum from Mare Nubium, you may be able to make out the irregular Class 5 AGATHARCHIDES, 31 by 34 miles across and 1600 feet deep (Fig. 20). The old, battered walls are bright but low, and they are broken in numerous places on the north and south. Its floor, still fully illuminated, is somewhat brighter than the maria surfaces, but it tends to merge with the walls, making the enclosure a test for binoculars. South of Bullialdus the Campanus-Mercator pair stands at the "estuary" which joins Mare Nubium and Palus Epidemiarum. As noted last night, Campanus is seen readily while Mercator is camouflaged. Halfway between the pair and Bullialdus you may detect a black dot with a bright edge. That is Class 1 KONIG, 14 by 16 miles across and 5400 feet deep. It is not easy to see since the surrounding plain is dark gray.

South of Palus Epidemiarum, a distance about equal to its length, the curious, triple crater Hainzel has become prominent. The bright-rimmed enclosure is illuminated fully with the exception of a small shadow at the west wall. Tangent to it on the south, the much larger Mee, washed out for the past week, has acquired enough contrast to make it visible again. East of Mee is the first of several huge, black excavations that continue south along the terminator. The first is separated from the dark side by a bright east inner wall, and the visible crests of partitional walls identify it as the triple crater Wilhelm-Montanari-Lagalla. An equally large, black void is tangent to Montanari on the south. It exhibits a brilliant east crest extending over the dark side and a light outer west wall, and that is all that can be seen of the great mountain-walled plain Longomontanus. On the terminator beyond Longomontanus, a shallow black elliptical arc or bite out of the disk marks the west wall of Clavius, all of which is dark. Tangent to it on the south is the black elliptical interior of Blancanus separated from the dark side by its conspicuous east wall. The northwest wall of Blancanus almost touches the lighted east wall of equally large Scheiner, the western floor of which is in shadow. Southwest to west of Scheiner may be seen an end-to-end row of three smaller, narrow black ellipses—Kircher, Bettinus, and Zucchius. At the terminator south of Blancanus the double crater

Klaproth-Casatus shows more prominently than last night, and it probably can be resolved into its separate components. There are more large craters visible in the south polar region, particularly if the libration in latitude has a large negative value, and some of them are labeled on the charts.

TWENTY-FIVE-DAY MOON

(Chart VII)

As soon as we set binoculars on the crescent, our attention is drawn to a striking feature large enough to be seen with the naked eye. Near the north horn the terminator cuts through the middle of what appears to be a huge crater with high and massive walls. Our initial astonishment at this unexpected sight gives way to realization that we are looking at the west half of Sinus Iridum. The north and west portions of the wall are the high Jura Mountains which cast broad, black shadows on the dim plain, giving us the impression of a dark crater floor. The south wall consists of the faintly-illuminated, shadow-dotted Heraclides Promontory supplemented by its own shadow which stretches eastward to the dark side.

The maria still can be distinguished easily from the brighter uplands even though the contrast has fallen off considerably over the past week. Mare Humorum stands out conspicuously like a super crater ringed by high mountains. It resembles Mare Crisium at the terminator nine nights ago, but it lacks in its wall the breadth and height which characterized that of Mare Crisium. North of Mare Humorum, the great Oceanus Procellarum covers a large portion of tonight's visible moon.

Near the west limb dark-floored Grimaldi stands out strongly in contrast to its bright surroundings. No indication of walls can be seen around it in this light, and it looks more like a tiny mare than a large crater. Just northwest of it a much smaller but easily-seen black spot marks a portion of the north floor of Riccioli, another crater which seems to have no walls tonight and which actually is ⅔ as large as Grimaldi in diameter. Near the midpoint of the terminator, the Kepler splash pattern on the Procellarum plain still can be traced, but its brightness is fading. The crater itself should not be difficult to resolve since its interior is nearly ⅔ black and ⅓ bright. Aristarchus remains prominent with more

than half its interior brilliant, and its bright plume to the west continues to wash out its unfortunately placed neighbor Herodotus. The latter may be visible as a dull spot, its floor a little darker than the plain and its thin ring a little lighter. Just northeast of Aristarchus lie the shabby Harbinger Mountains. They make no better showing than they did two weeks ago, but perhaps you can resolve them as two or three black-edged light specks.

In the uplands west of Heraclides Promontory Mairan may be seen, and where the terminator cuts the Jura Mountains, similar Bianchini resembles a break in the dark Iridum wall. Midway between the two is Sharp. Northeast of Sharp and northwest of Bianchini, near the center of Sinus Roris, Harpalus stands out. The last four craters are approximately the same size, and tonight each appears about half black and half bright.

One Grimaldi length north of Harpalus large, irregular Babbage can be made out, its low walls revealed by the narrow shadows they cast except on the northeast where the wall is bright. The intruder Babbage A on the sunlit floor near the south wall is at the half-and-half stage. Tangent to Babbage on the northeast is the younger and more prominent Pythagoras with higher, sharper walls, a bright floor, and a central mountain that may be seen with good conditions. Close to the terminator, east of Babbage, the larger and better-defined John Herschel shows well as the strongest crater in the area, its irregular perimeter traced in black except for bright east and northeast walls. North of it smaller, sharper Carpenter can be resolved with ⅔ of its interior bright. East of Carpenter look for Class 5 ANAXIMENES, 57 by 64 miles across and 8500 feet deep. Its old walls appear to be formed of four linear sections that mark out a narrow parallelogram or diamond, and the flat gray floor is outlined in black on the west and faint white on the east. Actually the crater is elliptical. It lies in the dim zone and barely can be distinguished from its rugged surroundings. Southeast of Anaximenes similar Philolaus is conspicuous as a black pit at the terminator but separated from the dark side by its bright east crest. A larger, well-defined crater north of Carpenter may be seen as a narrow bright ellipse if the libration is favorable. It is 62 by 70 miles across and 10,800 feet deep, and it recently has been named PASCAL.

Returning southward, let us pause for a moment on the Procellarum plain at the edge of the Kepler ray area one Grimaldi length south of the crater. Can you see Encke? It is almost washed out, and only the narrow shadows of its walls give it visibility. Farther south, between Oceanus Procellarum and Mare Humorum, Gassendi shows splendidly with its brilliant east wall bordered in black and its broader black west wall. The central peak and Gassendi A in the north wall can be resolved. North and a little west of Gassendi the equally large half-crater Letronne is visible, or, rather, its bright, shadow-bordered southeast wall is. The southwest wall of the bay presents so little contrast that it barely can be seen. Northeast of Letronne, about the same distance as Gassendi, a smaller half crater may be detected as a thin white semicircle on the dark plain. The southwest half of that nameless ancient crater, once 38 miles in diameter, has been melted away, and its southeast edge is marked by tiny WICHMANN, 7 miles in diameter and 3000 feet deep.

On the south shore of well-defined Mare Humorum the black-bordered white walls of Vitello can be seen, but its central peak probably is beyond our reach. Tangent to it on the west we see the more prominent and much larger half crater Lee M, of which the southwest portion is the half crater Lee. Overlapping Lee M on the northwest you may be able to make out the dim, gray semi-ellipse of the half crater Doppelmayer. If you do, push your skill a little further and see if you can resolve the bright central mountain.

At the terminator, one Humorum width southeast of Vitello, Hainzel stands out as a strong feature, a black oblong pit with a brilliant border extending over the dark side. Tangent to it on the south, the larger, irregular Mee shows well, its walls part dark and part light. Some of the features of its broad, rough floor may be detected through binoculars. Farther south, Schiller is prominent as a black gash in the disk near the terminator with a barely-resolved bright border on the northeast. Wedged between Schiller and the terminator is Class 1 BAYER, 30 by 33 miles across and 8200 feet deep. It is entirely dark except for a narrow bright east-wall crest, and it appears as a black speck. The long axis of Schiller extended to the terminator reaches another black speck as large as Bayer but not so sharply bounded. That is Class 2 ROST, 31 miles in diameter and 8200 feet deep. South of Schiller and Rost three black dashes may be seen in a parallel line, the middle one slightly displaced to the north. Those are the craters Zucchius, Bettinus, and Kircher. They are approximately at the half-and-half stage with bright east inner walls. With a favorable libration in latitude,

several other craters might be picked out here, including enormous Bailly, the center of which lies south of Zucchius and the length of which is greater than the three-crater line just mentioned.

TWENTY-SIX-DAY MOON

(Chart VII)

Five nights ago I discussed the difficulty of attempting to select the most prominent feature among several, each of which is outstanding in a different way. This morning no such problem confronts us because not a single prominent feature is left, and, indeed, very little of the moon's accessible surface remains visible. As the slender orange crescent clears the distant trees in the last hour before dawn, we are surprised at how little light it sends us. We have to look closely to find any indication of shadows cast by it. However, if the night is a good one, the orange soon brightens to gold, and the gold gives way to the more familiar silvery hue. At best we have a feeble luminary which sends us no more than 2 per cent as much light as does the full moon. Although we receive little light from the bright side, the "dark" side gives us much more than usual. This morning the earthshine is relatively strong on the large portion of the lunar surface turned toward us but away from the sun. Through binoculars we can spot easily the dark oval area near the east limb that marks Mare Crisium. South, west, and northwest of it a much larger irregular dark area can be identified as the connected Maria Fecunditatis, Tranquillitatis, and Serenitatis.

The dark floor of Grimaldi has brightened considerably, but still it is seen easily as a shaded ellipse near the midpoint of the bright limb. Its lesser companion on the northwest, the Riccioli dark spot, presents so little contrast that it is difficult to resolve. Near the terminator, northeast of Grimaldi, Aristarchus shows well, but it has lost its dazzling luster. The black interior contrasts with the bright western slope of a small mountain on the northwest. Just west of Aristarchus, shallow Herodotus has acquired enough shadow to be seen for a few hours before disappearing onto the dark side, but it is a very weak feature. If observing conditions are good, which would be most unusual in view of the moon's low altitude, you may want to try for Schroter's Valley—possible but difficult. Through a small telescope it appears as a weak, black arch north of Herodotus, but in a few hours, as the terminator crosses Aristarchus, the east half of the valley will be marked by brilliant walls visible through binoculars.

Along the terminator northeast of Aristarchus, about half the distance of Grimaldi in the opposite direction, a strip of mountainous territory may be recognized by slight irregularities in the terminator and a few bright peaks to the east on the dark side. It is the edge of the highlands west of Sinus Iridum. On the flat plain west of those mountains and north of Aristarchus, the low, lumpy, multiple-dome Rumker is about to make one of its brief and hesitant fortnightly appearances. Look for an oval area about half again as large as Aristarchus slightly darkened on the east and slightly brightened on the west. It is a difficult test in this light.

Farther north in the narrow horn, as far northeast of Aristarchus as Grimaldi is southwest of it, you may find some of the larger craters of the polar continent. All are now relatively weak features, and, if the libration is unfavorable, it will be difficult to resolve any of them in the needle-like cusp. The best delineated is Pythagoras with its high new walls and central peak. Tangent to it on the southwest is larger Babbage with low, irregular walls and a visible floor crater, Babbage A, south of center. Tangent to Babbage on the south and close to the terminator is the even more ancient and dilapidated SOUTH, a rectangular enclosure 63 by 67 miles across and 5200 feet deep. It seems rather inappropriately selected to memorialize the astronomer James South in view of its location in latitude 57 degrees north, but in fairness to its namer I must add that it is situated on the south edge of the polar continent. Just east of Pythagoras is vast, heart-shaped Anaximander. On its northeast wall, where the point should be, you may find, instead, sharp young Carpenter. Just east of Carpenter, the terminator cuts through Anaximenes which possibly may be visible as a small black bite.

Returning to the southern hemisphere, we cross Oceanus Procellarum and pause at its south shore. Near the terminator, southeast of Grimaldi, is a difficult unmatched pair—light Hansteen and dark Billy. North of Billy and east of Grimaldi the terminator cuts through small, inconspicuous Flamsteed, named for the first Astronomer Royal of England, John Flamsteed, who began his work at

the original Greenwich Observatory in 1675 charting more accurate stellar positions which greatly improved navigation on the high seas. The numbers assigned to faint, unnamed, naked-eye stars in his famous catalog of 3000 stars are still used today. About three Grimaldi lengths northwest of the Flamsteed area, where it once was believed the first lunar explorers might walk, Reiner makes one of its stronger appearances—a black speck on the Procellarum plain with the curious bright plume just west of it. It is equidistant from Hansteen and Aristarchus.

South of Billy, three times the Billy-Hansteen separation, larger Mersenius shows well with walls of black and white. It is one of the stronger features this morning as the terminator reaches toward its east wall. Just southwest of Mersenius may be seen smaller Class 3 CAVENDISH, 33 by 36 miles across and 12,100 feet deep (Fig. 45). You might even notice a slight irregularity on the southwest wall—much younger CAVENDISH E, 15 by 17 miles across and 8100 feet deep. South of Cavendish, a distance equal to that of Mersenius to the northeast, lies the larger but less well-defined Vieta, and almost tangent to Vieta on the southeast is Fourier, about the size of Cavendish. The fact that Cavendish has a sharper and more conspicuous outline than either Vieta or Fourier appears to be in conflict with Baldwin's classification (Class 2 for Vieta and Fourier; 3 for Cavendish). Another strong indication that Cavendish is younger than the other two is the fact that it has a brighter and more definite ring at full moon. How do you compare them?

Southeast of Vieta the great mountain-walled plain Schickard shows well near the terminator. It is slightly darker than the highlands between it and the limb, but its curiously stained floor has lost its mottled appearance. Just southeast of it the narrower and slightly shorter Phocylides-Nasmyth combination presents a similar appearance. A black notch in the west floor indicates the position of the unresolved dividing wall between the last two craters. One Schickard length southeast of the southeast rim of Phocylides we may find the light-bordered black ellipse of Zucchius close to the terminator, libration permitting. Its companions to the southeast, Bettinus and Kircher, are tangent to the sunset line and about to merge with the dark side. Look at the limb south of Zucchius for the gigantic circular mountain-walled plain Bailly. Half again as long as Schickard, Bailly is visible weakly at least in part, regardless of libration value, but only with a negative libration in latitude does it show reasonably well. If it is visible with appreciable width, look for the darkened floor crater Bailly B, as large as Cavendish, near the east wall. Also look behind Bailly for the Doerfel Mountains which may appear as tiny pimples scattered along the limb and extending beyond the crater limits in both directions.

TWENTY-SEVEN-DAY MOON

(Chart VII)

Unless we live in the mountains or on the desert, we might have a bit of a problem observing the narrow crescent this morning. The sky will be dark when it appears, ground fog and haze permitting, but the deep orange crescent will bear little resemblance to the moon with which we have become well acquainted. We wonder if our binoculars are out of focus because the crescent is not only dim but blank as we first view it. All this comes about from looking through a lot of water vapor, an experience very common on nice clear mornings when we observe objects at low altitude. As we watch, we note that conditions are getting better as the moon rises, but if we wait for it to rise high enough for good seeing and transparency, daylight will overtake us, and the moon's features will be washed out by the scattered rays of the sun. So we compromise and begin observing as soon as the moon loses the ruddy hue it so frequently exhibits when near the horizon. We look for the larger, stronger features first and pass over the smaller ones, making a mental note to return to them a little later when conditions improve.

As the moon gains altitude, we see that the crescent no longer is devoid of detail. As expected, the southern half now looks rugged and mountainous. The north horn presents a similar appearance, but the rest of the northern crescent remains smooth since it is part of Oceanus Procellarum. About ⅕ the way around the crescent from the south cusp, a broad but shallow bite out of it locates Schickard. The vast enclosure is more than half cut off by the terminator, and all of it is in darkness. Midway between Schickard and the south cusp the enormous bulk of Bailly stretches from terminator to limb, more or less. With large nega-

tive libration in latitude it actually extends over the dark side, and the rest of its boundary is marked by a weak black wall line. The floor crater Bailly B, a thin black ellipse near the northeast rim, shows much better than its host. With considerable positive libration Bailly narrows almost to a line along the limb. Just southeast of Schickard, Phocylides and Nasmyth also have joined the dark side and are indicated by a small bite out of the crescent. Between Schickard and the limb Inghirami may be seen as a weak black ellipse with a partial bright border.

Near the center of the crescent Grimaldi, long a conspicuous landmark, may be sought in vain. At this phase its dark floor shines almost as brightly as the surrounding territory, and its ancient, fragmentary walls produce no appreciable shadows. While Grimaldi is camouflaged almost completely, its location can be spotted easily. The upland of hills and small craters that lies along its east and southeast rim shows well at the terminator about halfway between the cusps. Midway between the points, where the invisible craters Grimaldi and Schickard should be, we look for Byrgius. The rays from Byrgius A on the east wall have faded away except in the immediate vicinity of the crater which appears a little brighter than its light environment and exhibits a weak black border. West and northwest of Byrgius the narrow, dark outline of the large multiple-crater Darwin-Lamarck stretches parallel to the limb. Unexpectedly it proves to be one of the stronger features this morning.

With favorable libration and good seeing we might resolve a few pimples along the limb behind Byrgius and Darwin-Lamarck. They are the CORDILLERA MOUNTAINS where heights up to 20,000 feet have been reported but where the measures of the Army Map Service reveal relative elevations no greater than ⅓ that amount. On the excellent plates of the *Rectified Lunar Atlas,* corrected by Kuiper and associates to show the surface as it would appear from directly above, the Cordilleras are seen to stretch some 600 miles in a circular arc resembling the Altai Scarp. This strongly suggests that they may be part of the outer shock wall of a circular mare sometimes glimpsed at the west limb—MARE ORIENTALE. The photograph taken on May 25, 1967, by Lunar Orbiter 4 proved the inference correct. Facing the moon from above the western limb, the wide-angle camera snapped a picture that showed most of the lunar surface in view (Fig. 52). The Cordillera Moun-

CHART VII

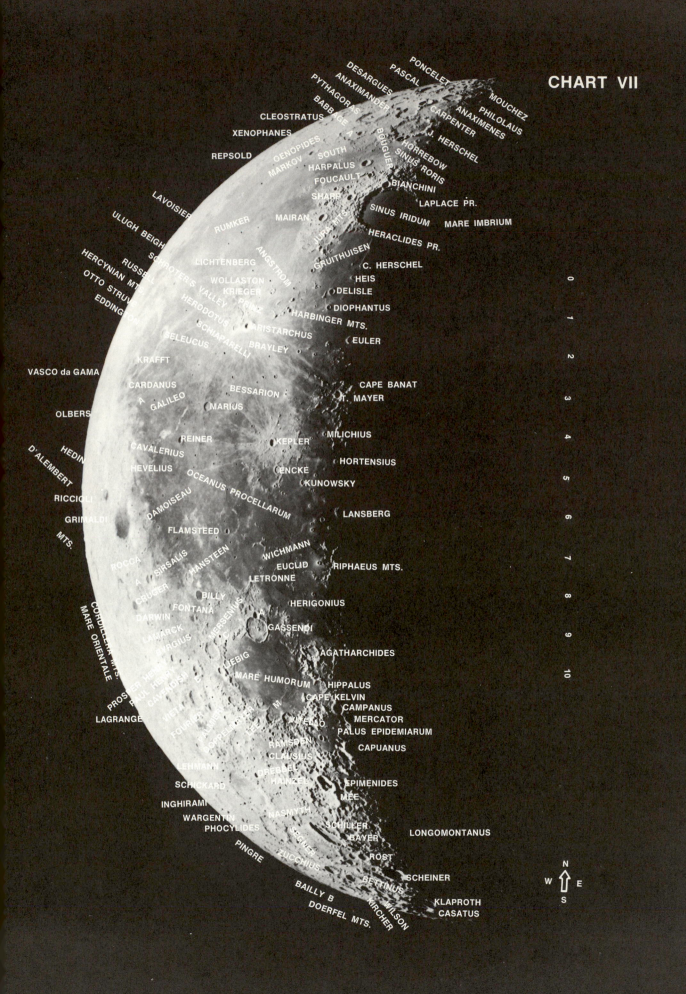

PONCELET
DESARGUES PASCAL
ANAXIMANDER MOUCHEZ
PYTHAGORAS ANAXIMENES PHILOLAUS
BABBAGE A CARPENTER
CLEOSTRATUS J. HERSCHEL
XENOPHANES HORREBOW
REPSOLD GENOPIDES SOUTH BOUGUER SINUS RORIS
MARKOV HARPALUS LAPLACE PR.
FOUCAULT BIANCHINI
LAVOISIER SHARP SINUS IRIDUM MARE IMBRIUM
RUMKER MAIRAN
ULUGH BEIGH JURA MTS. HERACLIDES PR.
HERCYNIAN MTS. RUSSELL ANGSTROM GRUITHUISEN HERACLIDES PR.
OTTO STRUVE SCHROTER'S LICHTENBERG C. HERSCHEL
EDDINGTON WOLLASTON KRIEGER HEIS
VALLEY PRINZ DELISLE
HERODOTUS DIOPHANTUS
SELEUCUS SCHIAPARELLI HARBINGER MTS.
ARISTARCHUS EULER
VASCO da GAMA KRAFFT BRAYLEY
CARDANUS BESSARION CAPE BANAT
A GALILEO T. MAYER
OLBERS MARIUS
MILICHIUS
HEDIN REINER KEPLER
D'ALEMBERT CAVALERIUS HORTENSIUS
HEVELIUS ENCKE
RICCIOLI KUNOWSKY
GRIMALDI OCEANUS PROCELLARUM LANSBERG
MTS. DAMOISEAU
FLAMSTEED
ROCCA WICHMANN
A SIRSALIS EUCLID RIPHAEUS MTS.
CRUGER HANSTEEN LETRONNE
BILLY
FONTANA HERIGONIUS
DARWIN MERSENIUS
LAMARCK A GASSENDI
BYRGIUS
CORDILLERA MTS. AGATHARCHIDES
MARE ORIENTALE LIEBIG HIPPALUS
PROSSER HENRY MARE HUMORUM CAPE KELVIN
PAUL HENRY CAMPANUS
CAVENDISH MERCATOR
LAGRANGE VIETA VITELLO PALUS EPIDEMIARUM
FOURIER CAPUANUS
DOPPELMAYER RAMSDEN
LEHMANN CLAUSIUS
SCHICKARD DREBBEL EPIMENIDES
HAINZEL
INGHIRAMI MEE
WARGENTIN NASMYTH
PHOCYLIDES SCHILLER LONGOMONTANUS
SEGNER BAYER
PINGRE ZUCCHIUS ROST
BETTINUS SCHEINER
BAILLY B KIRCHER WILSON KLAPROTH
DOERFEL MTS. CASATUS

N
W ⬆ E
S

0
1
2
3
4
5
6
7
8
9
10

tains form a ring more than 600 miles in diameter. The even more obscure ROOK MOUNTAINS mark a smaller concentric ring about 400 miles across, and the last of six such rings is the nearly circular shore of a mare plain 300 miles in diameter. Pimplelike limb markings occasionally seen west of Grimaldi are the D'Alembert Mountains. At the terminator and on the south edge of the rough region southeast of Grimaldi is a black speck marking the Siamese-twin craters Sirsalis and Sirsalis A.

Just north of Grimaldi is smaller but better-walled Hevelius, its light floor delicately outlined in black, and tangent to Hevelius on the north, is smaller Cavalerius, a short but strong black streak. Three of its lengths north of Hevelius we find Cardanus, a young crater similar to Cavalerius, and just north of Cardanus is its identical twin Krafft. Half bright and half dark, they appear as elongated black dots since the bright east halves are only slightly brighter than the Procellarum plain. Northeast of Krafft, nearly twice the distance to Cardanus, is the similar crater Seleucus. The rays that pass south of it, running from Cardanus toward Aristarchus beyond the terminator, still may be visible, but they are very weak. Just northwest of Seleucus is the bright wishbone—the large crater Eddington with its south wall melted away. Its west wall is also the displaced east wall of Otto Struve, which, with its large intruder Russell on the north, forms a double crater that extends twice as far north as does the wishbone. If you can resolve the long, irregular west walls of Otto Struve and Russell you will see the Hercynian Mountains, an alias by which those features are known. They have the strange property of appearing bright even when the sunset terminator is near, a property which casts some doubt on their mountainous character. With a large positive libration in longtitude, Eddington, Otto Struve, Russell, and the Hercynians are foreshortened into a single ridge close to the limb. If, in addition, there is a large negative libration in latitude, the north cusp of the crescent appears perfectly smooth and reveals none of the craters listed in the next paragraph.

The most prominent crater of the far northern continent is Pythagoras. Close to the terminator, with average libration, it has a broad bright east wall and an even broader black northwest wall. The bright central peak rises from the edge of the wall shadow and casts its own long, pointed shadow across the floor and up the east wall. On the southwest edge of Pythagoras may be seen a broad, narrow, darkened area or bite. That is the western half of Babbage through which the terminator passes this morning. Tangent to Babbage on the west is smaller OENOPIDES, 46 miles in diameter and 7200 feet deep, which at last has acquired enough wall contrast to be seen. Just southwest of the last crater is younger, deeper MARKOV, 25 by 27 miles across and 8900 feet deep. It appears as a black speck close to the terminator.

I have pushed along swiftly this morning with little pause for description and discussion, a procedure dictated by the short interval in which we are permitted to observe the moon at this advanced age. The 27-day moon is an unusual sight—a sight which most people probably never see during their entire lives.

TWENTY-EIGHT-DAY MOON

(Charts VII and IV)

What about the 28-day moon? This is an optional exercise for those few who like to try "the impossible." Obviously, we have here a situation similar to that of the one-day moon. The crescent is but a tiny sliver. When it becomes visible above the horizon, the sky around it no longer is dark. After brief but poor visibility, it fades into the bright background before attaining high enough altitude for good observation. Since the lunation period averages 29.53 days, we are theoretically a little better off here than at the other end of the cycle. At observation time, the average 28-day moon is farther from the sun than the average one-day moon by five degrees or 10 hours of travel time. However, that slight advantage probably is more than offset by the common humidity increase during the night which makes early-morning observing conditions inferior to those of early evening, particularly for low-altitude objects. So the problem is a tough one. Confidentially, I have searched the eastern horizon on several beautiful mornings from moonrise to sunrise without a single glimpse of the 28-day moon. However, the more significant fact is that on several other such occasions I have found it and observed its markings. The most favorable time of the year in the Northern Hemisphere is in

September or October. With the sun near the autumnal equinox, the 28-day moon is higher above the horizon at sunrise than in any other season.

Let us assume that through persistent search and the smile of fortune we have found the elusive crescent, thin, dim, and red near the horizon. What may we expect to see? As in the case of the one-day moon, the first impression is likely to be that of a blank disk. As it moves upward and assumes its normal color, the south half presents a jumble of craters seen nearly edge on, most of which remain unnamed. The north half continues without detail. Again, as in the case of the one-day moon, the line connecting the cusps may be inclined to the north-south line on the lunar disk by as much as 10 degrees or so. That would shift the midpoint of the crescent above or below the west point on the disk by a corresponding amount. In any case, the one fairly prominent feature of the entire crescent will be found near if not at the latter's midpoint. It resembles a rubber band with its midsection constricted slightly, and it extends 200 miles, or about 11 degrees approximately parallel to both terminator and limb. The south loop of the combination enclosure is Riccioli, and the north loop is HEDIN, 82 by 90 miles across and 6500 feet deep. The odd enclosure is shown best on Chart IV where it appears near the sunrise terminator and with lighting opposite to that of this morning. Riccioli's dark spot has looked small on recent mornings, compared with that of its neighbor Grimaldi, because only a small portion of the floor is stained. Don't let that impression cause you to underestimate the crater. Remember that Riccioli is a major crater 97 miles across and 7500 feet deep. Grimaldi, half again as large, is bisected by the terminator, and it shows nothing but a portion of the low dark west wall which it shares with Riccioli. The floor stains have lost their contrast in both craters.

At the terminator in front of Hedin the west wall of Hevelius may be noted as a black spot or small nick in the crescent. The sun has set on all that we can see of it, and its smaller companion on the north, Cavalerius, has joined the dark side. Just north of Hedin the west wall of Olbers presents a black spot, stronger at its north end behind which the recently conspicuous ray crater Olbers A is located. North of Hedin, a distance equal to the length of the Hedin-Riccioli structure, a small black spot close to the terminator is the interior of Cardanus. One-third that distance north of Cardanus is the almost identical but slightly stronger

Krafft. Near the limb west of Cardanus and Krafft you may see a group of craters, highly foreshortened, the largest of which is VASCO DA GAMA, 50 by 57 miles across and 6900 feet deep. North of Krafft, a little farther than Cardanus to the south, a strong black wall catches the eye. It is the west wall of the large half crater Eddington, the east wall of which is lost in the dim zone at the terminator. The double crater Otto Struve-Russell, which joins Eddington on the west and northwest, is difficult to resolve since its west wall, formed by the Hercynian Mountains, is so low that it casts very little shadow. To the north the crescent continues to look blank. About halfway from Otto Struve to the north cusp a bright ridge cuts across the Procellarum plain. West of it is the large irregular crater REPSOLD, 64 by 68 miles across and 8900 feet deep, difficult in this light to distinguish from the surrounding highlands. A little farther north XENOPHANES, 67 by 70 miles across and 10,500 feet deep, may show slightly better because of its higher black west wall. Introducing new features on a moon as difficult to observe as this morning's is like assigning a show television time between 4 and 5 A.M., but in this lunar longitude there is not much choice. The only other time these craters show well is just before full moon, when the competition is very strong. Look upon the new features as your reward for coming out and trying.

The south half of the crescent is so crowded that we find few features which stand out against the mottled background. South of Riccioli, a distance about equal to that of Olbers to the north, a dark spot more conspicuous than its neighbor marks the high outer east wall of Rocca—a slope that separates it from the ordinarily inconspicuous Mare Aestatis on the east. A little farther south the terminator cuts through the large enclosure Darwin-Lamarck which is marked by its long, dark west wall. Easier seen than Darwin is the broad, dark valley tangent to it on the northwest and directed toward Mare Orientale at the limb. It is 30 miles wide and about 120 miles long, and probably it was produced by the same event that dug the mare basin. Far down toward the cusp, as far south of Riccioli as the bright Repsold ridge is north of it, a vast elliptical excavation about 175 miles long is wedged between the terminator and the limb. It is PINGRE, and its most conspicuous part is a normal crater on its southeast floor 52 by 62 miles across and 9200 feet deep. Pingre offers little contrast and is not easy to pick out. Just beyond it the larger Bailly is bisected by the terminator and may reveal itself, if the libration is favorable, by a long, dark wall stretched across the crescent near the horn.

As you have seen, there are a great many small craters and other kinds of depressions along the south half of the limb this morning, but if you have located half of those I have listed you have done exceedingly well, and I am sure you have used up the brief observation period. Also you have used up the observable portion of the lunation.

Tomorrow the moon will be born again, and on the evening of the following day the skilled observer with good luck may find the one-day crescent. As again we pursue the moon through its phases and continue to gather observational knowledge firsthand, let us broaden our understanding of our satellite with some additional ideas, facts, and theories by reading some of the texts listed in the Reference Books section on page 205.

THE NEW WORLD OF TOMORROW

We have scrutinized the face of the moon through our binoculars. We have examined its features under varying illumination from the slanted rays of early-morning sun, through high noon, to the dim light of late evening. Every feature down to the order of five miles in dimensions has been revealed to us, and occasionally we have glimpsed bright spots that are much smaller. We have looked at the moon from afar just as men before us over the past several million years have gazed outward on occasion. We have seen much more than most of them but less than some who have used telescopes of small, moderate, or large aperture. We have shared with professional scientists the rich treasure of high-resolution pictures sent back in 1964 to 1973 by the marvelous moon probes created and operated by the National Aeronautics and Space Administration and its contractors.

The Rangers blazed the trail through 240,000 miles of hostile emptiness to report back subtelescopic lunar detail in surprising abundance. The Surveyors landed softly on that strange world and scouted representative areas. Surveyor 1 revealed protrusions and depressions down to

one-fiftieth of an inch in diameter. Surveyor 3 dug into the lunar soil and found it to be granular and cohesive, exhibiting the mechanical properties of fine, moist sea sand. Surveyor 5, with its alpha-particle spectrograph, analyzed the moon's material and showed it to be similar in chemical composition to the basaltic rock of the earth. The five Lunar Orbiters carried out their missions with remarkable efficiency. While following their elliptical courses as satellites of the moon, they reported on conditions, and they photographed the entire surface—both near and far sides—recording every detail down to objects of card-table size.

Altogether, approximately 100,000 exceptionally clear photographs had been obtained from lunar altitudes of 1000 miles down to a few feet. Physical and chemical tests had been performed on the lunar surface material. Selenographers had received data complete both in range and detail far exceeding the wildest dreams of earlier generations. The moon could be mapped more precisely than was possible for much of the earth.

But this was just a beginning. It was all preliminary to the main event. The challenge which inspired such unprecedented activity and accomplishment had been thrown down on May 25, 1961, when President John Kennedy in a message to the Congress declared that the United States should commit itself to the goal of landing a man on the moon and returning him safely to the earth before the end of the decade. Twenty days before that address, astronaut Alan Shepard had ridden the one-man Mercury "capsule" *Freedom 7* to an altitude of 116 miles on a 15-minute suborbital flight into space. Nine months later, astronaut John Glenn orbited the earth in *Friendship 7,* and three more Mercury missions of increasing length followed in the next 15 months. They proved that man could exist and function, when properly protected, in the strange new realm of zero gravity and zero atmosphere.

From March 1965 through November 1966, ten missions were flown in the larger, more maneuverable, two-man Gemini spacecraft. They lasted up to 14 days and included rendezvous and docking with other spacecraft, "walks" in space, numerous experiments, and all operations in and out of the spacecraft that might be required on lunar missions.

After two earth-orbital manned missions of the great three-man Apollo-Saturn space transport and two nonstop round trips to the moon in 1967–1969, Apollo 11 lifted off on July 16, 1969, carrying Neil Armstrong and Edwin Aldrin to explore a bit of

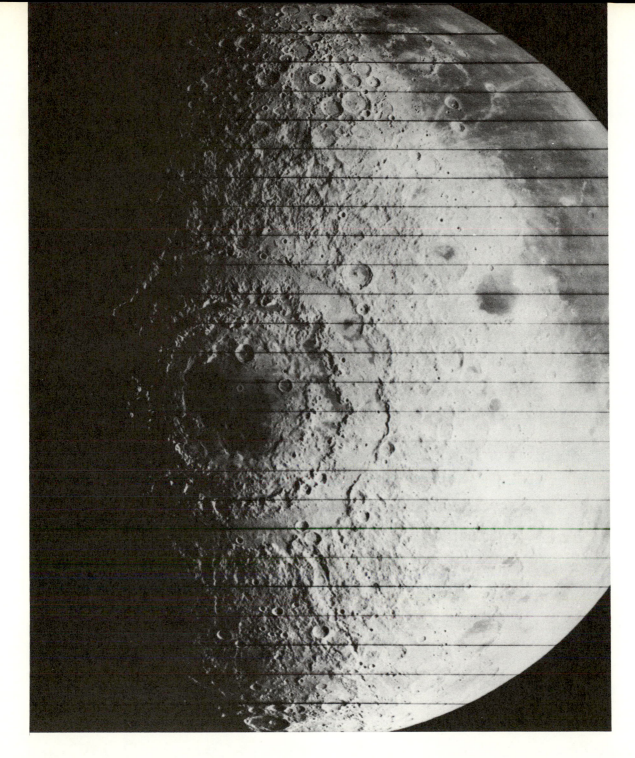

FIGURE 52.   Mare Orientale, the "bull's eye," where a minor planet some 40 miles in diameter crashed head-on into the moon as the latter sped around the earth at the orbital speed of 2300 miles per hour. The enormous energy of impact not only blasted out a 300-mile crater, into which lava flowed, but also generated shock waves that buckled and broke the surrounding surface to produce the ring-mountain ranges, three of which show conspicuously in this photograph by Lunar Orbiter 4. The explosion hurled out great blocks of stone and smaller debris which gouged and blanketed the outer surface to produce the obvious radial grooves extending at least 800 miles from the mare center. The largest sharp crater on the floor of Mare Orientale is 35-mile MAUNDER with a central peak. The next-largest is 25-mile peakless KOPFF. The smooth, darkened upper right section of the moon is the far western strip of Oceanus Procellarum. The large round dark spot at right on the eleventh line from the top is Grimaldi, and the equally large crater with a smaller floor stain above and to the left of Grimaldi is Riccioli. The smaller prominent two craters near the Procellarum shore below the fourth and fifth lines are Krafft and Cardanus, respectively. Above them is the "wishbone" of huge, ancient Eddington joined on the left to the equally old and larger double crater Russell—Otto Struve.

FIGURE 53. Candidate sites for the first manned landing on the moon. Five of the eight were photographed by Lunar Orbiter 1. Those represented by dark rectangles passed scrutiny by both Lunar Orbiters 2 and 3. Those marked by black dots looked good on the excellent pictures returned by Lunar Orbiter 3 and were confirmed by Lunar Orbiter 4 or 5. On July 20, 1969, the Apollo 11 lunar module touched down near site 6. Apollo 12 landed on November 19, 1969 near site 9.

Mare Tranquillitatis and Michael Collins to man the return ship parked in lunar orbit. Environmental monitors were set up to broadcast data back to the earth, numerous photographs were taken, and the effects of living, walking and working in a gravitational field only one-sixth as strong as that of the earth were experienced for the first time. Finally, 48 pounds of lunar rocks were brought back to anxiously waiting scientists. Apollo 12 astronauts Charles Conrad and Alan Bean visited Surveyor 3 on the plain 70 miles southeast of Lansberg on November 19, 1969. They deployed more recording instruments, explored craters up to 1000 feet across, and brought back Surveyor's camera which had been weathering in a small crater for 31 months. On February 5, 1971, Apollo 14 astronauts Alan Shepard and Edgar Mitchell studied the Fra Mauro highlands 115 miles east of the Apollo 12 camp. They broke out a sophisticated "shopping cart" to carry geology tools and samples for exploration of an area 1½ miles long. On July 30, 1971, David Scott and James Irwin of Apollo 15 drove the first Lunar Rover, an electric car, around the foothills of the northern Apennines 90 miles southeast of Autolycus. Meanwhile, Alfred Worden in the orbiting Command-Service Module operated a large package of geochemical exploration equipment and retrieved its records in the first spacewalk made by an astronaut without a backup associate inside the spacecraft. Apollo 16 landed April 21, 1972, in the Descartes highlands 130 miles south of Delambre and a mile above the nearest mare level. John Young and Charles Duke used explosive devices to gather seismic data. Onsite exploration ended with the intensive geological work of Eugene Cernan and Harrison Schmitt of Apollo 17 which landed December 11, 1972, in a valley of the Taurus Mountains 50 miles north of Vitruvius. In addition to a great many experiments and observations by the twelve moon walkers, a total of 778 pounds of lunar rock and soil, including several core samples, were brought back for laboratory investigation.

THE CHARTS

Since it is our purpose to observe the moon itself rather than merely to read about it indoors, it is essential that I provide you with the best and most conveniently arranged charts for the identification of the many surface features revealed through binoculars. I have selected for this purpose a sequence of seven photographs from the Lick Observatory collection of excellent plates. Each may be used on several nights, as indicated in the sections devoted to nightly observation, and the result is fully as satisfactory as if a separate chart were provided for each night. Actually it is more satisfactory to use such a sequence because the small variations of terminator location from one lunation to another would stand out much more conspicuously on a series of 28 nightly photographs.

The photographic charts, designated by Roman numerals to distinguish them from other illustrations, are made from plates taken with the great 36-inch refractor by the late Joseph Moore and the late Fred Chappell, two masters of their craft with whom it was my privilege to work some years ago. Both devoted their lives to research at the Lick Observatory—Moore as an astrophysicist and onetime director; Chappell as a photographer of exceptional skill and artistic talent. These photographs are not new. They are now a little over forty years old. With newer and larger telescopes better photographs have been taken of small sections of the lunar surface, but the Moore-Chappell collection remains unexcelled and unequaled among photographs of the whole moon. As every amateur photographer knows, taking a picture is a simple matter, taking a good picture is an exacting assignment, and taking a prize winner requires a touch of genius. There are three requisites for outstanding lunar photographs. The first is an excellent telescope of large aperture and long focal length. The second is exceptionally fine seeing. The third is knowledge and skill acquired through decades of experience at the telescope and in the darkroom. All three were combined in a remarkable way on those occasional and unpredictable nights when Moore and Chappell were called to photograph the moon.

Good definition and high resolving power in a photograph depend upon the degree to which the first two requisites are met, and their deficiencies alone are sufficient to condemn almost all lunar photographs to mediocrity or worse. But aside from such fundamental requirements, he who would portray the moon soon finds that his subject is a very difficult one indeed because of the large variation of brightness over the disk. As previously noted, the albedo or reflecting power varies over a wide range, rendering features such as the central peak of Aristarchus fully three times as bright as dark ones such as the floor of Grimaldi. In addition, the scattering properties of the surface cause the limb regions to appear bright while the zone along the terminator shines but dimly. We are well aware of such differences of illumination when we view the moon through binoculars, but the human eye has a wide range of adaptation. We can pick out detail on the dark plains if we look for it, and we can do the same in all but the very brightest of spots. The photographic plate, however, is much more limited in this respect. If properly exposed for detail in the brighter regions, it records nothing in the dark areas, and a large portion of the moon comes out completely black on the print. If properly exposed for detail on the maria, the intense light of the brighter regions "burns out" the plate, and the print shows a large portion of the moon completely white with no detail. Here is where the third requisite comes into play, and it brings us excellent pictures in spite of the severe limitations of the medium.

As you doubtless have noticed, two pages are devoted to each chart. The photographs on the two pages are identical, but the left one is more or less covered by identification labels. In some spots the density of labels per square inch runs rather high, but a strong effort has been made to find places for them so that each label begins or ends close to the small feature it identifies or passes near the center if the feature is a large one. The unlabeled photograph, of course, is the one to be compared with the moon as seen through binoculars (or telescope). Identification of objects so located then may be made by comparing the two photographs. For observation through binoculars rotate the book until the orientation of the photograph is the same as that of the moon seen with the naked eye. If a standard astronomical (inverting) telescope is used, rotate the book 180 degrees farther. If your refractor has a 90-degree diagonal between the objective and the eyepiece, or if your reflector has three reflecting surfaces, right and left may be interchanged in your view and no orientation of the photograph may match it with what you see. In that case, observe without the diagonal or bypass the third reflecting surface if

possible. If not, use a hand mirror to reflect the image of the photograph to your eyes instead of looking directly at it.

The compass rose near the lower right corner of the left photograph shows the modern or astronautical directions as used throughout this book. Compass directions here are used to locate a new object with reference to a known one. Consequently, they are apparent directions laid off on the flat lunar disk as we see it and not true directions as would be measured on the nearly spherical surface of the moon. Near the center of the disk the two systems of directions are essentially the same. For example, Julius Caesar is southeast of Manilius both apparently and truly (Charts II and V). The two systems practically are in coincidence also along the equator and along the central meridian. Large differences occur, however, in the general neighborhood of the northeast, southeast, southwest, and northwest limbs. For example, Phocylides appears southeast of Schickard (Charts IV and VII), and that is how we locate it, but on the lunar surface Phocylides actually is nearly southwest of Schickard. You need not be concerned with such differences unless you compare the photographs with rectified charts such as those produced by the Aeronautical Chart and Information Center and Army Map Service.

A scale consisting of small white numbers in line has been provided for the purpose of locating objects as explained in the next section. The units correspond to distances in hundreds of miles for the region around the center of the disk. Distances elsewhere may be measured approximately with the scale, provided measurement is made parallel or nearly parallel to the nearest section of the limb. Otherwise the measured distance is too small by an unknown factor owing to foreshortening on the curved surface. The "unknown factor" can be computed, of course, but that is a bit beyond the scope of our operation here. The scale may be used also to estimate approximate crater diameters if the longest dimension of a foreshortened crater is taken.

THE GAZETTEER

A majority of the objects listed alphabetically in this section have been discussed in the sections devoted to nightly observations, but they are repeated here for quick reference and for completeness. In addition I have included objects visible through binoculars which were not mentioned previously since they are generally less conspicuous than their neighbors. Finally, a number of small features have been added which are identified on the large-scale photographs throughout the book and which may be of interest to readers equipped with small telescopes. This section should serve the needs of those who come upon the name of a lunar surface feature in their reading and would like to identify and locate it. It should be useful also to those who spot a name on one of the charts and would like to find the basic data about the object. As you doubtless have noted, the origin of the name is obvious in the cases of many craters. Some of the less-famous scientists whose contributions have been discussed in this book also are memorialized on the moon. If you wish to follow up unfamiliar names, I refer you to a book which I have quoted several times—to H. Percy Wilkins and Patrick Moore, *The Moon* (1961). There you will find a clue to every name then in use, a clue which may be developed upon reference to an encyclopedia.

Each object is located by measurement from one of the 10 conspicuous Reference Features chosen because they are fairly well distributed over the disk and because they should be among the first you learn as the sunrise terminator passes over them. The Reference Features selected are: Aristarchus, Copernicus, Mare Crisium, Gassendi, Grimaldi, Langrenus, Plato, Posidonius, Theophilus, and Tycho. Most of the objects are labeled on more than one chart, but reference is made to one only—usually the chart on which the feature shows most prominently. Having located it, you should experience no difficulty in spotting it on the other charts that show it. The general direction of the object from the Reference Feature on the chart is given to the nearest 45 degrees by adding to the directions marked on the familiar compass rose the intermediate directions northeast, southeast, southwest, and northwest. The distance of the object from the Reference Feature is given in scale units to the nearest tenth of a unit. As usual, the distance between two features is measured center to center unless otherwise noted. You need not be precise in laying off the direction or the distance. Approximate values should suffice to bring to your attention the label of the feature you are seeking. If the object appears in one of the large-scale photographs in the book, a second item gives the figure number. All directions and distances are apparent directions and distances as we see them projected on the flat lunar disk. Consequently, when using the gazetteer you need give no thought to what the true direction and distance should be. Just lay off the coordinates from the Reference Feature as they are given, using two fingers as dividers separated by the given distance.

In case you wish to review the positions of the Reference Features you may locate them on Chart IV by measurement from the white dot at the center of the lunar disk just above the label "Herschel." Distances in chart scale units and directions from the dot are tabulated below. Directions are given first to the nearest 45 degrees. If more precise information is desired, the figures in parentheses give directions to the nearest degree measured clockwise around the disk from the north point.

ARISTARCHUS	NW	9.0 units	(300 degrees)
COPERNICUS	NW	4.8	(299)
MARE CRISIUM	NE	9.5	(65)
GASSENDI	SW	7.5	(246)
GRIMALDI	W	10.1	(264)
LANGRENUS	E	9.4	(95)
PLATO	N	9.1	(349)
POSIDONIUS	NE	7.6	(32)
THEOPHILUS	E	4.6	(106)
TYCHO	S	7.1	(192)

While all the Reference Features are shown and labeled on Chart IV, most of them are much more prominent on other charts. Note, for example, Gassendi on Chart VII and Theophilus on Chart II.

The general category of feature to which an object belongs is obvious from its name in the cases of the maria, sinus, paludes, lacus, rills, valleys, and ridges. Each other object belongs to one of the six following types of features indicated in the list by the symbol shown:

Crater	C.
Mountain	Mt.
Mountain Range	MR.
Promontory	Pr.
Cliff	Cl.
Dome	D.

For those craters which have central mountains, the distinction between a single peak and a multiple peak is indicated as follows:

Single Peak	Pk.
Multiple Peak	MuPk.

The horizontal dimensions of craters are given in miles, as in previous chapters. If the crater is approximately circular, a single figure is given which is the diameter. If it departs significantly from the circular, the minimum and maximum widths are given. Example: 25 × 28 mi. Remember that these dimensions refer to the actual or rectified crater shapes and not to the apparent shapes as we view them from the earth. All craters near the limb appear as narrow ellipses, but for those which are truly circular, a diameter only is given. Vertical dimensions such as crater depths (maximum) and central mountain heights are given in feet as before. Corresponding dimensions of mountains, promontories, cliffs, and domes are given in the same units as those of craters.

The following sample entry and subsequent explanation should make clear the interpretation of all data and abreviations:

MAUROLYCUS C. 73 mi. 16,700 ft. MuPk. 5200 ft. 2. (II Theophilus SW 5.4 and Fig. 51) 6 Day.

We read: Maurolycus is a crater. It is approximately circular in actual shape with a diameter of 73 miles. The crater depth is 16,700 feet measured from the highest point on the rim, and a multiple central mountain rises 5200 feet above the floor. Maurolycus belongs to age Class 2 according to Ralph Baldwin, *The Measure of the Moon* (1963). (See Three-Day Moon section for explanation of this classification.) Maurolycus may be found on Chart II by measuring from Theophilus southwest

5.4 scale units. It also may be examined in greater detail in Figure 51. The crater is introduced and discussed in the Six-Day Moon section.

Other abbreviations used occasionally in the list: AMS Lunar Maps produced by the Army Map Service, Corps of Engineers, U.S. Army. LAC Lunar Charts produced by the Aeronautical Chart and Information Center, U.S. Air Force.

ABENEZRA C. 25 x 28 mi. 10,500 ft. Pk. 4900 ft. 1. (II Theophilus SW 3.2) 21 Day.

ABULFEDA C. 40 mi. 10,500 ft. 1. (II Theophilus W. 2.3 and Fig. 50) 21 Day.

MARE AESTATIS 2 insignificant short dark streaks near W limb totaling about 65 x 10 mi. (V Grimaldi S 1.7) 13 Day.

SINUS AESTUUM Elliptical plain 140 x 180 mi. (III Copernicus E 2.2 and Fig. 7) 8 Day.

AGARUM Pr. Broad blunt headland jutting 40 mi. over Mare Crisium. 12,000 ft. (I Crisium SE shore) 16 Day.

AGASSIZ Pr. on Alpine crest above Mare Imbrium. 7500 ft. (VI Plato SE 1.8 and Fig. 9).

AGATHARCHIDES C. Irregular 31 x 34 mi. 1600 ft. 5. (VII Gassendi SE 1.5 and Fig. 20) 24 Day.

AGRIPPA C. 28 mi. 9800 ft. Pk. 2000 ft. 1. (II Theophilus NW 4.0) 7 Day.

AIRY C. Irregular, rectilinear 22 x 31 mi. Pk. 3400 ft. Very old. (II Theophilus W 4.0.)

ALBATEGNIUS C. 81 mi. 14,400 ft. Pk. off center to W appears to be remains of large central Mt. or C. 4500 ft. on Ranger 9 pictures (previously 2300 ft.). 5. (II Theophilus W 4.1.) 7 Day.

ALEXANDER C. Generally shallow elliptical depression 49 x 58 mi. Wall missing in rough territory on NE but rises to 10,100 ft. in Caucasus Mts. on W. Flooded. Very old. (II Posidonius NW 2.8 and Fig. 8.)

ALFRAGANUS C. 12 mi. 9200 ft. Pk. low off center to S. Young. (II Theophilus NW 1.6 and Fig. 50) 15 Day.

ALFRAGANUS C C. 7 mi. 3600 ft. Young. (II Alfraganus SW 0.2 and Fig. 50) 15 Day.

ALHAZEN C. 21 mi. 5200 ft. Young. (I Crisium just beyond E border).

ALIACENSIS C. 46 x 55 mi. 12,900 ft. Pk. small off center to NW appears to be remains of large central Mt. or C. 1. (II Theophilus SW 5.1) 7 Day.

ALMANON C. 31 mi. 6600 ft. 2. (II Theophilus SW 2.3 and Fig. 50) 21 Day.

ALPETRAGIUS C. 25 mi. 9800 ft. Pk. large conical 6200 ft. 1. (VI Tycho N. 4.2 and Fig. 26) 8 Day.

ALPHONSUS C. 64 x 73 mi. 10,500 ft. (6600 feet— Ranger 9). Pk. 3800 ft. 5. (VI Tycho N. 4.6 and Fig. 26) 8 Day.

ALPINE VALLEY Artificial looking, cut straight through heart of Alps perpendicular to crest and directed to Mare Imbrium. 110 mi. long and 1 to 13 mi. wide. (III Plato E. 1.5 and Fig. 9) 7 Day.

ALPS MR. covering approximately rectangular area 150 x 350 mi. rising progressively from Mare Frigoris on N to crest, part of which borders Mare Imbrium. 12,000 ft. (III Plato E 1.5 and Fig. 9) 7 Day.

ALTAI SCARP Cl. following circular arc from Piccolomini on S into Mare Tranquillitatis on N 600 mi. Averages about 1 mi. high but exceeds 2 mi. at some points. S half is stronger (II Theophilus SW 1.9 to S 2.9 and Figs. 48 and 5) 5 Day.

AMPERE Mt. of Apennine crest along Mare Imbrium. 8 x 15 mi. 10,600 ft. elevation reported but LAC gives 6400 ft. and AMS 8200 ft. (VI Copernicus NE 3.5 and Fig. 7).

ANAXAGORAS C. 33 mi. 8900 ft. Pk. 1. (III Plato N 1.9) 8 Day.

ANAXIMANDER C. 46 mi. 7200 ft. 5. (VI Plato NW 2.7) 12 Day.

ANAXIMENES C. 57 x 64 mi. 8500 ft. 5. (VI Plato N 2.4) 25 Day.

ANDEL C. Roughly rectangular 15 x 19 mi. 4900 ft. gap in S wall. Old. (II Theophilus W 2.5 and Fig. 50).

ANGSTROM C. 6 mi. 3800 ft. Young. (VII Aristarchus NE 1.5 and Fig. 44).

MARE ANGUIS Small irregular dark plain covering about 2500 square mi. (II Crisium just beyond NE border).

ANSGARIUS C. 58 mi. 10,000 ft. Old. (I Langrenus E 1.3) 1 Day.

APENNINES MR. Magnificent range forming SE wall of Mare Imbrium for 450 mi. (III Copernicus NE 1.4 to 5.7 and Fig. 7) 8 Day.

APIANUS C. 38 x 44 mi. 10,100 ft. 2. (II Theophilus SW 4.4) 21 Day.

APOLLONIUS C. 30 x 32 mi. 6400 ft. 5. (I Crisium S 2.4).

ARAGO C. 16 mi. 5900 ft. 1. (II Theophilus N. 3.4 and Fig. 10) 20 Day.

ARCHERUSIA Pr. separating Maria Serenitatis and Tranquillitatis on the W. 3600 ft. (III Posidonius S 2.7 and Fig. 11) 15 Day.

ARCHIMEDES C. 51 mi. 6800 ft. 5. (VI Plato S 3.4 and Fig. 42) 8 Day.

ARCHYTAS C. 20 x 22 mi. 7500 ft. MuPk. 1. (III Plato NE 1.6 and Fig. 9) 21 Day.

ARGAEUS Mt. or Pr. separating Maria Serenitatis and Tranquillitatis on the E. 4100 ft. (II Posidonius S 2.2 and Fig. 11).

ARGELANDER C. 20 x 25 mi. 8400 ft. Pk. 3200 ft. Old. (II Theophilus W 3.9).

ARIADAEUS C. 7 mi. 4600 ft. Young. (II Theophilus NW 3.3 unlabeled and Fig. 10).

ARIADAEUS RILL Runs 145 mi. W from point just N of Ariadaeus. Graben type depression two to three mi. wide. Appears quite straight but consists of several offset segments. (II Theophilus NW 4.0 and Fig. 10) 21 Day.

ARISTARCHUS C. 25 mi. 11,900 ft. Pk. Possibly youngest of the prominent craters—formed no more than 50,000,000 years ago. Conspicuous rays. 1. (Reference Feature: IV Dot NW 9.0 and Fig. 44) 11 Day.

ARISTILLUS C. 36 mi. 10,500 ft. MuPk. 1. (III Plato SE 2.8 and Fig. 42) 7 Day.

ARISTOTLE C. 55 mi. 12,000 ft. Two Pk. remnants on spine from S wall. 1. (II Posidonius NW 3.7 and Fig. 8) 6 Day.

ARNOLD C. Hexagonal 50 x 64 mi. 6600 ft. SW wall broken. Old. (II Posidonius N 5.1) 19 Day.

ARZACHEL C. 59 mi. 13,000 ft. Pk. 7 x 18 mi. 4600 ft. 3. (VI Tycho N 3.8 and frontispiece) 8 Day.

ASCLEPI C. Irregular 25 x 31 mi. 9200 ft. Very old. (II Theophilus S 6.4 unlabeled and Fig. 49).

ATLAS C. 54 mi. 10,000 ft. MuPk. 2600 ft. 5. (I Crisium NW 5.9) 4 Day.

MARE AUSTRALE Series of dark spots more or less connected but foreshortened to lines along SE limb. Some are probably flooded craters. Visible portion scattered over area some 300 x 500 mi. and more on far side. (II Theophilus SE 5.5) 11 Day.

AUTOLYCUS C. 25 mi. 11,100 ft. Several hills and spines on floor. 1. (III Plato SE 3.3 and Fig. 42) 7 Day.

AUWERS C. Roughly rectangular depression 11 x 15 mi. 2300 ft. Old. (II Theophilus N 5.2)

AUZOUT C. 18 x 21 mi. 10,700 ft. Pk. Young. (I Crisium just beyond SE border).

AZOPHI C. Irregular 29 x 34 mi. 11,200 ft. 1. (II Theophilus SW 3.2) 21 Day.

BABBAGE C. Roughly rectangular 85 x 95 mi. 6200 ft. Several breaks in wall. Extremely old. (VI Plato NW 3.4) 12 Day.

BABBAGE A C. 17 mi. 8900 ft. Young. (VI on floor of Babbage S of center) 12 Day.

BACON C. 43 mi. 12,000 ft. 1. (II Theophilus S 6.2 and Fig. 51) 6 Day.

BACON A C. 25 mi. 4600 ft. 1. (II tangent to Bacon on S and Fig. 51) 6 Day.

BACON B C. 28 mi. 4600 ft. 1. (II tangent to Bacon on NW and Fig. 51) 6 Day.

BAILLAUD C. 55 x 62 mi. 9500 ft. Breaks in wall. Old. (V Plato NE 3.6).

BAILLY C. 200 mi. 13,800 ft. 5. (IV Tycho SW 3.4) 14 Day.

BAILLY B C. 44 mi. 11,500 ft. Old. (IV on floor of Bailly E of center) 14 Day.

BAILY C. Roughly rectangular 16 x 20 mi. 1600 ft. Breaks in wall. Very old. (II Posidonius N 2.9).

BALL C. Irregular 25 mi. 5900 ft. Pk. 2600 ft. 2. (III Tycho N 1.2 and Fig. 6) 22 Day.

CAPE BANAT Pr. Broad shoulder of Carpathian Mts. extending 20 mi. onto Mare Imbrium. 5400 ft. (III Copernicus NW 1.8 and Fig. 4) 24 Day.

BAROCIUS C. 53 mi. 11,300 ft. 2. (V Tycho E. 3.9 and Fig. 51) 6 Day.

BAROCIUS B C. 25 mi. 4400 ft. 1. (V in NE wall of Barocius and Fig. 51) 6 Day.

BARROW C. Egg-shaped 53 x 63 mi. 7900 ft. Several breaks in wall. Very old. (II Posidonius NW 6.4) 7 Day.

BARROW A C. 20 mi. 4300 ft. Young. (II in SW wall of Barrow) 7 Day.

BAYER C. 30 x 33 mi. 8200 ft. 1. (VI Tycho SW 2.2) 25 Day.

BEAUMONT C. 33 x 38 mi. 6600 ft. N wall broken. 5. (II Theophilus S 1.2 and Fig. 48).

BEER C. 6 mi. 3400 ft. 1. (III Copernicus NE 3.7 and Fig. 42).

BEHAIM C. 33 mi. 11,000 ft. Pk. off center to W. Old. (I Langrenus SE 1.5) 1 Day.

BELKOVICH C. 100 x 120 mi. 8200 ft. (I Posidonius N 5.0).

BELLOT C. 12 mi. 7300 ft. Young. (II Langrenus SW 1.8 and Fig. 47).

BERNOUILLI C. 31 mi. 11,500 ft. Pk. 1. AMS finds S floor 1600 ft. higher than N floor. (I Crisium N 3.3) 16 Day.

BEROSUS C. 44 mi. 10,500 ft. 1. (I Crisium N 3.0) 15 Day.

BESSARION C. 7 mi. 3000 ft. Young. (VI Copernicus NW 2.8).

BESSEL C. 10 mi. 5200 ft. 1. (II Posidonius SW 2.4 and Fig. 11) 14 Day.

BETTINUS C. 49 mi. 12,500 ft. Pk. off center to N. 1. (IV Tycho SW 2.9) 11 Day.

BIANCHINI C. Irregular 25 mi. 10,000 ft. MuPk. off center to N. 1. (VI Plato W 2.9) 11 Day.

BIELA C. 48 mi. 6600 ft. Pk. 2600 ft., off center to N. Old. (II Theophilus S 5.9) 4 Day.

BIELA C C. 17 mi. 2300 ft. Young. (II in NE wall of Biela) 4 Day.

BILLY C. 30 mi. 4000 ft. Dark floor. 5. (V Grimaldi SE 2.4 and Fig. 45) 12 Day.

BIRMINGHAM C. Bell-shaped 55 x 57 mi. 5600 ft. Rough floor. Large gaps in wall. Extremely old. (VI Plato N 1.6 and Fig. 9).

BIRT C. 11 mi. 7700 ft. 1. (III Tycho N 3.4) 15 Day.

BLANC Mt. 16 x 19 mi. One of the higher Alpine peaks bordering Mare Imbrium. Heights to 14,000 ft. reported, but AMS finds no more than 9200 ft. (VI Plato SE 1.3 and Fig. 9).

BLANCANUS C. 63 x 66 mi. 12,500 ft. MuPk. 1. (III Tycho S 2.4) 9 Day.

BLANCHINUS C. 45 x 51 mi. 5200 ft. Numerous breaks in wall. Extremely old. (VI Tycho NE 3.1) 7 Day.

BODE C. 11 mi. 7200 ft. 1. (III Copernicus E 3.3) 15 Day.

BOGUSLAWSKY C. 61 x 66 mi. 11,500 ft. 2. (V Tycho SE 4.9).

BOHNENBERGER C. 21 mi. 7700 ft. MuPk. Gap in N wall. Old. (II Theophilus E 2.2) 19 Day.

GEORGE BOND C. Irregular 13 x 15 mi. 8200 ft. Young. (II Posidonius E 0.9).

WILLIAM BOND C. Nearly rectangular 90 x 107 mi. 9800 ft; Numerous gaps in wall. Very old. (III Plato NE 1.8 and Fig. 9) 21 Day.

BONPLAND C. 34 x 38 mi. 3300 ft. Most of wall less than 2000 ft. S wall missing. Flooded. Extremely old. (III Copernicus S 3.5 and Fig. 22) 9 Day.

BORDA C. Irregular 26 x 31 mi. 9600 ft. 5 x 10 mi. Pk. 2000 ft., off center to NW. Spine from S wall 2300 ft. Very old. (V Theophilus SE 3.5).

BOSCOVICH C. Roughly pentagonal 21 x 30 mi. 3000 ft. Dark floor. Numerous breaks in wall. Flooded. Extremely old. (V Posidonius SW 4.7).

BOUGUER C. 15 mi. 8200 ft. Young. (VI Plàto W 2.7) 11 Day.

BOUSSINGAULT C. 81 x 89 mi. 9500 ft. C. 50 mi. in diam. covers much of floor and between walls of the two the wall of a third can be traced in part. Very old. (V Tycho SE 5.5).

BRADLEY Mt. on Apennine crest 15 x 30 mi. Elevations to 16,000 ft. reported, but LAC gives 11,400 ft. for S Pk. and 10,900 ft. for N Pk. AMS shows 12,100 ft. (VI Copernicus NE 4.5 and Fig. 42).

BRAYLEY C. 10 mi. 7600 ft. Young (VI Aristarchus SE 1.3 and Fig. 44.)

BREISLAK C. 31 mi. 4900 ft. Old. (III Tycho E 3.8 and Fig. 51) 6 Day.

BRIGGS C. 24 mi. 3800 ft. Young. (V Aristarchus NW 1.7) 13 Day.

BROWN C. Roughly crescent-shaped 12 x 25 mi. Half destroyed by intruder on E 15 mi. diam. which is excluded from its host. Old. (VI Tycho SW 0.8 unlabeled on NE wall of Longomontanus and Fig. 6).

BUCH C. 28 x 34 mi. 5900 ft. Young. (II Theophilus SW 4.9 and Fig. 51).

BULLIALDUS C. 38 mi. 11,300 ft. MuPk. 1. (III Tycho NW 4.2 and Fig. 22) 9 Day.

BURCKHARDT C. 27 x 34 mi. 15,900 ft. Pk. 1300 ft., off center to W. 1. (I Crisium N 2.6) 16 Day.

BURG C. 24 mi. 12,000 ft. MuPk. 6200 ft. and W. of center. 1. (II Posidonius NW 2.3) 5 Day.

BUSCHING C. Irregular 29 x 34 mi. 7900 ft. 3. (II Theophilus S 4.6 and Fig. 51).

BYRD C. 51 x 57 mi. 8500 ft. Large breaks in S wall. Old. (II Posidonius NW 7.1).

BYRGIUS C. 45 x 53 mi. 15,000 ft. 3. (IV Gassendi SW 2.5 and Fig. 45) 13 Day.

BYRGIUS A C. 10 x 13 mi. 4000 ft. Young. (IV in NE wall of Byrgius and Fig. 45) 13 Day.

CALIPPUS C. Distorted pentagonal 19 x 21 mi. 9800 ft. 1. (III Plato SE 3.2 and Fig. 8) 21 Day.

CAMPANUS C. Irregular 31 mi. 6700 ft. Pk. has collapsed or has been removed almost completely by a C. three mi. across. 5. (III Tycho NW 3.7) 10 Day.

CAPELLA C. 31 mi. 10,500 ft. Pk. 3600 ft. Old. (II Theophilus NE 1.7) 5 Day.

CAPUANUS C. Fig-shaped 43 x 47 mi. 6200 ft. 5. (III Tycho NW 2.8) 10 Day.

CARDANUS C. 31 mi. 6600 ft. 2. (IV Grimaldi N 3.5) 13 Day.

CARLINI C. 8 mi. 5500 ft. 1. (IV Plato SW 3.6.)

CARPATHIANS MR. stretching 200 mi. and consisting chiefly of transverse ridges up to 65 mi. long. 6600 ft. (III Copernicus NW 1.3 and Fig. 4) 9 Day.

CARPENTER C. 38 mi. 10,200 ft. Pk. Old. (VI Plato NW 2.6) 12 Day.

CASATUS C. 66 x 78 mi. 12,800 ft. Dark floor. 2. (IV Tycho S 3.2) 23 Day.

CASSINI C. 36 mi. 3500 ft. 5. (VI Plato SE 2.3 and Fig. 9) 7 Day.

CASSINI A C. 10 x 12 mi. 5100 ft. 1. (VI on floor of Cassini and Fig. 9) 7 Day.

JACQUES CASSINI Valley 30 x 90 mi. 8200 ft. (VI Plato N 2.4).

CATHARINA C. 64 mi. 9200 ft. 4. (II Theophilus SW 1.3 and Fig. 5) 5 Day.

CAUCASUS MR. 75 x 310 mi. Highest pk., just W of Calippus, rises 17,400 ft. above Imbrium plain. (II Posidonius W 3.3 and Fig. 8) 6 Day.

CAUCHY C. 8 mi. 5900 ft. Between two nearly parallel rills running NW-SE; one over 200 mi. long, other 125 mi., visible through small telescope. Young. (II Posidonius SE 4.4) 19 Day.

CAVALERIUS C. 37 mi. 10,800 ft. MuPk. 3600 ft. 1. (IV Grimaldi N 2.0) 13 Day.

CAVENDISH C. Irregular 33 x 36 mi. 12,100 ft. Pk. 3. (VII Gassendi SW 1.9 and Fig. 45) 26 Day.

CAVENDISH E C. Irregular 15 x 17 mi. 8100 ft. Old. (VII in S wall of Cavendish and Fig. 45) 26 Day.

CAYLEY C. 9 mi. 8200 ft. Young. (V Theophilus NW 3.5 and Fig. 10).

CENSORINUS C. 1.5 mi. 1300 ft. Young. (IV Theophilus NE 2.4) 15 Day.

CEPHEUS C. 27 mi. 10,500 ft. MuPk. 1300 ft. 1. (I Posidonius NE 2.2) 4 Day.

CHACORNAC C. Irregular 32 x 36 mi. 6400 ft. Breaks in wall. Very old. (II Posidonius SE 0.5 and Fig. 11).

CHALLIS C. 37 mi. 9200 ft. N. wall missing. Old. (V Plato NE 2.6) 21 Day.

CHLADNI C. 8 mi. 6900 ft. Young. (III Copernicus SE 4.1) 15 Day.

CICHUS C. 25 mi. 7900 ft. Pk. 1. (III Tycho NW 2.2) 9 Day.

CICHUS C C. 8 mi. 4900 ft. Young. (III on SW crest of Cichus) 9 Day.

CLAIRAUT C. Irregular 37 x 59 mi. 9900 ft. Large breaks in wall. 2. (III Tycho E 3.3 and Fig. 51) 21 Day.

CLAIRAUT A C. 22 x 24 mi. 4900 ft. Old. (III in S wall of Clairaut and Fig. 51) 21 Day.

CLAUSIUS C. 13 x 16 mi. 5900 ft. Young. (VII Gassendi S 3.2).

CLAVIUS C. 132 x 152 mi. 16,100 ft. Floor and walls heavily battered. 2. (III Tycho S 1.8) 9 Day.

CLAVIUS B C. 30 mi. 10,200 ft. MuPk. 1. (III in NE wall of Clavius) 9 Day.

CLAVIUS D C. 19 mi. 9500 ft. 1. (III largest of the Clavius floor craters, E of center) 9 Day.

CLEOMEDES C. Irregular 81 x 92 mi. 14,300 ft. MuPk. 5100 ft., off center to NW. 5. (I Crisium N 2.0) 3 Day.

CLEOSTRATUS C. 36 x 40 mi. 7200 ft. Old. (VI Plato NW 3.5).

MARE COGNITUM Elliptical plain 120 x 200 mi. (VI Copernicus S 3.7 and Fig. 22) 9 Day.

COLUMBUS C. Square with rounded corners 56 mi. 7300 ft. Pk. 1700 ft., off center to N. 5. (I Crisium SW 2.5) 18 Day.

COLUMBUS A C. 28 mi. 6000 ft. 1. (I Tangent to Columbus on NW) 18 Day.

CONDAMINE C. Irregular 25 mi. 6600 ft. 5. (III Plato W 2.1) 10 Day.

CONDORCET C. 49 mi. 10,100 ft. 5. (I Crisium SE 1.5) 16 Day.

CONON C. 14 mi. 8900 ft. 1. (II Crisium W 9.0 and Fig. 42).

COOK C. 26 x 30 mi. 3800 ft. Flooded. Old. (I Langrenus SW 2.5) 18 Day.

COPERNICUS C. 60 mi. 12,600 ft. MuPk. 2000 ft. Conspicuous rays. 1. (Reference Feature: IV Dot NW 4.8 and Fig. 4) 9 Day.

CORDILLERAS MR. 600 mi. observable. Probably a scarp concentric with Mare Orientale on far side. Heights over 20,000 ft. reported (VI Gassendi W 2.6) 27 Day.

MARE CRISIUM Elliptical plain 270 x 350 mi. covering some 66,000 square mi. The long axis lies E-W and not N-S as it appears owing to foreshortening. Wall 14,400 feet on N. (Reference Feature: IV Dot NE 9.5) 2 Day.

CROZIER C. Irregular 13 x 16 mi. 4300 ft. Flooded. Old. (II Langrenus SW 1.7 and Fig. 47).

CRUGER C. 26 x 30 mi. 3300 ft. Dark floor. Flooded. Extremely old. (VI Grimaldi S 2.2 and Fig. 45) 13 Day.

CURTIUS C. 59 x 63 mi. 13,000 ft. Pk. 3000 ft., off center to NW. Breaks in wall. Very old. (III Tycho SE 3.1) 7 Day.

CUSANUS C. 40 mi. 11,800 ft. Old. (II Posidonius N 5.5) 16 Day.

CUVIER C. 47 x 53 mi. 12,100 ft. 1. (III Tycho SE 2.8 and Fig. 51) 7 Day.

CYRILLUS C. 62 mi. 11,800 ft. Pk. 6200 ft. 3. (II joins Theophilus on SW and Fig. 5) 5 Day.

CYRILLUS A C. Pear-shaped 10 mi. 3300 ft. Young. (IV in SW wall of Cyrillus and Fig. 5) 15 Day.

CYRILLUS F. C. Rectangular 20 x 30 mi. 8200 ft. Very old. (II joins Cyrillus on SE and Fig. 5) 5 Day.

CYSATUS C. Irregular 31 mi. 10,200 ft. 1. (III Tycho SE 2.7) 8 Day.

D'ALEMBERTS MR. of unknown extent on W limb. Heights to 20,000 ft. reported. (V Grimaldi W 0.7) 15 Day.

DAMOISEAU C. Irregular 32 mi. 3900 ft. Very old. (IV Grimaldi NE 0.5) 12 Day.

DANIELL C. 15 x 19 mi. 6200 ft. Young. (V Posidonius N 0.5 and Fig. 11).

DARNEY C. 10 mi. 7000 ft. Young. (VI Gassendi E 2.3 and Fig. 22).

D'ARREST C. 18 x 21 mi. 5200 ft. Breaks in wall. Old. (II Theophilus NW 3.2 and Fig. 10).

DARWIN C. Irregular 80 x 89 mi. 10,100 ft. S wall missing. Very old. (IV Grimaldi S 2.8 and Fig. 45) 13 Day.

DA VINCI C. Irregular 18 x 23 mi. 3600 ft. Breaks in wall. Very old. (V Theophilus NE 4.9).

DAVY C. 20 mi. 4900 ft. 5. (VI Gassendi E 5.0 and frontispiece) 23 Day.

DAVY A C. 9 mi. 6000 ft. Young. (VI in SE wall of Davy and frontispiece) 23 Day.

DAWES C. 11 mi. 5900 ft. 1. (V Posidonius S 2.5 and Fig. 11) 15 Day.

DEBES C. 20 mi. 7100 ft. S wall missing. Old. (I Crisium NW 2.5) 18 Day.

DE GASPARIS C. 18 x 22 mi. 10,200 ft. Breaks in wall. Old. (VII Gassendi SW 1.7 unlabeled between Vieta and Liebig, and Fig. 45).

DELAMBRE C. 32 mi. 11,500 ft. Pk. 2. (II Theophilus NW 2.3 and Figs. 24 and 10) 6 Day.

DE LA RUE C. Irregular 67 x 91 mi. 11,800 ft. Gaps in wall. Very old. (I Posidonius N 4.2).

DELAUNAY C. Irregular 28 x 30 mi. 6300 ft. Divided into three main parts by walls up to 4800 ft. high. Very old. (II Theophilus SW 4.8).

DELISLE C. Irregular 17 mi. 7900 ft. 1. (III Copernicus NW 4.0).

DEMOCRITUS C. Irregular 20 x 25 mi. 8900 ft. Pk. 3600 ft. 1. (V Copernicus NE 10.2) 18 Day.

DEMONAX C. 72 x 79 mi. 9800 ft. Old. (V Tycho SE 5.3).

DE MORGAN C. Egg-shaped 6 x 8 mi. 4900 ft. Old. (V Theophilus NW 3.4 unlabeled between Cayley and D'Arrest and Fig. 10).

DESARGUES C. 47 x 58 mi. 8900 ft. E wall destroyed. Old. (VII Aristarchus NE 7.0).

DESCARTES C. Poorly-defined enclosure 28 x 30 mi. 3000 ft. Extremely old. (IV Theophilus W 1.8 and Fig. 50).

DESCARTES C C. 2 mi. Young. (IV Theophilus W 1.8 and Fig. 50) 15 Day.

DE SITTER C. 36 x 42 mi. 12,100 ft. SW wall missing. Spine crosses floor SW–NE with highest point near center. Old. (V Plato NE 3.2).

DESLANDRES C. 136 x 152 mi. 11,500 ft. The vast enclosure, which has gone by the names Horbiger and Hell Plain, is probably the C. drawn by Galileo (Fig. 16). Cassini's Bright Spot on floor E of center is ray pattern 35 mi. across about C. three mi. in diameter. 4. (III Tycho NE 1.8) 8 Day.

DE VICO C. 13 x 14 mi. 8200 ft. Young. (VII Gassendi W 2.5 unlabeled and Fig. 45).

DEVILLE Pr. Mt. mass 15 x 25 mi. rising to Alpine crest 7200 ft. above Mare Imbrium (VI Plato SE 1.6 and Fig. 9).

DIONYSIUS C. 12 mi. 8300 ft. High albedo. Young. (II Theophilus NW 3.0 and Fig. 10) 15 Day.

DIOPHANTUS C. 12 mi. 8500 ft. Pk. 1. (VII Aristarchus E 2.2).

DOERFELS MR. of unknown extent, a few peaks of which can be seen on SW limb with favorable libration. Heights to 26,000 ft. reported. (IV Tycho SW 3.5) 14 Day.

DOLLOND C. 7 mi. 5100 ft. Young. (IV Theophilus W 2.2 and Fig. 50).

DOLLOND E C. 3 mi. 1100 ft. Bright spindle connects Dollond E and Descartes C. Young. (IV Theophilus W 1.9 and Fig. 50) 15 Day.

DONATI C. Irregular 22 mi. 9000 ft. Pk. 5200 ft. Very old. (II Theophilus SW 4.2).

DOPPELMAYER C. 37 x 44 mi. 2600 ft. MuPk. 2300 ft. NE wall missing. Flooded. Old. (IV Grimaldi SE 5.4) 11 Day.

DOVE C. Irregular depression consisting of several parts 20 x 27 mi. 3300 ft. Very old. (II Theophilus S 5.3 unlabeled and Fig. 49).

DRAPER C. 6 x 7 mi. 3700 ft. Old. (III Copernicus N 1.5 unlabeled and Fig. 4).

DREBBEL C. Irregular 20 x 23 mi. 6900 ft. Young. (VII Gassendi S 3.7 and Fig. 46).

DUNTHORNE C. 11 mi. 4900 ft. Young. (VI Gassendi SE 2.7) 15 Day.

EDDINGTON C. 80 x 90 mi. 4300 ft. S wall missing. Flooded. Old. (VI Aristarchus W 1.8) 14 Day.

EGEDE C. Polygonal 19 x 25 mi. 2300 ft. 5. (II Posidonius NW 4.1 and Fig. 8).

EIMMART C. Irregular 25 x 28 mi. 11,400 ft. 1. (I Crisium, near N edge).

ENCKE C. Hexagonal 18 mi. 2300 ft. MuPk. 5. (VII Gassendi N 4.1) 11 Day.

ENDYMION C. 77 mi. 16,100 ft. Dark floor. 5. (II Posidonius N 3.7) 3 Day.

PALUS EPIDEMIARUM Triangular plain 110 x 175 mi. broken by many C. up to 45 mi. diameter and numerous rills. (III Tycho NW 3.2) 10 Day.

EPIGENES C. 35 mi. 8900 ft. Pk. 2600 ft. off center to SE. 5. (VI Plato NE 2.1 and Fig. 9) 22 Day.

EPIMENIDES C. Irregular 14 x 16 mi. 5200 ft. Pk. Old. (VII Gassendi SE 4.5).

ERATOSTHENES C. 37 mi. 12,300 ft. MuPk. 6600 ft. 1. (III Copernicus NE 1.9 and Fig. 4) 8 Day.

EUCLID C. 8 mi. 4400 ft. Brilliant. 1. (III Copernicus SW 3.7 and Fig. 22) 15 Day.

EUCTEMON C. 40 mi. 8200 ft. Old. (V Plato NE 3.3).

EUDOXUS C. 40 x 43 mi. 14,300 ft. MuPk. 2500 ft. 1. (II Posidonius NW 3.1 and Fig. 8) 6 Day.

EULER C. 17 mi. 7200 ft. MuPk. 2200 ft. 1. (III Copernicus NW 2.7) 10 Day.

FABRICIUS C. 48 mi. 14,400 ft. MuPk. 2600 ft. 2. (II Theophilus S 4.8) 4 Day.

FARADAY C. Irregular 33 x 44 mi. 7500 ft. MuPk. 2. (III Tycho E 2.8 and Fig. 51) 7 Day.

FARADAY E C. Irregular 20 mi. 4900 ft. MuPk. Young. (III in S wall of Faraday and Fig. 51) 7 Day.

FARADAY K C. Irregular 25 mi. 5600 ft. MuPk. Old. (III in SW wall of Faraday and Fig. 51) 7 Day.

FAUTH C. 8 x 9 mi. 6100 ft. Young. (III Copernicus S 0.6 and Fig. 4).

FAYE C. Irregular 23 x 27 mi. 7800 ft. Pk. 4600 ft. Large gaps in wall. Extremely old. (II Theophilus SW 4.5).

MARE FECUNDITATIS Relatively craterless plain in form of square 350 mi. on side with smaller rectangular appendage on S corner. Area about 150,000 square mi. (I Crisium S 4.2) 3 Day.

FERMAT C. 24 mi. 7400 ft. 2. (II Theophilus SW 2.4 and Fig. 48) 20 Day.

FERNELIUS C. 41 mi. 6200 ft. 3. (II Theophilus SW 5.8 and Fig. 51) 21 Day.

FERNELIUS A C. 18 mi. 3000 ft. Old. (II in W wall of Fernelius) 21 Day.

FEUILLEE C. 7 mi. 3600 ft. 1. (III Copernicus NE 3.8 and Fig. 42).

FIRMICUS C. 34 x 39 mi. 8400 ft. 5. (I Crisium S 2.0.)

FLAMMARION C. 44 x 50 mi. 8200 ft. Gaps in wall. Flooded. Extremely old. (III Copernicus SE 3.9) 22 Day.

FLAMSTEED C. 13 mi. 5500 ft. 1. On S floor of ghost C. 70 mi. marked by ridges up to 2100 ft. (VII Grimaldi E 2.7) 11 Day.

FONTANA C. Irregular 19 x 21 mi. 8000 ft. NE wall broken. Very old. (VII Gassendi W 2.3 and Fig. 45) 12 Day.

FONTENELLE C. 25 mi. 9200 ft. Pk. 5. (III Plato NW 1.4 and Fig. 9) 23 Day.

FOUCAULT C. Irregular 15 mi. 7200 ft. Young. (VI Plato W 3.2) 11 Day.

FOURIER C. 30 x 35 mi. 12,200 ft. 2. (VII Gassendi SW 2.3) 12 Day.

FRACASTORIUS C. 73 mi. 5800 ft. MuPk. 1000 ft. N wall missing. 5. (II Theophilus SE 1.9 and Fig. 48) 5 Day.

FRA MAURO C. 60 mi. 2400 ft. W wall missing. Gaps in N and E walls. Flooded. Extremely old. (III Copernicus S 3.0 and Fig. 22) 9 Day.

FRANKLIN C. 34 mi. 10,700 ft. Pk. 2000 ft. 2. (II Posidonius NE 2.3) 4 Day.

FRAUNHOFER C. 33 mi. 7900 ft. 3. (I Langrenus S 5.5) 16 Day.

MARE FRIGORIS Longest, narrowest, and least dark of the plains. Over 700 mi. long, but shrinks to width of 45 mi. (IV Plato N 0.8) 6 Day.

FURNERIUS C. Hexagonal. 81 mi. 11,000 ft. 3. (I Langrenus S 4.9) 2 Day.

FURNERIUS A C. 9 mi. 3300 ft. Young. (IV on outer N wall of Furnerius) 15 Day.

FURNERIUS C C. 12 mi. 4600 ft. Young. (IV on outer N wall of Furnerius) 15 Day.

GALILEO C. 10 mi. 6600 ft. 1. (IV Grimaldi N 3.0) 13 Day.

GALILEO A   C. 8 mi. 4600 ft. 1. (IV just N of Galileo) 13 Day.

GALLE   C. 15 mi. 6600 ft. Young. (II Posidonius NW 4.1 and Fig. 8) 20 Day.

GAMBART   C. Polygonal 16 mi. 3600 ft. Old. (III Copernicus S 1.8 and Fig. 4).

GARTNER   C. 66 mi. 4300 ft. S wall missing. Flooded. Extremely old. (II Posidonius N 4.1) 19 Day.

GASSENDI   C. 70 mi. 6600 ft. MuPk. 3600 ft. 5. (Reference Feature: IV Dot SW 7.5 and Fig. 20) 11 Day.

GASSENDI A   C. 20 x 24 mi. 8500 ft. MuPk. 1600 ft. 1. (IV on N wall of Gassendi and Fig. 20) 11 Day.

GASSENDI E   C. 5 mi. 2600 ft. Young. (IV Gassendi SW 0.4 and Fig. 20) 15 Day.

GAUDIBERT   C. Irregular 18 x 20 mi. 5200 ft. MuPk. 2600 ft. Old. (V Theophilus E 1.9.)

GAURICUS   C. 46 x 52 mi. 8900 ft. Walls heavily cratered. Very old. (III Tycho N 1.5) 9 Day.

GAUSS   C. 112 mi. 13,000 ft. MuPk. 4300 ft. Very old. (I Crisium N 3.7) 1 Day.

GAY-LUSSAC   C. 16 x 17 mi. 5600 ft. Spines across floor. Walls broken on N and S. Very old. (III Copernicus N 0.8 and Fig. 4.)

GEBER   C. 27 mi. 9600 ft. 1. (II Theophilus SW 2.8 and Fig. 50) 21 Day.

GEMINUS   C. 53 mi. 9400 ft. MuPk. 4800 ft. 1. (I Crisium N 3.3) 3 Day.

GEMINUS C   C. 9 mi. 4200 ft. Young. (I on E rim of Geminus) 15 Day.

GEMMA FRISIUS   C. 56 mi. 17,100 ft. 3. (II Theophilus SW 4.6 and Fig. 51) 6 Day.

GIOJA   C. 27 mi. 9800 ft. Young. (VI Plato NE 3.3.)

GOCLENIUS   C. Double; 34 x 46 mi. 6100 ft. Pk. near center of W section. Flooded. Very old. (V Theophilus E 2.9) 15 Day.

GODIN   C. 22 mi. 10,500 ft. Pk. 3600 ft. 1. (II Theophilus NW 3.7) 7 Day.

GOLDSCHMIDT   C. 71 x 77 mi. 7500 ft. Breaks in wall. Flooded. Very old. (VI Plato NE 2.6) 8 Day.

GOODACRE   C. 28 mi. 10,500 ft. Pk. Old. (II Theophilus SW 4.3) 6 Day.

GRIMALDI   C. 140 x 145 mi. 10,500 ft. Dark floor. 5. (Reference Feature: IV Dot W 10.1) 13 Day.

GRIMALDI G   C. 8 mi. 4100 ft. One of many rim craters, near center of bright area. Young. (IV on SE wall of Grimaldi) 15 Day.

GROVE   C. 16 x 17 mi. 7800 ft. Pk. 1. (V Posidonius N 1.3.)

GRUEMBERGER   C. 58 mi. 13,500 ft. 2. (III Tycho S 2.6.)

GRUITHUISEN   C. 10 mi. 5200 ft. Young. (VII Aristarchus NE 2.2.)

GUERICKE   C. 36 x 39 mi. 3100 ft. Numerous gaps in wall. 5. (III Copernicus S 4.2 and frontispiece) 9 Day.

GUTENBERG   C. Irregular 45 mi. 7500 ft. MuPk. 1600 ft. E and S walls destroyed. 5. (I Crisium S 5.5) 4 Day.

GUTENBERG E   C. 18 mi. 2600 ft. SE wall missing. Flooded. Old. (I on E wall of Gutenberg) 4 Day.

HADLEY   Mt. on Apennine crest near NE end of range, 15 x 28 mi. Elevations around 15,000 ft. reported, but LAC and AMS agree on about 9500 ft. above Imbrium plain. (VI Aristarchus E 7.8 and Fig. 7.)

HAEMUS   MR. of ruined Mts. stretching 300 x 60 mi.

Highest pk. 10,000 ft. (AMS), 8000 ft. (LAC) above Serenitatis plain. (II Posidonius SW 3.7) 6 Day.

HAGECIUS   C. 50 mi. 7500 ft. Very old. (II Theophilus S 6.4) 18 Day.

HAHN   C. 50 mi. 10,200 ft. Pk. 2300 ft., off center to W. Very old. (I Crisium N 2.7) 15 Day.

HAINZEL   C. Triple; 44 x 58 mi. 10,500 ft. N and SE sections have Pk. Very old. (VI Tycho W 3.0) 10 Day.

HALL   C. Irregular 23 x 26 mi. 5900 ft. W wall missing. Flooded. Old. (II Posidonius NE 1.0.)

HALLEY   C. 22 mi. 9000 ft. 2. (II Theophilus W 3.7) 7 Day.

HANNO   C. 34 x 36 mi. 6600 ft. Old. (I Langrenus S 7.8.)

HANSEN   C. 23 x 25 mi. 4900 ft. Pk. off center to W. Young. (I Crisium E 1.6).

HANSTEEN   C. 29 mi. 4100 ft. MuPk. 5. (VII Gassendi NW 2.2 and Fig. 45) 12 Day.

HARBINGERS   MR. scattered over 40 x 100 mi. area. 5800 ft. (VII Aristarchus E 1.1 and Fig. 44) 11 Day.

HARPALUS   C. 26 mi. 9800 ft. Spine on S floor. 1. (VI Plato W 3.3) 11 Day.

HASE   C. 50 mi. 9000 ft. MuPk. 3800 ft. 2. (I Langrenus S 3.7) 3 Day.

HAYN   C. 48 x 50 mi. 11,800 ft. MuPk. Old. (II Posidonius N 5.1.)

HECATAEUS   C. 87 mi. 15,700 ft. Pk. 1000 ft. Young. (I Langrenus SE 2.1) 1 Day.

HEDIN   C. 82 x 90 mi. 6500 ft. MuPk. Gaps in wall. Very old. (IV Grimaldi N 1.6) 28 Day.

HEINSIUS   C. 43 mi. 6900 ft. SW wall destroyed. 3. (III Tycho NW 1.2 and Fig. 6) 9 Day.

HEINSIUS A   C. 13 mi. 6200 ft. Young. (III on floor of Heinsius and Fig. 6) 9 Day.

HEINSIUS B   C. 14 x 18 mi. 4900 ft. Young. (III on SW wall of Heinsius and Fig. 6) 9 Day.

HEINSIUS C   C. 15 mi. 3200 ft. Young. (III on S wall of Heinsius and Fig. 6) 9 Day.

HEIS   C. 9 mi. 4900 ft. Young. (VII Aristarchus E 3.0.)

HELICON   C. 16 mi. 7400 ft. 1. (III Plato SW 2.5) 24 Day.

HELL   C. 22 mi. 6600 ft. Pk. Young. (VI Tycho N 1.5) 8 Day.

PAUL HENRY   C. Irregular 24 x 26 mi. 6800 ft. Old. (VII Gassendi SW 2.1 and Fig. 45).

PROSPER HENRY   C. Irregular 25 mi. 12,800 ft. Spine across floor N−S. Old. (VII Gassendi SW 2.3 and Fig. 45).

HERACLIDES   Pr. 35 x 50 mi. rising 5900 ft. above Imbrium plain. (III Plato W 3.7) 10 Day.

HERACLITUS   C. 31 mi. 9500 ft. Two rectangular excavations run NE from the lowered wall with dimensions of 20 x 30 and 25 x 40 mi. Very old. (II Theophilus SW 6.9 and Fig. 51) 7 Day.

HERCULES   C. 45 mi. 12,500 ft. 1. (II Posidonius N 2.5) 4 Day.

HERCULES A   C. 20 mi. 4900 ft. Young. (II Hercules N 0.6) 18 Day.

HERCULES D   C. 10 mi. 2600 ft. Young. (II on floor of Hercules) 15 Day.

HERCYNIANS   MR. 125 mi. long. 4300 ft. W walls of craters Otto Struve and Russell constitute the range. (V Aristarchus W 2.2) 14 Day.

HERIGONIUS   C. 10 mi. 5600 ft. Young. (VII Gassendi NE 1.1 and Fig. 20).

HERODOTUS   C. 20 x 22 mi. 4700 ft. 5. (IV Aristarchus W 0.3 and Fig. 44) 11 Day.

HERSCHEL   C. 22 x 25 mi. 12,800 ft. MuPk. 4900 ft. 1. (III Copernicus SE 4.4) 8 Day.

CAROLINE HERSCHEL   C. 8 mi. 4800 ft. 1. (VII Aristarchus NE 3.3.)

JOHN HERSCHEL   C. Irregular 105 mi. 5600 ft. Extremely old. (VII Aristarchus NE 6.7) 10 Day.

HESIODUS   C. 26 mi. 4600 ft. Flooded. Very old. (III Tycho NW 2.4.)

HEVELIUS   C. 66 x 69 mi. 7000 ft. Pk. 3500 ft. 5. (IV Grimaldi N 1.4) 13 Day.

HIND   C. 17 mi. 9200 ft. 1. (II Theophilus W 3.5) 7 Day.

HIPPALUS   C. 33 x 37 mi. 3000 ft. SW wall missing. Flooded. Very old. (VII Gassendi SE 2.2.)

HIPPARCHUS   C. 83 x 89 mi. 7500 ft. MuPk. 1900 ft. 5. (II Theophilus W 4.1) 7 Day.

HIPPARCHUS C   C. 11 mi. 8700 ft. Young. (II Hipparchus SE 0.7) 7 Day.

HIPPARCHUS G   C. 9 mi. 7700 ft. Young. (II on E wall of Hipparchus) 15 Day.

HIPPARCHUS L   C. 8 mi. 7100 ft. Young. (II Hipparchus E 0.8) 7 Day.

HOLDEN   C. 28 x 33 mi. 13,500 ft. Old. (I Langrenus S 1.8) 16 Day.

HOMMEL   C. 75 x 90 mi. 9200 ft. Pk. 3. (II Theophilus S 6.2 and Fig. 49) 5 Day.

HOMMEL A   C. 33 mi. 5900 ft. Very old. (II on E floor of Hommel and Fig. 49) 5 Day.

HOMMEL C   C. 33 mi. 5900 ft. Very old. (II on W floor of Hommel and Fig. 49) 5 Day.

HOOKE   C. Irregular 21 x 25 mi. 3400 ft. Very old. (I Posidonius NE 3.0) 18 Day.

HORREBOW   C. 15 x 17 mi. 6600 ft. Young. (VII Aristarchus NE 6.2.)

HORROCKS   C. 17 x 19 mi. 9200 ft. Pk. 1. (II Theophilus W 3.9) 7 Day.

HORTENSIUS   C. 9 mi. 7800 ft. 1. (III Copernicus W 1.6 and Fig. 4) 24 Day.

HUGGINS   C. 45 mi. 8500 ft. Very old. (III Tycho E 1.4 and Fig. 6) 8 Day.

HUMBOLDT   C. 125 mi. 15,400 ft. MuPk. 37 x 40 mi., 5200 ft. 2. (IV Langrenus S 3.5) 1 Day.

MARE HUMBOLDTIANUM   Elliptical plain 145 x 170 mi. About ¼ of surface stained dark. Peaks on E and W rims rise 9800 ft. above floor (AMS) although elevations of 16,000 ft. reported. (I Posidonius NE 4.5) 11 Day.

MARE HUMORUM   Elliptical plain 250 x 275 mi. covering some 50,000 square mi. Long axis E–W rather than NW–SE as it appears because of foreshortening. Walls relatively low, rising to 4900 ft. above wrinkled floor at one point. (VII Gassendi SE 1.3 and Fig. 20) 11 Day.

HUYGENS   Mt. on Apennine crest. 10 x 30 mi. Elevations up to 21,000 ft. reported, but modern measures vary from 16,000 ft. (AMS) down to 7500 ft. (LAC) above Imbrium plain. (VI Plato S 5.2 and Fig. 7.)

HYGINUS RILL   Graben-type depression divided near midpoint into two approximately linear sections by Hyginus (6 mi., 2900 ft.). Angle between sections 140 degrees. Length 140 mi., width up to 2 mi. Many small craters within. (II Theophilus NW 5.1) 21 Day.

HYPATIA   C. Irregular 17 x 25 mi. 4600 ft. Resembles Indian arrowhead. MuPk. 2. (II Theophilus N 1.5.)

IDELER   C. 21 x 25 mi. 5600 ft. Very old. (II Theophilus S 5.9 unlabeled and Fig. 49).

MARE IMBRIUM   Elliptical plain 670 x 750 mi. covering some 340,000 square mi. and bounded by several major and minor MR. Exceptionally dark. Surface declines some 8000 ft. from SE to NW edge. (VI Copernicus N 4.5 and Figs. 4 and 9) 7 Day.

INGHIRAMI   C. 62 mi. 10,800 ft. Low central hill on ridge across floor. 2. (IV Tycho W 5.1 and Fig. 46) 13 Day.

SINUS IRIDUM   Circular plain 160 mi. Jura Mts. rising 12,700 ft. form over half the boundary. Surface declines some 1300 ft. from SE to NW edge continuing the slope of Mare Imbrium. (III Plato W 3.0) 10 Day.

ISIDORUS   C. 25 mi. 9900 ft. 3. (V Theophilus NE 1.4) 5 Day.

JACOBI   C. 43 mi. 12,500 ft. 2. (V Tycho SE 3.3) 21 Day.

JANSSEN   C. Hexagonal 122 x 153 mi. 9300 ft. Breaks in wall. Extremely old. (V Tycho E 6.7) 4 Day.

JULIUS CAESAR   C. Irregular 45 x 55 mi. 4000 ft. Much of SE wall missing. Flooded. Very old. (II Theophilus NW 4.3 and Fig. 10) 6 Day.

JURAS   MR. extending 300 mi. along perimeter of circle 160 mi. in diameter to form boundary of Sinus Iridum from which foothills radiate 90 to 160 mi. 12,700 ft. (III Plato W 3.2) 10 Day.

KAISER   C. Rectangular 31 x 34 mi. 6500 ft. 3. (II Theophilus SW 5.5 and Fig. 51.)

KANE   C. 34 mi. 4300 ft. S wall broken. Old. (II Crisium NW 10.0) 20 Day.

KANT   C. 19 x 22 mi. 9400 ft. Pk. 2900 ft., off center to NW. 1. (II Theophilus W 1.1 and Fig. 5) 15 Day.

CAPE KELVIN   Mt. on Humorum plain near SE edge. 13 x 30 mi. 6200 ft. (VII Gassendi SE 2.1.)

KEPLER   C. 20 mi. 7500 ft. Conspicuous ray pattern. 1. (VI Copernicus W 2.8) 11 Day.

KIRCH   C. 8 mi. 5600 ft. Young. (VI Plato S 1.8 unlabeled and Fig. 42.)

KIRCHER   C. 49 mi. 15,400 ft. 1. (IV Tycho S 3.0) 11 Day.

KLAPROTH   C. 65 x 77 mi. 10,200 ft. S wall removed. Very old. (III Tycho S 2.8) 23 Day.

KLEIN   C. 30 mi. 6600 ft. Pk. 1500 ft. Flooded. Old. (II Theophilus W 4.3) 7 Day.

KONIG   C. 14 x 16 mi. 5400 ft. 1. (III Tycho NW 3.9) 24 Day.

KRAFFT   C. 31 mi. 6600 ft. Old. (IV Grimaldi N 4.1) 13 Day.

KRIEGER   C. Polygonal 15 mi. 3600 ft. 5. (VII Aristarchus NE 1.1 and Fig. 44).

KUNOWSKY   C. Hexagonal 12 mi. 1600 ft. Pk. 2. (III Copernicus SW 2.5.)

LACAILLE   C. 40 x 44 mi. 9100 ft. 3. (VI Tycho NE 3.2) 7 Day.

LADE   C. Polygonal 35 x 40 mi. 3100 ft. S wall missing. Flooded. Very old. (II Theophilus NW 3.5.)

LAGALLA   C. Remnant; shaped like a slice of pie 50 mi. across. 4300 ft. Originally about 75 mi. diameter. Extremely old. (III Tycho W 1.6 and Fig. 6) 9 Day.

LAGRANGE   C. Irregular 90 x 95 mi. 7900 ft. MuPk. Very old. (IV Gassendi SW 3.3.)

LAHIRE   Mt. isolated on Mare Imbrium. 9 x 13 mi. 2600 ft. (LAC), 4900 ft. (AMS). May be part of original Imbrium crater wall. (III Plato SW 4.5.)

LALANDE   C. 14 x 15 mi. 9600 ft. Pk. 3300 ft. 1. (VI Copernicus SE 3.4.)

LAMARCK   C. Irregular 65 x 80 mi. 9200 ft. N wall missing. Very old. (IV Grimaldi S 3.3 and Fig. 45) 13 Day.

LAMBERT   C. 19 mi. 7900 ft. Pk. 1000 ft. 1. (III Copernicus N 2.9) 9 Day.

LAME   C. 55 mi. 11,800 ft. Pk. 2200 ft. Breaks in wall. Very old. (I Langrenus S 1.1 and Fig. 47) 2 Day.

LAMECH   C. 9 mi. 5200 ft. Young. (II Posidonius NW 3.3 unlabeled and Fig. 8.)

LAMONT   C. 39 x 55 mi. Walls melted down to ridges a few hundred feet high visible near terminator. Flooded. Very old. (II Theophilus N 3.0 unlabeled and Fig. 10.)

LANGRENUS   C. 85 mi. 16,200 ft. MuPk. 6400 ft. 1. (Reference Feature: IV Dot E 9.4 and Fig. 47) 2 Day.

LANGRENUS B   C. 21 mi. 8500 ft. Very old. (I Langrenus N 0.8 and Fig. 47) 16 Day.

LANGRENUS F   C. 30 mi. 8400 ft. Very old. (I Langrenus NW 0.7 and Fig. 47) 16 Day.

LANGRENUS K   C. 19 mi. 8200 ft. Pk. 2700 ft. Very old. (I Langrenus NW 0.6 and Fig. 47) 16 Day.

LANSBERG   C. 24 mi. 10,900 ft. MuPk. 2900 ft. 1. (III Copernicus SW 2.2 and Fig. 4) 10 Day.

LAPEYROUSE   C. 46 mi. 10,000 ft. Pk. Old. (I Langrenus E 1.2) 1 Day.

LAPLACE   Pr. Blunt headland extending 35 mi. over Mare Imbrium. 9900 ft. above Sinus Iridum (III Plato W 2.2) 10 Day.

LASSELL   C. 14 mi. 3100 ft. Flooded. Breaks in wall. Old. (III Tycho N 4.6) 23 Day.

LAVINIUM   Pr. Triangular Mt. 12 x 20 mi. 5600 ft. (I Crisium, in SW wall) 4 Day.

LAVOISIER   C. Egg-shaped 45 x 58 mi. 7200 ft. Pk. Very old. (VII Aristarchus NW 2.9.)

LEE   C. 24 x 27 mi. 5900 ft. NE wall missing. Flooded. Very old. (VII Gassendi S 2.3) 11 Day.

LEE M   C. 45 mi. 1600 ft. N and W walls largely destroyed. Flooded. Very old. (VII Gassendi S 2.2) 11 Day.

LEHMANN   C. 29 x 35 mi. 6600 ft. S wall broken by passes into Schickard. Very old. (VII Gassendi S 3.5 and Fig. 46).

LEIBNITZ   MR. extending some 600 mi. along S limb. Visible with favorable libration. Heights to 33,000 ft. reported. (III Tycho SE 4.1) 14 Day.

LEMONNIER   C. 36 mi. 7300 ft. W wall reduced to ridges. Flooded. Old. (V Posidonius S 0.9 and Fig. 11) 20 Day.

LETRONNE   C. 75 mi. 3000 ft. MuPk. 1300 ft. N ⅓ wall missing. 5. (VII Gassendi NW 1.5 and Fig. 20) 11 Day.

LEVERRIER   C. 13 mi. 8000 ft. 1. (III Plato SW 2.3) 24 Day.

LEXELL   C. 36 mi. 7200 ft. Pk. N wall missing. Very old. (VI Tycho NE 1.3) 8 Day.

LICETUS   C. 47 x 50 mi. 9800 ft. Pk. S wall broken. 1. (III Tycho E 2.5 and Fig. 51) 7 Day.

LICHTENBERG   C. 9 x 10 mi. 3900 ft. Young. (V Aristarchus NW 1.8)

LIEBIG   C. 22 x 25 mi. 4900 ft. 1. (VII Gassendi SW 1.4 and Fig. 45).

LILIUS   C. 38 x 42 mi. 7900 ft. Pk. 1300 ft. 1. (III Tycho SE 2.5) 21 Day.

LILIUS A   C. 23 x 26 mi. 6200 ft. 1. (III Lilius SE 0.2) 21 Day.

LINDENAU   C. 29 x 33 mi. 8900 ft. MuPk. 1. (II Theophilus S 3.5 and Fig. 48) 5 Day.

LINNE   C. 1 mi. 1600 ft. on white patch 7 mi. diameter. Young. (V Plato SE 4.6) 15 Day.

LITTROW   C. 17 x 19 mi. 3900 ft. S wall broken. Flooded. Old. (II Posidonius SE 1.9. unlabeled and Fig. 11.)

LOHSE   C. 26 mi. 9700 ft. Pk. 5000 ft. Old. (I Langrenus S 0.9 and Fig. 47) 16 Day.

LONGOMONTANUS   C. Pentagonal 107 mi. 12,500 ft. MuPk. W of center. N wall broken. 2. (III Tycho SW 1.4 and Fig. 6) 9 Day.

LUBINIEZKY   C. 26 mi. 2000 ft. SE wall reduced to hills. Flooded. Old. (IV Gassendi E 2.3 and Fig. 22.)

LUTHER   C. 6 x 7 mi. 3000 ft. 1. (V Posidonius W 0.9 unlabeled and Fig. 11.)

LYELL   C. Irregular 20 x 23 mi. 4900 ft. SW wall reduced to hills. Flooded. Old. (I Crisium W 2.3.)

MACLAURIN   C. 32 x 34 mi. 3600 ft. SE wall broken. Old. (I Langrenus NE 1.6.)

MACLEAR   C. Polygonal 13 mi. 2300 ft. Flooded. Old. (II Theophilus N 4.2 unlabeled and Fig. 10) 20 Day.

MACROBIUS   C. 42 mi. 11,800 ft. MuPk. 3600 ft. 1. (I Crisium NW 1.7) 4 Day.

MADLER   C. 17 mi. 7900 ft. Central ridge 2500 ft. 1. (V Theophilus E 0.6 and Fig. 5) 15 Day.

MAGELLAN   C. Irregular 23 x 25 mi. 6400 ft. 5. (V Theophilus E 2.7) 18 Day.

MAGELLAN A   C. Irregular 21 mi. 5700 ft. Young. (V Magellan SE 0.2) 18 Day.

MAGINUS   C. 100 x 115 mi. 16,400 ft. MuPk. 3. (III Tycho SE 1.2 and Fig. 6) 8 Day.

MAGINUS G   C. 28 mi. 7200 ft. Breaks in wall. Old. (III on SW wall of Maginus and Fig. 6) 8 Day.

MAIN   C. Irregular 32 mi. 9200 ft. S wall missing. Old. (II Posidonius NW 6.9) 21 Day.

MAIRAN   C. 26 mi. 11,200 ft. MuPk. 1. (VII Aristarchus NE 3.4) 11 Day.

MANILIUS   C. 25 mi. 9300 ft. MuPk. 1. (II Posidonius SW 4.5) 7 Day.

MANNERS   C. 10 mi. 5900 ft. 1. (II Theophilus N 3.1 and Fig. 10) 20 Day.

MANZINUS   C. 55 x 63 mi. 9800 ft. 1. (V Tycho SE 4.4) 5 Day.

MARALDI   C. Irregular 24 mi. 4300 ft. Dark floor; several passes through walls. Flooded. Very old. (V Posidonius SE 2.5) 19 Day.

MARE MARGINIS   Irregular plain 210 x 290 mi. Resembles flooded complex valley with numerous inlets and isolated pools. (II Crisium E 1.9) 11 Day.

MARIUS   C. 27 mi. 5500 ft. 5. (VII Aristarchus S 2.3) 12 Day.

MARKOV   C. Irregular 25 x 27 mi. 8900 ft. Pk. Old. (VII Aristarchus N 4.8) 27 Day.

MASKELYNE   C. 15 mi. 8200 ft. Pk. 1300 ft. Young. (V Theophilus NE 2.7) 20 Day.

MASON   C. 20 x 25 mi. 6300 ft. N wall broken. 5. (V Posidonius N 1.7) 20 Day.

MAUPERTUIS   C. Approximately rectangular 25 x 29 mi. 5600 ft. Very old. (III Plato W 2.2) 10 Day.

MAUROLYCUS   C. 73 mi. 16,700 ft. MuPk. 5200 ft. 2. (II Theophilus SW 5.4 and Fig. 51) 6 Day.

MAURY C. 11 x 13 mi. 4600 ft. Young. (V Posidonius NE 1.3)

CHRISTIAN MAYER C. 24 x 26 mi. 5900 ft. Pk. 1300 ft. 2. (II Posidonius NW 5.2.)

TOBIAS MAYER C. 20 x 22 mi. 5200 ft. Old. (III Copernicus NW 1.8 and Fig. 4) 24 Day.

TOBIAS MAYER A C. 11 mi. 8200 ft. Young. (III Tobias Mayer E 0.2 and Fig. 4) 24 Day.

MC CLURE C. Irregular 15 x 16 mi. 4900 ft. Old. (I Langrenus SW 2.0 unlabeled and Fig. 47.)

SINUS MEDII Irregular plain 100 x 200 mi. containing center of lunar disk for zero libration. SW half dark; NE half covered with light ejecta from Triesnecker. (V Theophilus NW 5.5) 7 Day.

MEE C. Polygonal 85 mi. 7500 ft. Pk. W of center. Breaks in wall. Extremely old. (VII Gassendi SE 4.6) 10 Day.

MENELAUS C. 16 x 19 mi. 8700 ft. MuPk. 1. (II Posidonius SW 3.5) 6 Day.

MERCATOR C. Irregular 29 mi. 4300 ft. 5. (III Tycho NW 3.3) 10 Day.

MERCURY C. 40 mi. 9800 ft. Pk. 2000 ft., W of center. 2. (I Posidonius NE 3.9) 16 Day.

MERSENIUS C. 49 x 58 mi. 7500 ft. Wall broken on N and S. 3. (VII Gassendi SW 1.3 and Fig. 45) 11 Day.

MERSENIUS C C. 9 mi. 4900 ft. Young. (VII Mersenius NE 0.4 and Fig. 20) 15 Day.

MESSALA C. 72 mi. 8900 ft. N wall linear for 30 mi. Very old. (I Posidonius NE 3.4) 2 Day.

MESSALA B C. 10 mi. 3600 ft. Young. (I on S wall of Messala) 15 Day.

MESSIER C. 6 x 10 mi. 4100 ft. Young. (IV Langrenus NW 2.1) 15 Day.

METIUS C. 50 x 55 mi. 18,600 ft. MuPk. 1500 ft. 2. (V Theophilus S 5.1) 4 Day.

METON C. 66 x 79 mi. 8200 ft. Flooded. Largest of four C. forming clover-leaf complex 110 x 135 mi. Very old. (V Plato NE 2.9) 20 Day.

MILICHIUS C. 7 mi. 5200 ft. 1. (VII Aristarchus SE 3.4 and Fig. 4.)

MILLER C. 36 x 40 mi. 7500 ft. Pk. 1. (III Tycho E 1.8 and Fig. 6) 22 Day.

MILLER A C. 21 x 25 mi. 4900 ft. Pk. 1. (III Miller NE 0.4) 22 Day.

MITCHELL C. 21 mi. 3300 ft. Pk. Old. (V Plato E 3.5 and Fig. 8.)

MOLTKE C. 4 mi. 2400 ft. Young. (II Theophilus N 2.0 unlabeled and Fig. 10.)

MONGE C. 23 mi. 8500 ft. MuPk. 2000 ft. S wall broken. Old. (I Langrenus SW 2.8) 18 Day.

MONTANARI C. Polygonal 46 x 57 mi. 5200 ft. Extremely old. (III Tycho W 1.2 and Fig. 6) 9 Day.

MORETUS C. 75 mi. 13,800 ft. Pk. 6600 ft. 1. (III Tycho S 2.9) 8 Day.

LACUS MORTIS Polygonal plain 100 x 110 mi. Probably a true mountain-walled plain. Flooded. E wall largely melted away. (II Posidonius N 2.3) 5 Day.

MOSTING C. 16 mi. 9800 ft. Pk. 5500 ft. 1. (VI Copernicus SE 3.2) 22 Day.

MOSTING A C. 8 mi. 7800 ft. Young. (VI Mosting S 0.5) 15 Day.

MOUCHEZ C. 60 mi. 6600 ft. S wall missing. Very old. (VI Plato N 2.7.)

MURCHISON C. Irregular 35 x 40 mi. 3000 ft. Wall broken on N and SE. Very old. (VI Copernicus E 3.7) 8 Day.

MUTUS C. 50 mi. 9500 ft. 2. (V Tycho SE 4.7 and Fig. 49) 5 Day.

NANSEN C. 70 mi. Appears to have MuPk. off center to E. (I Posidonius N 6.5.)

NASIREDDIN C. 34 mi. 8900 ft. 1. (VI Tycho E 1.5 and Fig. 6) 8 Day.

NASMYTH C. Polygonal 34 x 54 mi. 4300 ft. Very old. (IV Tycho W 4.2 and Fig. 46) 12 Day.

NEANDER C. 27 x 32 mi. 9900 ft. Pk. 2000 ft. 2. (I Langrenus SW 5.5) 19 Day.

NEARCH C. 47 mi. 6200 ft. 1. (V Theophilus S 7.3 and Fig. 49) 4 Day.

NEARCH A C. 26 mi. 2000 ft. Young. (V Breaks S wall of Nearch and Fig. 49) 4 Day.

MARE NECTARIS Elliptical plain 220 x 265 mi. Long axis lies NW–SE. (II Theophilus SE 1.3 and Fig. 5) 4 Day.

NEISON C. 31 x 35 mi. 7900 ft. Old. (II Crisium NW 10.1) 20 Day.

NEWCOMB C. 23 x 26 mi. 9100 ft. MuPk. 4600 ft. 1. (V Posidonius E 2.0) 18 Day.

NEWTON C. Triangular 66 x 85 mi. 21,300 ft. Multiple C. consisting of at least five parts. 1. (IV Tycho SE 3.6.)

NICOLAI C. 24 x 28 mi. 7000 ft. Young. (V Tycho E 5.2 and Fig. 49) 20 Day.

NICOLLET C. 10 mi. 4600 ft. 1. (III Tycho N 3.5) 15 Day.

NOGGERATH C. 22 x 24 mi. 3600 ft. Very old. (VII Gassendi S. 4.8 unlabeled and Fig. 46.)

NONIUS C. Pentagonal 39 x 47 mi. 5200 ft. Extremely old C. remnant with composite inner W wall 22 mi. wide. (II Theophilus SW 5.7) 21 Day.

MARE NUBIUM Irregular plain 350 x 400 mi. Exceptionally dark. (VI Gassendi E 3.9) 9 Day.

OENOPIDES C. 46 mi. 7200 ft. Pk. off center to E. Wall breaks on SE. Very old. (VII Aristarchus N 5.3) 27 Day.

OKEN C. Irregular 45 x 48 mi. 10,500 ft. Flooded. Old. (IV Tycho E 9.5) 15 Day.

OLBERS C. 45 mi. 9800 ft. Pk. off center to S. 1. (IV Grimaldi N 2.3) 14 Day.

OLBERS A C. 26 mi. 8200 ft. Young. (IV unlabeled on NW wall of Olbers) 14 Day.

OLIVIUM Pr. Broad blunt Mt. mass. 6900 ft. (I Crisium in SW wall) 4 Day.

MARE ORIENTALE Well observed for first time July 20, 1965, by Russian lunar probe Zond 3, it appears to be a nearly circular dark plain about 330 mi. across with a broad peninsula from NE extending to center and bays on E and SW. (VI Gassendi W 2.7) 27 Day.

ORONTIUS C. Deformed 65 x 80 mi. 9200 ft. Walls heavily damaged. 3. (VI Tycho NE 0.9 and Fig. 6) 8 Day.

PALITZSCH C. Irregular funnel-shaped 33 x 93 mi. 11,200 ft. Valley formed by at least 3 connected C. (I Langrenus S 3.2) 16 Day.

PALLAS C. Irregular 30 mi. 5800 ft. Pk. 2600 ft. Many wall breaks on N and E. Very old. (VI Copernicus E 3.4) 8 Day.

PALMIERI C. 25 x 27 mi. 3900 ft. Wall breaks on N and S. Flooded. Old. (VII Gassendi S 1.9.)

PARROT C. 40 x 42 mi. 3600 ft. Heavily battered. Floor largely covered by ridges and wall fragments. Extremely old. (VI Copernicus SE 6.4.)

PARRY C. 28 x 30 mi. 3800 ft. 5. (III Tycho N 6.1 and Fig. 22) 9 Day.

PASCAL C. 62 x 70 mi. 10,800 ft. Old. (VII Aristarchus NE 7.6) 25 Day.

PEARY C. Egg-shaped 46 x 57 mi. 8900 ft. Old. (VI Plato NE 3.4.)

PEIRCE C. 12 mi. 5600 ft. Pk. 1. (I Crisium, on NW floor) 4 Day.

PENTLAND C. 36 mi. 10,200 ft. Pk. 2600 ft., NW of center. 1. (V Tycho SE 3.5) 21 Day.

PENTLAND A C. 25 mi. 10,800 ft. Young. (V Pentland S 0.2) 21 Day.

PETAVIUS C. 99 x 110 mi. 13,800 ft. MuPk. 8200 ft. 5. (I Langrenus S 3.0) 2 Day.

PETAVIUS B C. Irregular 21 x 23 mi. 11,800 ft. Young. (I Petavius N 0.9) 4 Day.

PETERMANN C. 48 mi. 12,100 ft. Old. (I Posidonius N 5.7) 16 Day.

PHILLIPS C. 80 mi. 10,500 ft. Very old. (IV Langrenus S 3.2) 1 Day.

PHILOLAUS C. 48 x 51 mi. 10,800 ft. MuPk. 1. (VI Plato N 2.3) 10 Day.

PHOCYLIDES C. Turnip-shaped 76 x 89 mi. 6900 ft. 3. (VII Gassendi S 5.1 and Fig. 46) 12 Day.

PIAZZI SMITH C. 9 mi. 5500 ft. 1. (VI Plato S 1.5 unlabeled and Fig. 9.)

PICARD C. 16 mi. 7600 ft. Pk. 1. (I Crisium, on SW floor) 4 Day.

PICCOLOMINI C. 54 mi. 11,800 ft. MuPk. 5600 ft. 1. (II Theophilus S 3.1 and Fig. 48) 5 Day.

EDWARD PICKERING C. 10 mi. 7800 ft. Young. (IV Langrenus W 8.6) 15 Day.

WILLIAM PICKERING C. 7 x 11 mi. 5700 ft. 1. (IV Langrenus NW 2.1) 15 Day.

PICO Mt. 10 x 18 mi. Height measures range from 7000 to 9600 ft., averaging about 8000 ft. except for the recent 4600 ft. (AMS). May be part of original Imbrium crater wall. (VI Plato S 0.8 and Fig. 9) 8 Day.

PICTET C. 35 x 43 mi. 7500 ft. MuPk. 3. (VI Tycho E 0.5 and Fig. 6).

PINGRE C. 140 x 175 mi. 4600 ft. Old. (IV Tycho SW 4.2) 28 Day.

PITATUS C. 60 x 69 mi. 2900 ft. Pk. 1600 ft. Flooded. Very old. (VI Tycho NW 2.1) 8 Day.

PITISCUS C. 50 mi. 10,200 ft. Pk. 5900 ft. 2. (II Theophilus S 5.8 and Fig. 49) 5 Day.

PITON Mt. 11 x 17 mi. 8200 ft. (VI Plato SE 1.7 and Fig. 9) 7 Day.

PLANA C. Irregular 26 x 31 mi. 6700 ft. Pk. 3400 ft. Walls broken. Flooded. Old. (V Posidonius NW 1.7) 20 Day.

PLATO C. 64 x 67 mi. 8000 ft. Floor very dark. 5. (Reference Feature: IV Dot N 9.1 and Fig. 9) 8 Day.

PLAYFAIR C. 27 x 30 mi. 7800 ft. 2. (II Theophilus SW 3.9) 21 Day.

PLINY C. 27 mi. 10,500 ft. MuPk. 3300 ft. 1. (V Posidonius S 2.9) 5 Day.

PLUTARCH C. Irregular 41 mi. 9200 ft. Pk. Old. (I Crisium NE 2.0) 1 Day.

POLYBIUS C. Irregular 25 mi. 7700 ft. Old. (II Theophilus S 1.9 and Fig. 48) 20 Day.

PONCELET C. 43 x 48 mi. 4300 ft. Broken walls. Very old. (VII Aristarchus NE 8.1.)

PONS C. Deformed 25 x 27 mi. 7500 ft. Heavily battered; SE wall destroyed. Old. (II Theophilus SW 2.7 and Fig. 48.)

PONTANUS C. Approximately rectangular 35 x 38 mi. 6900 ft. MuPk. 2000 ft. N wall broken. Very old. (II Theophilus SW 3.7) 21 Day.

POSIDONIUS C. 52 x 61 mi. 8500 ft. MuPk. 3000 ft. 5. (Reference Feature: IV Dot NE 7.6 and Fig. 11) 5 Day.

PRINZ C. 30 mi. 3600 ft. Pk. remnant. SW wall gone. Flooded. Old. (VII Aristarchus NE 0.6 and Fig. 44.)

OCEANUS PROCELLARUM Largest of the great lunar plains extending from Oenopides on N some 1200 mi. to Letronne on S and from Eratosthenes W almost to limb—an equal but foreshortened distance. Irregular outline. (VI Copernicus all directions) 11 Day.

PROCLUS C. 18 mi. 11,900 ft. Pk. 1600 ft. 1. (II Posidonius SE 4.1) 4 Day.

PROCTOR C. 32 x 35 mi. 4300 ft. Curious chain of rim C., 2 to 6 mi. in diameter, covers all but E third of rim. Very old. (VI Tycho SE 1.0 and Fig. 6) 8 Day.

PROTAGORAS C. 12 x 14 mi. 6900 ft. Pk. Young. (V Plato E 2.0 and Fig. 8) 21 Day.

PTOLEMY C. 93 mi. 9800 ft. 5. (VI Copernicus SE 4.9) 8 Day.

PTOLEMY A C. 6 mi. 3900 ft. Young. (VI on NE floor of Ptolemy) 8 Day.

PURBACH C. 62 x 73 mi. 9800 ft. MuPk. 1600 ft. 5. (VI Tycho N 2.7) 8 Day.

PALUS PUTREDINIS Irregular plain 45 x 75 mi. Wall height varies from zero in numerous places to 12,000 ft. near E corner. (VI Copernicus NE 5.0 and Fig. 42) 7 Day.

PYRENEES MR. 45 x 165 mi. (not including Gutenberg). Heights of 12,000 ft. on N and 6500 ft. on S reported, but LAC gives max. of 6100 ft. near N end, and AMS gives max. of 10,800 at S end. (I Langrenus W 2.9) 4 Day.

PYTHAGORAS C. 81 x 90 mi. 16,400 ft. MuPk. 6000 ft. 1. (VI Plato NW 3.2) 13 Day.

PYTHEAS C. 12 x 15 mi. 7100 ft. Spine across floor. 1. (III Copernicus N 2.0 and Fig. 4) 9 Day.

RABBI LEVI C. Distorted pentagonal 43 x 50 mi. 12,000 ft. Walls and floor broken by numerous small C. Extremely old. (II Theophilus S 3.9 labeled "Levi") 6 Day.

RAMSDEN C. 13 x 16 mi. 5900 ft. 1. (VII Gassendi SE 3.2) 15 Day.

REAUMUR C. 30 x 33 mi. 4400 ft. Walls broken in several places and serrated throughout by grooves directed toward Mare Imbrium. Very old. (VI Copernicus SE 4.4.)

REGIOMONTANUS C. Misshapen 53 x 76 mi. 9700 ft. MuPk. 2600 ft. on spine from N wall. Extremely old. (VI Tycho NE 2.4) 8 Day.

REICHENBACH C. Hexagonal 45 mi. 11,200 ft. Heavily battered. 1. (V Theophilus SE 4.0) 18 Day.

REINER C. 18 mi. 8500 ft. MuPk. 3300 ft. 1. More conspicuous is the curious diamond-shaped white spot just W of Reiner. Spot is 27 x 50 mi., without relief, strangely patterned, and apparently associated with no crater unless it be Kepler 360 mi. E. (VII Grimaldi NE 2.6) 12 Day.

REINHOLD C. 28 mi. 9000 ft. MuPk. 1. (III Copernicus SW 1.3 and Fig. 4) 9 Day.

REPSOLD C. 64 x 68 mi. 8900 ft. Pk. Old. (VII Aristarchus N 4.4) 28 Day.

RHAETICUS C. 25 x 32 mi. 5200 ft. Central ridge. Breaks in wall. Flooded; floor 3000 ft. above Sinus Medii. Very old. (II Theophilus NW 4.4) 22 Day.

RHEITA C. 44 mi. 14,500 ft. Pk. 3700 ft., off center to N. 1. (V Tycho E 8.1) 18 Day.

RHEITA VALLEY One of several valleys radial to Mare Nectaris. Main section 15 x 230 mi. and 8400 ft. deep. An extension 9 x 100 mi. continues S at angle of 15 degrees to original direction. (V Tycho E 7.8) 4 Day.

RICCIOLI C. 97 mi. 7500 ft. Breaks in wall. S half of floor very rough with peak near center. Dark section of N floor definitely flooded. 3(?). (IV Grimaldi NW 0.6) 14 Day.

RICCIUS C. Irregular 45 mi. 5900 ft. MuPk. Heavily battered. Extremely old. (II Theophilus S 4.2) 20 Day.

RIPHAEUS MR. 30 x 110 mi. rectangle includes most of the Mts. 4100 ft. The largest remaining section of Mare Cognitum C. wall. (III Copernicus S 3.4 and Fig. 22) 10 Day.

RITTER C. 18 mi. 4300 ft. Small linear hill W of center. 5. (II Theophilus N 2.7 and Fig. 10) 20 Day.

ROCCA C. 48 x 52 mi. 7200 ft. Large breaks in wall. Very old. (IV Grimaldi S 1.5) 13 Day.

ROMER C. 25 mi. 11,100 ft. Pk. 6200 ft. 1. (II Posidonius SE 1.6) 4 Day.

SINUS RORIS Poorly-defined plain of indefinite extent. The E portion, which appears to be part of Mare Frigoris, is roughly elliptical 130 x 175 mi. The W section is a rectilinear part of Oceanus Procellarum. (V Plato W 3.7) 11 Day.

ROSENBERGER C. 55 x 58 mi. 7200 ft. Pk. 1300 ft. 3. (V Theophilus S 6.8) 18 Day.

ROSS C. 17 mi. 5900 ft. Pk. 1. (II Posidonius S 3.8 and Fig. 10) 20 Day.

ROSSE C. 8 mi. 5000 ft. Young. (V Theophilus SE 1.7.)

ROST C. 31 mi. 8200 ft. 2. (VI Tycho SW 2.0) 25 Day.

ROTHMANN C. 26 x 28 mi. 9900 ft. 1. (II Theophilus S 3.2 and Fig. 48.) 20 Day.

RUMKER D. Exceptionally large dome composed of numerous small ones. Visible only when near terminator. Irregular 38 x 48 mi. Heights of 200 to 2500 ft. reported. Less than 1600 ft. (AMS). (IV Aristarchus N 2.6) 12 Day.

RUSSELL C. 62 mi. 3300 ft. S third of wall missing where floor joins that of Otto Struve. Flooded. Very old. (V Grimaldi N 5.9) 14 Day.

RUTHERFURD C. 30 x 33 mi. 7200 ft. Pk. off center to N. Old. (III Tycho S 2.1) 9 Day.

SABINE C. 18 mi. 4600 ft. Hill. 5. (II Theophilus N 2.5 and Figs. 10 and 25) 20 Day.

SACROBOSCO C. Irregular. 64 mi. 9200 ft. 2300-ft. ridge across floor. Heavily battered. 3. (II Theophilus SW 2.9) 6 Day.

SANTBECH C. 40 mi. 13,100 ft. Pk. 2300 ft., N of center. 1. (V Theophilus SE 2.9) 4 Day.

SASSERIDES C. Irregular 58 x 76 mi. 5900 ft. Battered

remnant. Extremely old. (VI Tycho N 0.6 unlabeled and Fig. 6.)

SAUSSURE C. 35 x 37 mi. 6900 ft. MuPk. 2. (VI Tycho E 1.0 and Fig. 6) 8 Day.

SCHEINER C. 71 mi. 15,100 ft. 2. (III Tycho SW 2.3) 10 Day.

SCHIAPARELLI C. 15 x 16 mi. 6900 ft. Hill SE of center. 1. (VII Aristarchus W 1.3 and Fig. 44.)

SCHICKARD C. 135 x 150 mi. 9500 ft. The two stained areas are 700 ft. lower than central part of floor. 5. (IV Tycho W 4.6 and Fig. 46) 12 Day.

SCHILLER C. 48 x 113 mi. 12,500 ft. Consists of one circular and two elliptical C. in line. SE floor 1600 ft. higher than middle section. 3. (VII Gassendi S 5.5 and Fig. 46) 11 Day.

SCHMIDT C. 8 mi. 5200 ft. Young. (II Theophilus NW 2.6 unlabeled and Fig. 10.)

SCHOMBERGER C. 53 x 56 mi. 11,200 ft. 1. (V Tycho SE 4.2) 20 Day.

SCHROTER C. 21 x 28 mi. 3400 ft. S wall gone; N wall broken and low. 5. (VI Copernicus SE 2.7 unlabeled and Fig. 4.)

SCHROTER'S VALLEY 125 mi. long, 2–5 mi. wide, 500–4500 ft. deep. Actually begins not at Cobra Head but at inner wall of Herodotus, winds N, NW, W, SW, and opens on a plain 65 mi. NW of Herodotus. (VI Aristarchus NW 0.1 to 0.5 and Fig. 44) 23 Day.

SCORESBY C. 35 mi. 11,800 ft. MuPk. S of center. 1. (V Plato NE 2.8) 21 Day.

SCOTT C. 70 mi. 9800 ft. Old. (III Tycho SE 4.1.)

SECCHI C. 14 x 15 mi. 5200 ft. Wide gaps in wall on N and S. Old. (I Langrenus NW 2.8)

SEGNER C. 45 mi. 6600 ft. Ridge across floor. 3. (VII Gassendi S 6.2) 11 Day.

SELEUCUS C. 28 mi. 7500 ft. Central hill. 1. (V Aristarchus W 1.9) 13 Day.

SERAO Mt. on Apennine crest. 5 x 16 mi. 7900 ft. (AMS) or 4900 ft. (LAC). (III Plato S 5.2 unlabeled and Fig. 7.)

MARE SERENITATIS Roughly elliptical plain 360 x 420 mi. Goodacre determined perimeter 1850 mi. and area 125,000 square mi. Possibly the oldest of the great maria. (V Posidonius SW 1.8) 5 Day.

SERPENTINE RIDGE Apparent length 250 mi. Actual length along convolutions and including weak end portions 350 mi. Widths to 7 mi. Max. height 1000 ft. over 3 mi. stretch near center. Max. slope across ridge 4 degrees. (V Posidonius SW 0.8 and Fig. 11) 5 Day.

SHARP C. Irregular 24 mi. 10,500 ft. MuPk. off center to E. 1. (VI Plato W 3.8) 11 Day.

SHEEPSHANKS C. Irregular 16 x 21 mi. 7200 ft. Young. (V Plato E 2.9.)

SHORT C. Egg-shaped 39 x 46 mi. 13,800 ft. Walls broken on N and S. 1. (III Tycho S 3.2.)

SHUCKBURGH C. Heart-shaped 24 x 33 mi. 4300 ft. Walls broken on N and S. Very old. (I Posidonius NE 2.8) 18 Day.

SILBERSCHLAG C. 9 mi. 7200 ft. Young. (II Theophilus NW 4.1 unlabeled and Fig. 10.)

SIMPELIUS C. Hexagonal 43 x 54 mi. 10,800 ft. Pk. 1000 ft., off center to N. 1. (III Tycho SE 3.5) 21 Day.

SIMPELIUS E C. Roughly square 30 x 33 mi. 12,800 ft. Joins Simpelius through wide valley made by C. between them. Young. (III Simpelius N 0.2) 21 Day.

SIRSALIS C. 23 x 28 mi. 8900 ft. Pk. 1. (IV Grimaldi SE 1.4 and Fig. 45) 12 Day.

SIRSALIS A C. 22 x 29 mi. 6900 ft. Old. (IV joins Sirsalis on SW and Fig. 45) 12 Day.

SIRSALIS F C. 9 mi. 5300 ft. Young. (IV just S of Sirsalis and Fig. 45) 15 Day.

SIRSALIS RILL Begins 70 mi. NE of Sirsalis and runs SW almost straight 210 mi. to De Vico A, through which it passes, curves W, and continues SW another 75 mi. Width 1–3 mi. (Fig. 45.)

MARE SMYTHII Roughly elliptical plain 165 x 225 (N–S) mi. (II Langrenus NE 2.5) 11 Day.

SNELLIUS C. Irregular 49 x 56 mi. 11,700 ft. Pk. 1. (I Langrenus SW 3.9) 3 Day.

SOMMERING C. 17 x 19 mi. 3300 ft. N wall broken; S wall gone. 5. (VI Copernicus SE 2.9 unlabeled and Fig. 4.)

PALUS SOMNI Roughly octagonal region of mountains and small plains 110 x 145 mi. Two-thirds of area very rough. Elevations outside craters up to 3000 ft. (plains) and 9800 ft. (Mts.) above Mare Tranquillitatis. (III Crisium W 1.5) 19 Day.

LACUS SOMNIORUM Irregular polygonal plain about 90 x 180 mi. interrupted by numerous hills, Mts., small C. Bordering highlands rise to 8200 ft. on E. (III Posidonius NW 1.1) 15 Day.

SOSIGENES C. 11 mi. 6600 ft. Flooded. Old. (II Theophilus N 4.1 unlabeled and Fig. 10.)

SOUTH C. Rectangular 63 x 67 mi. 5200 ft. Mu.hills. S half of wall reduced to few ridges and hills, one rising 2300 ft. Extremely old. (VII Aristarchus NE 5.6) 26 Day.

SPALLANZANI C. Approximately square 20 mi. on side; 5200 ft. Very old. (II Theophilus S 5.5 unlabeled and Fig. 49.)

SPITZBERGEN MR. 18 x 50 mi. 4600 ft. May be part of original Imbrium crater wall. (VI Plato S 2.6 and Fig. 42) 8 Day.

SPORER C. 15 x 17 mi. 2300 ft. Flooded. Very old. (VI Copernicus SE 4.3) 22 Day.

MARE SPUMANS Irregular plain 90 x 105 mi. resembling a reservoir in hilly country. Relatively C. free. Walls rise to 5600 ft. (I Langrenus N 2.0) 11 Day.

STADIUS C. 42 mi. 1200 ft. Most of wall reduced to low ridges or C. chains. A wall rising to 2500 ft. follows C. outline on NE for 22 mi., but it is part of Mt. mass extending S from W wall of Eratosthenes and may be an extrusion feature much younger than the C. 5. (VI Copernicus E 1.1 unlabeled and Fig. 4.)

STEINHEIL C. Hexagonal 35 x 41 mi. 6900 ft. Flooded. Old. (V Tycho E 6.9) 4 Day.

STEVINUS C. Irregular 48 mi. 15,400 ft. Pk. 3000 ft. 1. (II Theophilus SE 4.4) 3 Day.

STEVINUS R C. Square 16 mi. 7500 ft. Not the ray center. Old. (II Stevinus NW 0.3) 15 Day.

STIBORIUS C. Deformed 27 x 31 mi. 11,200 ft. Pk. 2300 ft. E wall nearly linear. 1. (II Theophilus S 3.7) 6 Day.

STOFLER C. 68 x 85 mi. 7500 ft. SE wall removed. N–S ridge rising 5600 ft. crosses floor just E of original center. 2. (II Theophilus SW 6.1 and Fig. 51) 7 Day.

STRABO C. 34 x 40 mi. 12,500 ft. Old. (I Posidonius N 4.5.)

STRAIGHT RANGE MR. Rectangular 15 x 55 mi. 6600 ft. Extremely artificial in appearance. May be part of original Imbrium crater wall. (III Plato W 1.4 and Fig. 9) 9 Day.

STRAIGHT WALL Cl. 75 mi. 1200 ft. Slopes upward to E at 41-degree angle; one of steepest known lunar slopes. Best seen on 23-Day Moon when wall reflects sunlight most efficiently. (VI Tycho N 3.2) 8 Day.

STREET C. 37 x 40 mi. 5600 ft. Very old. Outline scalloped by intruders along wall. (III Tycho S 0.5 unlabeled and Fig. 6.)

MARE STRUVE Small, irregular, indented dark plain 37 x 56 mi. Wall rises to 6900 ft. on E. Flooded. (I Posidonius NE 3.6.)

OTTO STRUVE C. Tomato-shaped 97 x 115 mi. 5600 ft. N and S walls missing. Flooded. Very old. (V Aristarchus W 2.3) 14 Day.

SYLVESTER C. 38 x 42 mi. 8900 ft. Old. (VI Plato N 2.9).

TACITUS C. Irregular 28 x 30 mi. 9300 ft. 1200-ft. ridge across floor. 1. (II Theophilus SW 1.6 and Fig. 5) 20 Day.

TANNERUS C. 18 x 20 mi. 5600 ft. Young. (II Theophilus S 6.6 unlabeled and Fig. 49.)

TARUNTIUS C. 36 mi. 3000 ft. MuPk. 2300 ft. 5. (I Langrenus NW 3.0) 4 Day.

TAURUS MTS. Irregular upland area some 200 x 300 mi. 9800 ft. Densely populated with C. from 30 mi. diameter on down. Numerous linear valleys and irregular depressions. The name does not appear on LAC 43 of the region. (I Posidonius SE 1.4) 4 Day.

TAYLOR C. Irregular 21 x 31 mi. 8100 ft. Pk. 1600 ft., off center to W. Wall broken on N and S. 1. (II Theophilus NW 2.0 and Fig. 50) 21 Day.

TAYLOR A C. Irregular 20 x 25 mi. 10,000 ft. Wall broken on SW. Old. (II Taylor NW 0.3 and Fig. 50) 21 Day.

TENERIFFES MR. Group of four multiple peak masses scattered over triangular section of Imbrium plain 35 x 65 mi. Heights of 8000 ft. reported, but LAC shows 6200 ft. max. May be part of original Imbrium crater wall. (VI Plato SW 0.8 and Fig. 9) 9 Day.

THALES C. 22 mi. 5900 ft. Young. (II Posidonius N 4.4) 15 Day.

THEAETETUS C. Irregular 16 mi. 7700 ft. 1. (V Plato SE 2.8 and Fig. 42) 15 Day.

THEBIT C. 32 x 36 mi. 10,400 ft. Pk. 3600 ft. on spine from N wall, off center to W. 2. (VI Tycho N 3.2) 8 Day.

THEBIT A C. 13 mi. 9700 ft. Young. (VI in NW wall of Thebit) 8 Day.

THEBIT L C. 6 mi. 2900 ft. Pk. Young. (VI in NW wall of Thebit A) 8 Day.

THEON JUNIOR C. Irregular 12 mi. 8900 ft. Young. (II Theophilus NW 2.5 and Figs. 10 and 24) 15 Day.

THEON SENIOR C. 13 mi. 10,500 ft. Young. (II Theophilus NW 2.7 and Figs. 10 and 24) 15 Day.

THEOPHILUS C. 65 mi. 22,300 ft. MuPk. 7500 and 5900 ft. 1. (Reference Feature: IV Dot E 4.6 and Fig. 5) 5 Day.

THEOPHILUS B C. 5 mi. 3900 ft. Young. (II or IV in NW wall of Theophilus and Fig. 5) 15 Day.

TIMAEUS C. 20 mi. 7400 ft. Pk. 2. (III Plato NE 1.5 and Fig. 9) 21 Day.

TIMOCHARIS C. 22 mi. 9400 ft. Pk. 2600 ft. 1. (VI Copernicus NE 3.5 and Fig. 7) 8 Day.

TISSERAND C. 22 mi. 9500 ft. 2. (I Crisium NW 1.5) 18 Day.

TORRICELLI C. Top-shaped 12 x 19 mi. 6900 ft. Multiple C. Flooded. Old. (V Theophilus N 1.4) 20 Day.

TRALLES C. Irregular 27 x 28 mi. 11,000 ft. Ridge across floor rises 2600 ft. 2. (I Crisium NW 2.2) 18 Day.

MARE TRANQUILLITATIS Roughly elliptical plain 400 x 550 mi. with extension to S 120 x 200 mi. Boundaries are upland slopes rather than Mt. walls. (II Theophilus N 4.1) 4 Day.

TRIESNECKER C. Irregular 16 mi. 9500 ft. Pk. 2. (VI Copernicus E 4.4) 15 Day.

TROUVELOT C. 7 x 8 mi. 5000 ft. Young. (III Plato E 1.8 unlabeled and Fig. 8.)

TURNER C. 8 mi. 6000 ft. Young. (VI Copernicus SE 2.4 unlabeled and Fig. 4.)

TYCHO C. 56 mi. 13,800 ft. Pk. 5200 ft. Around Full, most spectacular C., its conspicuous and numerous rays extending many hundred mi. Possibly no more than 50,000,000 years old and youngest of the prominent C. 1. (Reference Feature: IV Dot S 7.1 and Fig. 6) 8 Day.

UKERT C. Roughly triangular 14 x 16 mi. 9500 ft. Pk. 2600 ft. Young. (III Copernicus E 4.0) 15 Day.

ULUGH BEIGH C. Irregular 36 mi. 5600 ft. Pk. Wall broken on N and W. Flooded. Old. (VII Aristarchus NW 2.7.)

MARE UNDARUM Irregular plain with separated parts spread over approximately rectangular area 110 x 160 mi. Resembles cluster of flooded valleys. Surrounding uplands rise to 6900 ft. (II Crisium SE 2.3) 11 Day.

MARE VAPORUM Roughly elliptical plain 115 x 160 mi. joined on SE by an equally dark region half as large containing black horseshoe 50 x 75 mi. Latter area extremely rough with several peaks rising between 2000 and 2600 ft. including Schneckenberg Mt. Probably extruded features of black magma. (III Copernicus E 4.5 and Fig. 7) 7 Day.

VASCO DA GÁMA C. 50 x 57 mi. 6900 ft. Old. (VII Grimaldi N 3.9) 28 Day.

VENDELINUS C. Irregular 92 x 100 mi. 14,700 ft. Walls broken. 5. (I Langrenus S 1.3 and Fig. 47) 2 Day.

VIETA C. Irregular 51 x 54 mi. 17,400 ft. Pk. 1900 ft., off center to N. 2. (VII Gassendi SW 2.4) 12 Day.

VITELLO C. 25 x 29 mi. 4600 ft. Pk. 2300 ft. 5. (VII Gassendi S 2.4) 11 Day.

VITRUVIUS C. Irregular 17 mi. 7600 ft. 2600-ft. N–S ridge across floor. 5. (V Posidonius S 2.5) 15 Day.

VITRUVIUS A C. Irregular 11 mi. 8100 ft. 2000-ft. N–S ridge across floor. Young. (V Vitruvius E 0.4) 15 Day.

VLACQ C. 59 mi. 9800 ft. Pk. 3900 ft. 1. (V Theophilus S 6.6 and Fig. 49) 4 Day.

VOGEL C. Key-shaped multiple C. 18 x 39 mi. 7900 ft. Pk. on spine from N wall, off center to E. N crater, irregular 13 mi. 4600 ft., now designated Vogel B. Old. (II Theophilus W 3.8.)

WALLACE C. Irregular 15 x 17 mi. 1300 ft. SE wall missing. Flooded. Very old. (VI Copernicus NE 2.9 unlabeled and Fig. 7.)

WALTER C. 78 x 88 mi. 12,000 ft. Pk. 4000 ft., off center

to NE on complex spine from N wall. 4. (VI Tycho NE 2.0) 7 Day.

WARGENTIN C. 59 mi. Filled to within 500 ft. of rim presumably with lava. Circular plateau 1000 ft. above surroundings. 5. (IV Tycho W 4.4 and Fig. 46) 12 Day.

WATT C. Deformed 32 x 41 mi. 6500 ft. MuPk. 1 (V Tycho E 7.0) 4 Day.

WERNER C. 43 mi. 15,000 ft. MuPk. 2400 ft. 1. (III Tycho NE 3.3) 7 Day.

WHEWELL C. 9 mi. 7200 ft. Young. (V Theophilus NW 3.7 unlabeled and Fig. 10.)

WICHMANN C. 7 mi. 3000 ft. Young. Marks SE extremity of ghost C., 38 mi. 2900 ft., over half melted away. Latter conspicuous but nameless. (VII Gassendi N 1.8) 25 Day.

WILHELM C. 57 x 68 mi. 9500 ft. 3. (III Tycho W 1.3 and Fig. 6) 9 Day.

WILHELM A C. 12 mi. 3900 ft. Young. (III stronger of two craters on S wall of Wilhelm and Fig. 6) 23 Day.

WILHELM B C. 10 x 12 mi. 3600 ft. Old. (III on W wall of Wilhelm and Fig. 6) 23 Day.

WILKINS C. Irregular 34 x 37 mi. 3900 ft. Severely battered. Wall broken in several places. Very old. (II Theophilus SW 3.4 unlabeled and Fig. 48.)

WILSON C. 41 x 47 mi. 12,500 ft. Very old. (IV Tycho S 3.0) 11 Day.

WOLF C. Pear-shaped 15 mi. 2300 ft. Flat floor enclosed by clumps of hills. Flooded. Very old. (VI Copernicus S 6.1 and Fig. 22) 15 Day.

WOLFF Mt. on Apennine crest; triangular 22 mi. on side. Elevations 11,000 to 12,000 ft. reported; modern measures range from 11,200 ft. (AMS) down to 5600 ft. (LAC). (III Plato S 5.2 unlabeled and Fig. 7.)

WOLLASTON C. 6 x 7 mi. 3200 ft. Young. (VII Aristarchus N 1.3 and Fig. 44.)

WROTTESLEY C. 37 x 43 mi. 11,300 ft. MuPk. 1600 ft., off center to NW. 1. (I Langrenus S 2.9) 16 Day.

WURZELBAUER C. 52 x 60 mi. 7200 ft. 5. (VI Tycho NW 1.7) 9 Day.

XENOPHANES C. 67 x 70 mi. 10,500 ft. MuPk. 10 x 20 mi. 3000 ft. Old. (VI Plato NW 3.9) 28 Day.

YOUNG C. Irregular 40 x 50 mi. 14,100 ft. Rheita Valley cuts through it. Extremely old. (V Tycho E 8.0.)

ZACH C. 44 mi. 11,800 ft. Pk. 2000 ft., near N wall. 2. (V Tycho SE 2.9) 21 Day.

ZAGUT C. Irregular 50 x 55 mi. 9800 ft. Heavily battered. Very old. (II Theophilus S 3.6 and Fig. 48) 6 Day.

ZAGUT E C. 23 mi. 7300 ft. Old. (II on NE floor of Zagut and Fig. 48) 6 Day.

ZENO C. 35 x 43 mi. 8500 ft. Low, broad Pk. Heavily battered and deformed. Extremely old. (I Posidonius NE 4.1.)

ZOLLNER C. Peach stone-shaped 23 x 32 mi. 8900 ft. MuPk. 2300 ft., off center to N. Wall broken on N and S. Very old. (II Theophilus NW 1.4 and Fig. 50.)

ZUCCHIUS C. Irregular 45 mi. 10,500 ft. MuPk. 1600 ft. 1. (IV Tycho SW 3.2) 11 Day.

ZUPUS C. Irregular depression 21 x 28 mi. 4400 ft. Walls low except for Mt. mass on E; broken in many places. Floor very dark. Flooded. Old. (V Gassendi W 1.5 and Fig. 45.)

REFERENCE BOOKS

It is the author's hope that some who read this book may choose to continue their investigation of the New World of Tomorrow. For their guidance the list that follows has been prepared. It is by no means complete, but it is representative. Contemporary atlases have been included since most of them are instructive when used at the desk as well as at the telescope. They differ considerably in the extent of their textual content. Additional references will be found in most of the contributions here listed.

Alter, Dinsmore, *Pictorial Guide to the Moon*. Thomas Y. Crowell Co., New York, 1963.

Baldwin, Ralph B., *The Face of the Moon*, University of Chicago Press, Chicago, 1949.

———, *The Measure of the Moon*, University of Chicago Press, Chicago, 1963.

———, in *Annual Review of Astronomy and Astrophysics*, Goldberg, Leo, ed., Vol. 2, Annual Reviews, Inc., Palo Alto, 1964.

———, *A Fundamental Survey of the Moon*, McGraw-Hill Book Co., New York, 1965.

Barabashov, N. P., Mikhailov, A. A., and Lipsky, Yu. N., *An Atlas of the Moon's Far Side*, Rodman, Richard B., translator, Interscience Publishers, New York, and Sky Publishing Corp., Cambridge, 1961.

Both, Ernst E., *A History of Lunar Studies*, Buffalo Museum of Science, Buffalo, 1961.

Branley, Franklyn, M., *Exploration of the Moon*, American Museum of Natural History, New York, 1963.

Caidin, Martin, *The Moon: New World for Man*, Bobbs-Merrill Co., Inc. Indianapolis, 1963.

de Callatay, Vincent, *Atlas of the Moon*, St. Martin's Press, New York, 1964.

Cooper, Henry S. F., *Moon Rocks*, The Dial Press, New York, 1970.

Cortright, Edgar M., *Exploring Space With a Camera*, NASA SP-168, Washington, 1968.

———, ed., *Apollo Expeditions to the Moon*, NASA SP-350, Washington, 1975.

Darwin, George Howard, *The Tides*, (paperback reprint), W. H. Freeman & Co., San Francisco, 1962.

Eades, J. B., ed., *Proceedings of the Conference on Lunar Exploration*, Virginia Polytechnic Institute, Blacksburg, 1963.

Elger, T. G., *The Moon*, George Philip & Son, London, 1895.

———, *Map of the Moon*, (diameter approximately 18 inches), Sky Publishing Corp., Cambridge.

Farmer, Gene, and Hamblin, Dora, *First on the Moon*, Little, Brown and Company, Boston, 1970.

Fauth, Philipp, *Mondatlas*, Olbers-Gesellschaft, Bremen, 1964.

Fielder, G., *Structure of the Moon's Surface*, Pergamon Press, New York, 1961.

Firsoff, V. A., *Moon Atlas*, Viking Press, New York, 1962.

———, *The Old Moon and the New*, A. S. Barnes and Company, New York, 1969.

French, Bevan M., *The Moon Book*, Penguin Books, Baltimore, 1977.

Gold, T., in *Monthly Notices of the Royal Astronomical Society*, Vol. 115, 1955.

Goodacre, W., in *Splendour of the Heavens*, Phillips, T. E. R.; and Steavenson, W. H., eds., Robert M. McBride & Co., New York, 1925.

———, *The Moon*, Pardy & Son, Bournemouth, 1931.

Heacock, R. L.; Kuiper, G. P.; Shoemaker, E. M.; Urey, H. C.; and Whitaker, E. A.; *Technical Report No. 32–700, Ranger VII, Part II. Experimenters' Analyses and Interpretations*, Jet Propulsion Laboratory, Pasadena, 1965.

———, *Technical Report 32–800. Ranger VIII and IX, Part II. Experimenters' Analyses and Interpretations*, Jet Propulsion Laboratory, Pasadena, 1966.

Hess, Wilmot N.; Menzel, Donald H.; and O'Keefe, John A.; eds., *The Nature of the Lunar Surface: Proceedings of the 1965 IAU-NASA Symposium*, Johns Hopkins Press, Baltimore, 1966.

Jacobs, J. A., ed., *Proceedings of the Conference on Artificial Satellites*, Virginia Polytechnic Institute, Blacksburg, 1964.

Jet Propulsion Laboratory, *Ranger VII Photographs of the Moon, Part I: Camera "A" Series*, National Aeronautics and Space Administration, Washington, 1964.

———, *Ranger VII Photographs of the Moon, Part II: Camera "B" Series*, 1965.

———, *Ranger VII Photographs of the Moon, Part III: Camera "P" Series*, 1965.

———, *Ranger VIII Photographs of the Moon*, 1966.

———, *Ranger IX Photographs of the Moon*, 1966.

Kopal, Zdenek, *The Moon*, Academic Press, New York, 1964.

———, *An Introduction to the Study of the Moon*, D. Reidel Publishing Company and Gordon and Breach Publishers, New York, 1966.

———, *A New Photographic Atlas of the Moon*, Taplinger Publishing Company, Boston, 1974.

———, ed., *Physics and Astronomy of the Moon*, Academic Press, New York, 1962.

———, and Carder, Robert W., *Mapping of the Moon*, D. Reidel Publishing Company, Boston, 1974.

———, and Mikhailov, Zdenka K., eds., *The Moon*, Academic Press, New York, 1962.

———, *et al., Photographic Atlas of the Moon,* Academic Press, New York, 1965.

Kuiper, Gerard P., and Middlehurst, Barbara M., eds., *Planets and Satellites,* University of Chicago Press, Chicago, 1961.

———, *et al.,* eds., *Orthographic Atlas of the Moon,* University of Arizona Press, Tucson, 1960.

———, *et al.,* eds., *Photographic Lunar Atlas,* University of Chicago Press, Chicago, 1960.

Mailer, Norman, *Of a Fire on the Moon,* Little, Brown and Company, Boston, 1970.

Markov, A. V., *The Moon: A Russian View,* University of Chicago Press, Chicago, 1962.

Masursky, Harold; Colton, C. W.; and El-Baz, Farouk, eds.; *Apollo Over the Moon: A View from Orbit,* NASA SP-362, Washington, 1978.

Middlehurst, Barbara M., and Kuiper, Gerard P., eds., *The Moon, Meteorites, and Comets,* University of Chicago Press, Chicago, 1963.

Moore, Patrick, *A Guide to the Moon,* W. W. Norton & Co., New York, 1953.

———, *A Survey of the Moon,* W. W. Norton & Co., New York 1963.

———, *Space in the Sixties,* Penguin Books, Baltimore, 1963.

———, *The Moon,* Rand McNally & Company, New York, 1981.

Mutch, Thomas A., *Geology of the Moon,* Princeton University Press, Princeton, 1970.

Neison, Edmund, *The Moon,* Longmans, Green & Co., London, 1876.

Ronan, Colin A., *The Practical Astronomer,* Macmillan Publishing Co., Inc., New York, 1981.

Salisbury, John W., and Glaser, Peter E., eds., *The Lunar Surface Layer,* Academic Press, New York, 1964.

Shoemaker, Eugene M., in *National Geographic,* November, 1964.

———, in *Scientific American,* December, 1964.

Sidgwick, J. B., *Observational Astronomy for Amateurs* (paperback reprint), Dover Publications, Inc., New York, 1980.

———, *Amateur Astronomer's Handbook* (paperback reprint), Dover Publications, Inc., New York, 1981.

U.S. Air Force, *Lunar Charts,* Aeronautical Chart and Information Center, St. Louis.

U.S. Army, *Lunar Maps,* Army Map Service, Washington.

Verne, Jules, *From the Earth to the Moon, and All Around the Moon,* Roth, Edward, translator, (paperback reprint), Dover Publications, Inc., New York.

Webb, T. W. *Celestial Objects for Common Telescopes* (paperback reprint with revisions, Mayall, Margaret W., ed.), Dover Publications, Inc., New York, 1962.

Whitaker, E. A.; Kuiper, G. P.; Hartmann, W. K.; and Spradley, L. H.; *Rectified Lunar Atlas,* University of Arizona Press, Tucson, 1963.

Wilford, John Noble, *We Reach the Moon,* Bantam Books, New York, 1969.

Wilkins, H. Percy, and Moore, Patrick, *The Moon,* The Macmillan Company, New York, 1961.

INDEX

Abenezra Crater, 162, 192
Abulfeda Crater, 162, 192
Academy of Sciences, U.S.S.R., 72
Aeronautical Chart and Information
    Center, 43, 44, 45, 109, 110,
    111, 119, 120, 174, 190, 192
Aestatis Mare, 134, 143, 185, 192
Aestuum Sinus, 104, 192
Agarum Promontory, 145, 192
Agatharchides Crater, 177, 192
Age of the moon, 6, 9, 80
Agrippa Crater, 99, 141, 162, 192
Air Force, U.S., 43, 45, 85, 109, 120,
    138, 192
Albategnius Crater, 41, 78, 95, 143, 162,
    172, 192
Albedo, 37, 189
Aldrin, Edwin, 189, 229
Alfraganus Crater, 142, 192
Aliacensis Crater, 99, 163, 172, 192
Almagestum Novum, 41
Almanon Crater, 162, 192
Alpetragius Crater, 106, 172, 192
Alphonsus Crater, 55, 105, 172, 173, 192
Alpine Valley, 21, 99, 192
Alps Mountains, 21, 40, 99, 169, 173,
    192
Altai Scarp, 89, 158, 193
Alter, Dinsmore (1888–1968), 105, 120,
    125, 131, 172
Anaxagoras Crater, 104, 139, 148, 155,
    169, 193
Anaximander Crater, 126, 180, 193
Anaximenes Crater, 178, 180, 193
Anders, William, 137
Angular diameter, 37
Ansgarius Crater, 81, 138, 193
Apennine Mountains, 20, 40, 101, 119,
    148, 154, 159, 168, 173, 174, 189, 193
Apianus Crater, 163, 193
Apogee, 36, 78
Apollo Program, 1, 13, 43, 70, 73, 77,
    124, 142, 149, 154, 155, 162, 172,
    175, 180, 186, 187, 188
Apollo 8, 137
Apollo 11, 29, 54, 155, 186
Apollo 12, 15, 189
Apollo 14, 189
Apollo 15, 23, 103, 189
Apollo 16, 165, 175, 189
Apollo 17, 31, 189

Arago Crater, 155, 193
Archerusia Promontory, 141, 193
Archimedes Crater, 101, 139, 168, 173,
    193
Archytas Crater, 159, 193
Ariadaeus Rill, 21, 162, 193
Aristarchus Crater, 37, 120, 126, 140,
    148, 154, 158, 168, 174, 178, 179,
    189, 191, 193
Aristillus Crater, 98, 139, 159, 168, 169,
    193
Aristotle Crater, 94, 139, 148, 154, 159,
    193
Armstrong, Neil, 124, 189, 229
Army Map Service, U.S., 45, 48, 85,
    118, 130, 149, 159, 181, 190, 192
Arnold Crater, 153, 193
Arthur, D. W. G., 43
Arzachel Crater, 105, 172, 193
Ashbrook, Joseph (1918–1980), 106
Ashen light, 10
Astrogeology Branch, U.S. Geological
    Survey, 45, 109, 119
Atlas-Agena rocket, 66, 72
Atlas-Centaur rocket, 66
Atlas Crater, 87, 139, 145, 149, 153,
    193
Atmosphere, 20, 173
Australe Mare, 119, 138, 193
Autolycus Crater, 98, 139, 159, 168,
    169, 189, 193
Azophi Crater, 162, 193

Babbage Crater, 126, 178, 180, 184, 193
Baco, 49
Bacon Crater, 49, 94, 158, 193
Bacon, Roger (1214–1294), 49
Bailly Crater, 13, 113, 137, 179, 180,
    181, 186, 193
Baldwin, Ralph B. (1912–   ), 74, 85,
    88, 93, 98, 99, 104, 105, 106, 108,
    109, 113, 114, 118, 121, 125, 126,
    135, 138, 141, 152, 153, 157, 172,
    177, 180
Ball Crater, 172, 193
Banat Cape, 176, 193
Barabashov, Nikolai P. (1894–1971),
    65, 72, 118
Barnard, Edward E. (1857–1923), 139
Barocius Crater, 94, 154, 163, 193
Barr, Edward, 120

Barrow Crater, 99, 155, 162, 193
Bayer Crater, 179, 193
Bean, Alan, 15, 189
Beer, William B. (1797–1850), 104, 115
Behaim Crater, 81, 193
Bernouilli Crater, 145, 194
Berosus Crater, 138, 194
Bessel Crater, 159, 194
Bettinus Crater, 125, 177, 179, 180, 194
Bianchini Crater, 121, 177, 178, 194
Bible, The, 137
Biela Crater, 87, 152, 194
Billy Crater, 127, 143, 180, 194
Binoculars, care of, 4
    focusing, 4
    selection of, 2
    testing, 3
Birt Crater, 143, 172, 194
Blancanus Crater, 113, 176, 177, 194
Blanchinus Crater, 100, 172, 194
Bode Crater, 141, 169, 194
Bohnenberger Crater, 153, 194
Bond, William Crater, 159, 169, 173,
    194
Bonpland Crater, 112, 194
Borman, Frank, 137
Both, Ernst E., 41
Bouguer Crater, 121, 194
Brahe (*see* Tycho Brahe)
Breislak Crater, 94, 158, 194
Briggs Crater, 134, 194
Bullialdus Crater, 112, 114, 143, 175,
    177, 194
Burckhardt Crater, 145, 194
Burg Crater, 92, 155, 194
Byrgius Crater, 127, 134, 143, 148,
    181, 194

Caesar (*see* Julius Caesar)
Calendar, 134
California Institute of Technology, 55
Calippus Crater, 159, 194
Cambridge University Observatory, 148
Campanus Crater, 118, 143, 175, 177,
    194
Cape, 21
    (*see also* Kennedy, Cape)
Capella Crater, 92, 153, 194
Capuanus Crater, 118, 175, 194
Carbon, gaseous emission of, 105
Cardanus Crater, 131, 184, 185, 194

Carpathian Mountains, 109, 176, 194
Carpenter Crater, 126, 178, 180, 194
Casatus Crater, 176, 178, 194
Cassini Crater, 99, 169, 194
Cassini, Giovanni D. (1625–1712), 41, 107, 194
Cassini, Jacques J. (1677–1756), 49
Cassini's Bright Spot, 107, 143
Catharina Crater, 89, 95, 142, 154, 194
Caucasus Mountains, 20, 40, 93, 101, 148, 154, 159, 168, 194
Cauchy Crater, 153, 194
Cavalerius Crater, 65, 131, 184, 185, 194
Cavendish Crater, 180, 194
Celestial Objects for Common Telescopes, 20
Censorinus Crater, 142, 195
Cepheus Crater, 87, 149, 195
Cernan, Eugene, 31, 189
Cetus, 80
Challis Crater, 162, 195
Changes on lunar surface, 37, 38, 105, 114, 120
Chappell, James Frederick (1891–1964), 189
Charboneaux, 139
Charts, 38, 41, 43, 45, 77, 189
Chart I, 82–83
Chart II, 90–91
Chart III, 96–97
Chart IV, 116–117
Chart V, 146–147
Chart VI, 170–171
Chart VII, 182–183
Chart, directions for use, 189–192
Chart scale, 190, 191, 192
Cheese, The Thin, 130
Chladni Crater, 141, 195
Cichus Crater, 113, 175, 195
Clairaut Crater, 163, 195
Clavius Crater, 6, 20, 40, 41, 108, 112, 173, 175, 177, 195
Cleomedes Crater, 85, 86, 145, 149, 195
Cloud, 177
Cobra Head, 174
Cognitum Mare, 55, 112, 175, 177, 195
Coincidences in nature, 33
Collins, Michael, 189
Columbus Crater, 149, 195
Compass directions, 77, 190, 191
Condamine Crater, 115, 177, 195
Condorcet Crater, 145, 195
Conrad, Charles, 15, 189
Continental areas, 21
Cook Crater, 152, 195
Copernicus Crater, 40, 41, 73, 108, 114, 119, 126, 140, 148, 154, 158, 168, 174, 176, 191, 195
Copernicus, Nicolaus (1473–1543), 8, 34, 35, 131
Cordillera Mountains, 181, 195
Crater, 13, 20, 178
    summit, 32

Crater class, 85, 192
Crater origin, 54, 62, 63, 114, 126, 144, 145, 152, 172
Crater size, 13, 66, 85, 190
Crescent moon, 10, 173, 178, 179, 181, 185
Crimean Observatory, 105
Crisium Mare, 36, 81, 85, 86, 145, 149, 179, 191, 195
Cruger Crater, 134, 143, 195
Curtius Crater, 100, 168, 195
Cusanus Crater, 145, 195
Cusps of the moon, 11
Cuvier Crater, 100, 163, 195
Cyrillus Crater, 89, 95, 142, 154, 195
Cysatus Crater, 108, 195

D'Alembert Mountains, 137, 184, 195
Damoiseau Crater, 127, 195
Darwin Crater, 135, 181, 185, 195
Darwin, George H. (1845–1912), 33
Davy Crater, 175, 195
Dawes Crater, 141, 195
Debes Crater, 149, 195
Delambre Crater, 51, 95, 155, 162, 195
Democritus Crater, 149, 153, 195
Descartes Crater, 142, 189, 195
Deslandres Crater, 41, 49, 106, 172, 173, 196
Dimensions, lunar, 6
Dionysius Crater, 141, 155, 196
Dioptrice, 38
Dipper, Big, 81
Doerfel Mountains, 101, 136, 181, 196
Dollfus, Audouin, 112, 118
Dollond Crater, 142, 196
Dome, 32, 119, 126
Doppelmayer Crater, 124, 179, 196
Duke, Charles, 165, 189
Dunthorne Crater, 143, 196
Dust, 70, 124, 131

Eagle, 124
Earth, selenographic latitude of, 136
    viewed from moon, 76
Earth-moon system model, 6
Earthshine, 10, 70, 176, 179
Eclipse, 9, 33, 81, 130, 137
Eddington Crater, 136, 184, 185, 196
Elongation of the moon, 12
Encke Crater, 119, 140, 178, 196
Endymion Crater, 86, 119, 138, 145, 196
Epidemiarum Palus, 118, 175, 177, 196
Epigenes Crater, 169, 196
Equator, 130, 137, 190
Eratosthenes Crater, 101, 119, 168, 174, 196
Erosion, 66, 124
Euclid Crater, 140, 177, 196
Eudoxus Crater, 94, 139, 148, 154, 159, 196
Euler Crater, 115, 140, 176, 196
Everest, Mount, 136

Fabricius Crater, 87, 152, 154, 196
Face of the Moon, The, 85
Faraday Crater, 100, 163, 196
Far side of the moon, 32, 72, 73, 77, 137, 138, 218, 219, 224, 225
Fault, 54, 106, 161, 165, 167
Feature classification, 13, 70, 192
Fecunditatis Mare, 85, 86, 149, 153, 179, 187, 196
Fermat Crater, 158, 196
Fernelius Crater, 163, 196
Fiocco, Giorgio (1931–   ), 95
First quarter moon, 11
Flammarion Crater, 172, 196
Flamsteed Crater, 66, 124, 180, 196
Flamsteed, John, 180
Fontana Crater, 127, 196
Fontenelle Crater, 174, 196
Foucault Crater, 121, 196
Fourier Crater, 127, 180, 196
Fracastorius Crater, 92, 154, 196
Fra Mauro Crater, 112, 175, 189, 196
Frame, television, 51
Franklin Crater, 87, 149, 196
Fraunhofer Crater, 148, 196
Freedom 7, 186
Friendship 7, 186
Frigoris Mare, 36, 94, 153, 196
Full moon, 11, 130, 135
Furnerius Crater, 84, 86, 144, 148, 196

Galileo Crater, 131, 196. 197
Galileo Galilei (1564–1642), 1, 8, 13, 32, 38, 39, 41, 70, 106, 113, 131, 173, 174
Galle Crater, 155, 197
Gartner Crater, 153, 197
Gassendi Crater, 121, 143, 179, 191, 197
Gassendi, Pierre (1592–1655), 39, 41
Gauricus Crater, 112, 175, 197
Gauss Crater, 81, 138, 197
Geber Crater, 162, 197
Gemini Program, 186
Geminus Crater, 86, 139, 145, 197
Gemma Frisius Crater, 94, 163, 197
General Biographical Dictionary, 134
Geological Survey, U.S., 45, 109, 119
Ghost crater, 106, 112, 115, 124
Gibbous moon, 11
Glenn, John, 186
Goclenius Crater, 141, 149, 197
Godin Crater, 99, 141, 162, 197
Gold, Thomas (1920–   ), 124
Goldschmidt Crater, 104, 155, 169, 197
Goldstone Deep Space Network, 70
Goodacre Crater, 94, 163, 197
Goodacre, Walter (1856–1938), 100, 108, 118, 124, 125, 131, 134, 149, 159,
Graben, 54
Greenacre, James A., 120
Grimaldi Crater, 131, 136, 143, 148, 154, 158, 168, 174, 178, 179, 181, 185, 190, 191, 197

Grimaldi, Francesco M. (1618–1663), 41
Cuericke Crater, 51, 55, 112, 175, 197
Gutenberg Crater, 87, 153, 197

Hackman, R. J., 119
Haemus Mountains, 93, 155, 159, 197
Hagecius Crater, 152, 197
Hahn Crater, 138, 197
Hainzel Crater, 118, 175, 177, 179, 197
Hale, George Ellery (1868–1938), 1
Hall, John S. (1908–   ), 120
Halley Crater, 95, 142, 162, 197
Hansteen Crater, 127, 180, 197
Harbinger Mountains, 119, 178, 197
Harpalus Crater, 121, 139, 178, 197
Harriot, Thomas (1560–1621), 13
Hartmann, William K., 49
Hase Crater, 86, 148, 197
Hecateaus Crater, 81, 197
Hedin Crater, 185, 197
Heinsius Crater, 113, 197
Helicon Crater, 177, 197
Heliometer, 135
Hell Crater, 107, 172, 197
Heraclides Promontory, 114, 178, 197
Heraclitus Crater, 100, 163, 197
Hercules Crater, 87, 139, 149, 153, 197
Hercynian Mountains, 136, 184, 185, 197
Herodotus Crater, 120, 121, 140, 174, 178, 179, 198
Herring, Alika, 127
Herschel, Caroline L. (1750–1848), 49
Herschel Crater, 106, 169, 198
Herschel, John F. W. (1792–1871), 49
Herschel, John, Crater, 115, 178, 198
Herschel, William (1738–1822), 49
Hevelius Crater, 131, 184, 185, 198
Hevelius, John (1611–1687), 41, 42
Hind Crater, 98, 142, 162, 198
Hipparchus Crater, 95, 98, 142, 143, 162, 172, 198
History of Lunar Studies, 41
Holden Crater, 148, 198
Hommel Crater, 89, 95, 154, 158, 198
Hooke Crater, 149, 198
Horrocks Crater, 95, 162, 198
Hortensius Crater, 176, 198
Huggins Crater, 107, 172, 198
Humboldt Crater, 81, 138, 198
Humboldtianum Mare, 119, 138, 198
Humidity, 5, 173, 181
Humorum Mare, 121, 143, 168, 174, 178, 179, 198
Huygens, Christian (1629–1693), 174
Hyginus Rill, 21, 98, 162, 198

Imbrium Mare, 40, 98, 101, 114, 169, 173, 177, 198
Inghirami Crater, 135, 181, 198
Institutio Astronomica, 39

International Astronomical Union, 49, 55, 77
Iridum Sinus, 114, 115, 174, 177, 178, 198
Irwin, James, 103, 189
Isidorus Crater, 92, 153, 198
Isostatic adjustment, 126

Jacobi Crater, 168, 198
Janssen Crater, 87, 144, 152, 154, 198
Jet Propulsion Laboratory, 51, 55, 70, 97
Julius Caesar Crater, 95, 162, 190, 198
Jupiter's satellites, 39
Jura Mountains, 21, 114, 174, 177, 178, 198

Kane Crater, 155, 198
Kant Crater, 142, 158, 198
Kennedy, Cape, 51, 55, 66, 72, 187
Kennedy, John F. (1917–1963), 65, 186
Kepler Crater, 45, 119, 126, 140, 148, 158, 168, 174, 176, 178, 198
Kepler, John (1571–1630), 8, 34, 35, 38, 131
  first law, 8, 35, 36
  second law, 35, 36, 78
Kipling, Rudyard (1865–1936), 77
Kircher Crater, 125, 177, 179, 180, 198
Klaproth Crater, 176, 178, 198
Klein Crater, 95, 162, 172, 198
Konig Crater, 177, 198
Kopal, Zdenek (1914–   ), 114, 119
Kozyrev, Nikolai A. (1908–1983), 105, 120
Krafft Crater, 131, 184, 185, 198
Kuiper, Gerard P. (1905–1973), 1, 43, 49, 51, 54, 59, 60, 63, 85, 109, 123, 151, 181

LAC (Lunar Aeronautical Chart), 43
Lacaille Crater, 100, 172, 198
Lacus (lake), 13
Legalla Crater, 113, 177, 198
Lamarck Crater, 135, 181, 185, 199
Lambert Crater, 109, 174, 199
Lame Crater, 84, 148, 199
Langley Research Center, 73
Langrenus Crater, 84, 86, 118, 138, 145, 148, 191, 199
Langrenus, Michel F. (1600–1675), 41
Lansberg Crater, 115, 176, 189, 199
Lapeyrouse Crater, 81, 138, 199
Laplace Promontory, 114, 176, 177, 199
Laser signals, 95, 107
Lassell Crater, 175, 199
Last quarter moon, 11
Latinizing names, 49
Lava drainage channels, 123
Lavinium Promontory, 88, 149, 199
Lee Crater, 125, 179, 199
Leibnitz Mountains, 136, 199
LEM (Lunar excursion module), 66, 124

Lemonnier Crater, 155, 199
Letronne Crater, 124, 179, 199
Leverrier Crater, 177, 199
Lexell Crater, 107, 172, 199
Libration, diurnal, 34, 35, 135
Libration in latitude, 33, 34, 78, 99, 100, 125, 152, 158, 162, 168, 178, 179, 181, 184
Libration in longitude, 34, 36, 37, 78, 126, 138, 169, 184
Licetus Crater, 100, 163, 199
Lick Observatory, 39, 106, 139, 189
Light-gathering power, 3
Lilius Crater, 168, 199
Limb of the moon, 11
Lindenau Crater, 89, 158, 199
Lineament, 161, 165, 167
Linne Crater, 38, 139, 199
Lohse Crater, 148, 199
Longomontanus Crater, 113, 175, 177, 199
Lovell, James, 137
Lowell Observatory, 120
Lowell, Percival (1855–1916), 120
Luminescence, 119
Luna 9, 65, 131
Luna 10, 72
Lunar far side, 32, 72, 73, 77, 137, 138, 218, 219, 224, 225
Lunar module (LEM), 66, 124
Lunar Orbiter, 70, 72, 73, 77, 119, 138, 186
Lunar Orbiter 1, 72, 73, 74, 75, 76, 188
Lunar Orbiter 2, 73, 188, 214, 215
Lunar Orbiter 3, 73, 77, 188, 217, 218, 221
Lunar Orbiter 4, 73, 77, 181, 187, 188, 220, 222, 223, 224
Lunar Orbiter 5, 77, 188, 216, 225, 226, 227
Lunar Reference Mosaic, 45
Lunar Rover, 103, 189
Lunation, 33, 78
Lunik 2, 98, 103
Lunik 3, 32, 72, 218

Maclear Crater, 155, 199
Macrobius Crater, 88, 141, 149, 199
Madler Crater, 142, 154, 199
Madler, John H. (1794–1874), 104, 115
Magellan Crater, 152, 199
Maginus Crater, 38, 40, 108, 173, 199
Magnifying power, 1, 2, 3
Main Crater, 162, 199
Mairan Crater, 121, 178, 199
Man in the moon, 13
Manilius Crater, 99, 141, 148, 159, 190, 199
Manned Flight Space Network, 73
Manners Crater, 155, 199
Manzinus Crater, 89, 95, 154, 158, 199
Mapping, early, 38, 41
  geologic, 45
  rectified, 43, 45

Mapping, relief, 45
Maraldi Crater, 153, 199
Mare (sea), 13, 118, 124, 126, 131, 159, 168, 174, 178
  circular, 126
Marginis Mare, 119, 199
Maria (see Mare)
Marius Crater, 127, 199
Marius, Simon (1570–1624), 13
Markov Crater, 184, 199
Maskelyne Crater, 154, 199
Mason Crater, 155, 199
Massachusetts Institute of Technology, 95
Maupertuis Crater, 115, 199
Maurolycus Crater, 94, 154, 163, 199
Mayer, Tobias, Crater, 176, 200
McMath-Hulbert Observatory, 20
Measure of the Moon, The, 85
Medii Sinus, 37, 98, 140, 162, 187, 200
Mee Crater, 118, 178, 179, 200
Menelaus Crater, 93, 141, 155, 159, 200
Mercator Crater, 118, 143, 175, 177, 200
Mercury Crater, 145, 200
Mercury Program, 186
Meridian, celestial, 10, 77
Mersenius Crater, 124, 143, 180, 200
Messala Crater, 84, 86, 139, 145, 200
Messier Crater, 37, 142, 200
Metius Crater, 87, 152, 154, 200
Meton Crater, 155, 200
Michigan, University of, 20
Miller Crater, 172, 200
Mitchell, Edgar, 189
Monge Crater, 152, 200
Montanari Crater, 113, 177, 200
Moon, far side of the, 32, 72, 73, 77, 137, 138, 218, 219, 224, 225
  manned landing on, 1, 13, 124, 189, 229
  new, 9, 78
  Pictorial Guide to the, 131
  shape of the, 73
  waning, 11, 80
  waxing, 10, 80
Moon, The, 115, 191
Moonrise daily delay of, 11, 144
Moore, E., 43
Moore, Joseph H. (1878–1949), 189
Moore, Patrick (1923–   ), 49, 92, 99, 100, 106, 108, 120, 121, 125, 130, 136, 139, 145, 148, 149, 168, 172, 174, 191
Moretus Crater, 36, 108, 168, 173, 200
Mortis Lacus, 93, 155, 200
Mosting Crater, 140, 172, 200
Mount Wilson Observatory, 105
Mountain, 20, 127, 168, 174, 175, 176, 179
Mountain-walled plain, 84, 118, 176
Murchison Crater, 104, 169, 200

Mutus Crater, 89, 95, 154, 158, 200

NASA (National Aeronautics and Space Administration), 55, 186, 187
Nasireddin Crater, 107, 172, 200
Nasmyth Crater, 130, 137, 180, 181, 200
Nature, 120
Neander Crater, 154, 200
Nearch Crater, 87, 154, 200
Nebularum Palus, 98
Nectaris Mare, 45, 86, 88, 152, 200
Neison Crater, 155, 200
Neison, Edmund (1851–1940), 43, 118, 125
New moon, 9, 78
Newcomb Crater, 149, 200
Newton Crater, 136, 200
Newton, Isaac (1642–1727), 36
Nicholson, Seth B. (1891–1963), 98
Nicolai Crater, 158, 200
Nicollet Crater, 143, 200
Nimbus, 177
Nininger, Harvey H. (1887–   ), 142
Nonius Crater, 163, 200
North pole, 104, 162
North Star, 9
Nubium Mare, 37, 51, 55, 112, 143, 168, 200

Objective lens or mirror, 3
Observatory, 5,
Observing light, 5
Observing screen, 5
Oenopides Crater, 184, 200
Oken Crater, 138, 200
Olbers Crater, 136, 140, 185, 200
Olivium Promontory, 88, 149, 200
Orbit, of planet or satellite, 8, 35, 130
  orbit, polar, 77
Orbital motion, 35
Orbiter (see Lunar Orbiter)
Orientale Mare, 181, 185, 200
Orontius Crater, 107, 172, 200
Outgassing, 105, 120

Palitzsch Crater, 148, 200
Pallas Crater, 104, 169, 200
Palus (swamp), 13
Paris Observatory (Meudon), 139
Parry Crater, 55, 112, 175, 201
Pascal Crater, 178, 201
Peirce Crater, 88, 201
Pentland Crater, 168, 201
Perigee, 36, 78
Perkins Observatory, 120
Petavius Crater, 84, 86, 144, 148, 201
Petermann Crater, 145, 201
Pettit, Edison (1889–1962), 98
Phase angle, 173
Phases of the moon, 9, 10, 12
Phillips Crater, 81, 201
Philolaus Crater, 36, 115, 169, 177, 178, 201

Phocylides Crater, 130, 137, 180, 181, 190, 201
Photographic Lunar Atlas, 1, 43, 49, 142, 201
Photometry, 173, 189
Physics and Astronomy of the Moon, 114
Pic du Midi Observatory, 119
Picard Crater, 88, 149, 201
Piccolomini Crater, 89, 154, 201
Pickering, Edward, Crater, 143, 201
Pickering, William, Crater, 37, 49, 142, 201
Pickering, William Hayward (1910–   ), 51
Pickering, William Henry (1858–1938), 32, 38, 49, 142
Pico Mountain, 104, 139, 169, 174, 201
Pie cuts, 35
Pingre Crater, 185, 201
Pitatus Crater, 106, 112, 175, 201
Pitiscus Crater, 89, 95, 154, 158, 201
Piton Mountain, 98, 169, 201
Plana Crater, 155, 201
Planet, minor, 114
Plato (427–347 B.C.), 131
Plato Crater, 40, 101, 138, 148, 154, 158, 168, 174, 191, 201
Playfair Crater, 163, 201
Pliny Crater, 92, 95, 141, 155, 201
Plutarch (46–120 A.D.), 159
Plutarch Crater, 81, 138, 201
Polybius Crater, 158, 201
Pontanus Crater, 163, 201
Posidonius Crater, 92, 139, 155, 191, 201
Postmare feature, 85
Premare feature, 85
Procellarum Oceanus, 43, 51, 65, 66, 70, 73, 118, 131, 174, 178, 180, 181, 185, 187, 201
Proclus Crater, 88, 95, 126, 141, 201
Proctor Crater, 108, 201
Promontory, 21
Protagoras Crater, 159, 201
Ptolemaeus, 49
Ptolemaic system, 8, 34
Ptolemy, Claude (2nd cent. A.D.), 49, 131
Ptolemy Crater, 41, 93, 105, 143, 169, 201
Purbach Crater, 107, 172, 201
Putredinis Palus, 98, 169, 201
Pyrenees Mountains, 87, 152, 153, 201
Pythagoras Crater, 134, 178, 180, 184, 201
Pytheas Crater, 109, 140, 174, 201

Rabbi Levi Crater, 94, 158, 201
Rackham, Thomas W., 119
Railway, The, 106
Ramsden Crater, 143, 201
Ranger Program, 55, 60, 65, 70, 124, 186
  Ranger 6, 93, 155
  Ranger 7, 45, 49, 50, 52, 112, 175, 177
  Ranger 8, 45, 51, 53, 54, 93, 155, 187

Ranger 9, 45, 55, 56, 57, 58, 59, 60, 61, 62, 63, 64, 105, 106, 172
Ray (streak), 20, 134, 135, 136, 140, 141, 144, 176, 184
Rectified Lunar Atlas, 49, 85, 125, 181
Reference Feature, 191
Regiomontanus Crater, 107, 172, 201
Reichenbach Crater, 152, 201
Reiner Crater, 65, 127, 140, 180, 201
Reinhold Crater, 109, 176, 202
Repsold Crater, 185, 202
Resolving power, 3
Revolution, 7, 8, 9, 10, 32, 34, 36, 37, 78
Rhaeticus Crater, 169, 202
Rheita Crater, 152, 202
Rheita Valley, 21, 87, 152, 202
Riccioli Crater, 136, 178, 179, 185, 202
Riccioli, Giovanni B. (1598–1671), 41
Riccius Crater, 158, 202
Ridge, 21, 92, 109, 114, 127, 155, 159, 169
Rill (crack), 21, 55, 141, 143, 148, 153, 174
Riphaeus Mountains, 55, 115, 140, 175, 177, 202
Ritter Crater, 51, 155, 202
RLC (Ranger Lunar Chart), 45
Rocca Crater, 134, 185, 202
Rocks, lunar, 66
Romer Crater, 88, 141, 153, 202
Rook Mountains, 184
Roris Sinus, 121, 178, 202
Rosenberger Crater, 152, 202
Ross Crater, 155, 202
Rosse (Laurence Parsons), Fourth Earl of , 98
Rost Crater, 179, 202
Rotation, 7, 8, 9, 10, 32, 34, 37, 78, 135
Rothmann Crater, 158, 202
Rumker Dome, 32, 126, 180, 202
Russell Crater, 49, 136, 184, 185, 202
Russell, Henry Norris (1877–1957), 49
Rutherfurd Crater, 113, 176, 202

Sabine Crater, 55, 124, 155, 202
Sacrobosco Crater, 95, 163, 202
Santbech Crater, 87, 152, 153, 202
Saturn rocket, 187
Saussure Crater, 108, 172, 202
Scarp (cliff), 87
Scheiner Crater, 118, 148, 177, 202
Schickard Crater, 130, 137, 180, 181, 190, 202
Schiller Crater, 125, 137, 179, 202
Schlesinger, Frank (1871–1943), 32
Schmidt, J. F. Julius (1825–1884), 13, 38, 175
Schmitt, Harrison, 31, 189
Schneckenberg Mountain, 49
Schomberger Crater, 158, 202
Schroter, John H. (1745–1816), 114, 174
Schroter's Valley, 174, 180, 202

Schurmeier, Harris M., 51
Scoresby Crater, 162, 202
Scott, David, 103, 189
Seasons, 33
Segner Crater, 125, 202
Selenographia, 41
Selenography, 13, 38, 186
Seleucus Crater, 134, 140, 184, 202
Serenitatis Mare, 88, 93, 155, 159, 179, 202
Serpentine Ridge, 92, 155, 202
Shakespeare, William (1564–1616), 6, 154, 181
Sharp Crater, 121, 178, 202
Shepard, Alan, 186, 189
Shoemaker, Eugene M. (1928–   ), 32, 45, 54, 62, 105, 109
Shuckburgh Crater, 149, 202
Sidereus Nuncius, 1, 38
Simpelius Crater, 168, 202
Sinus (bay), 13
Sirsalis Crater, 127, 143, 184, 203
Sirsalis Rill, 127, 143, 203
Sky and Telescope, 105, 119, 120, 127, 142
Smullin, Louis D. (1916–   ), 95
Smythii Mare, 73, 119, 203
Snellius Crater, 86, 148, 203
Soft landing, 65, 66, 124
Soil analysis, 65, 70, 93, 98, 115, 124, 186
Solar cell, 70
Somni Palus, 152, 203
Somniorum Lacus, 139, 203
South Crater, 180, 203
South, James (1785–1867), 180
South pole, 108, 152, 168
Space Age, 72, 124
Spacewalks, 186, 189
Spectrogram, 105
Spectrograph, 105
Spine (crater floor ridge), 172
Spitzbergen Mountains, 104, 139, 169, 203
Sporer Crater, 169, 203
Spradley, L. Harold, 49
Spumans Mare, 118, 203
Steinheil Crater, 87, 152, 203
Stevinus Crater, 86, 144, 148, 203
Stiborius Crater, 94, 154, 203
Stofler Crater, 100, 163, 203
Straight Range, 109, 139, 174, 203
Straight Wall, 106, 172, 175, 203
Stratovolcano, 109
Streak (*see* ray)
Struve, Otto, Crater, 49, 136, 184, 185, 203
Struve, Otto Wilhelm (1819–1905), 49
Sun, 7, 9, 35, 81
Surface structure, 65, 66, 70, 124, 131, 186.
Surveyor Program, 55, 65, 66, 70, 124, 186
Surveyor 1, 66, 67, 68, 69, 71, 124, 186, 187

Surveyor 3, 115, 186, 189
Surveyor 5, 93, 155, 186
Surveyor 6, 98
Surveyor 7, 107, 228

Tacitus Crater, 158, 203
Tapscott, J. W., 43
Taruntius Crater, 86, 141, 149, 203
Taurus Mountains, 88, 149, 153, 189, 203
Taylor Crater, 162, 203
Telescope, 2
Temperature, 70, 93, 98
Teneriffe Mountains, 109, 139, 169, 174, 203
Terminator, lunar, 20, 78, 135, 137, 148
Terminator, terrestrial, 73, 76
Thales Crater, 135, 139, 203
Theaetetus Crater, 139, 203
Thebit Crater, 106, 143, 172, 203
Theon Junior Crater, 142, 203
Theon Senior Crater, 142, 203
Theophilus (died 412 A.D.), 134
Theophilus Crater, 20, 55, 88, 95, 134, 142, 148, 154, 191, 203
Tidal theory, 33
Timaeus Crater, 159, 203
Timocharis Crater, 104, 139, 169, 174, 204
Tisserand Crater, 149, 204
Torricelli Crater, 154, 204
Tralles Crater, 149, 204
Tranquillitatis Mare, 51, 86, 88, 93, 124, 152, 154, 155, 159, 179, 187, 189, 204
Transparency, atmospheric, 173, 181
Triesnecker Crater, 141, 169, 204
Tsiolkovsky Crater, 73
Tycho Brahe (1546–1601), 131
Tycho Crater, 20, 40, 73, 107, 112, 114, 119, 126, 135, 144, 148, 154, 158, 163, 168, 172, 175, 191, 204

Ukert Crater, 141, 169, 204
Undarum Mare, 118, 204
United Nations, 77
Universal time, 79
Urey, Harold C. (1893–1981), 59, 103, 114, 130, 172

Valley, 21, 125, 127, 152, 162, 172, 174, 185
Vaporum Mare, 98, 159, 204
Vasco da Gama Crater, 185, 204
Vendelinus Crater, 84, 86, 148, 204
Venus, 37
Vieta Crater, 127, 180, 204
Vitello Crater, 121, 143, 179, 204
Vitruvius Crater, 141, 153, 189, 204
Vlacq Crater, 87, 89, 95, 152, 154, 204
Volcano, 58, 105, 107, 109, 174
    maar, 105, 109
    shield, 109

Walter Crater, 99, 163, 172, 204

Waning moon, 11, 80
Wargentin Crater, 130, 135, 137, 204
Water, 38, 101
Watt Crater, 87, 152, 204
Waxing moon, 10, 80
Webb, Thomas W. (1807–1855) 20, 101, 107, 108, 112, 114, 118, 120, 127, 131, 134, 159, 169, 176, 177
Werner Crater, 100, 163, 172, 204
Whitaker, Ewen A., 43, 49
Wichmann Crater, 179, 204
Wilhelm Crater, 113, 175, 177, 204

Wilkins, H. Percy (1896–1960), 49, 92, 99, 100, 106, 108, 120, 121, 125, 130, 136, 139, 145, 148, 149, 168, 172, 174, 191
Wilson Crater, 125, 204
Wolf Crater, 143, 204
Wood, Robert W. (1868–1955), 6
Worden, Alfred, 189
Wrottesley Crater, 148, 204
Wurzelbauer Crater, 112, 175, 204

XB-70A, 55

Xenophanes Crater, 185, 204

Young, John, 165, 189

Zach Crater, 168, 204
Zagut Crater, 94, 158, 204
Zenith, 10
Zond 3, 72
Zucchius Crater, 125, 177, 179, 180, 204

LUNAR PHOTOGRAPHS SUPPLEMENT

The supplement of sixteen photographs of lunar features in the 1983 edition of this book appears courtesy of the National Aeronautics and Space Administration (NASA). Taken by the cameras of Lunar Orbiters 2, 3, 4, and 5, Surveyor 7 and by astronaut Neil Armstrong of the Apollo 11 mission, the photographs show in exceptional detail the valleys, rills, domes, terraces on crater walls, mountains, double and ghost craters, ridges, maria, and, as of 1969, footprints that mark the lunar landscape. Areas not observable from the earth such as the farside crater Tsiolkovsky are revealed in varying degrees of resolution. Where possible, the notation used in the Gazetteer for determining the location of lunar features has been added to the captions of the photographs.

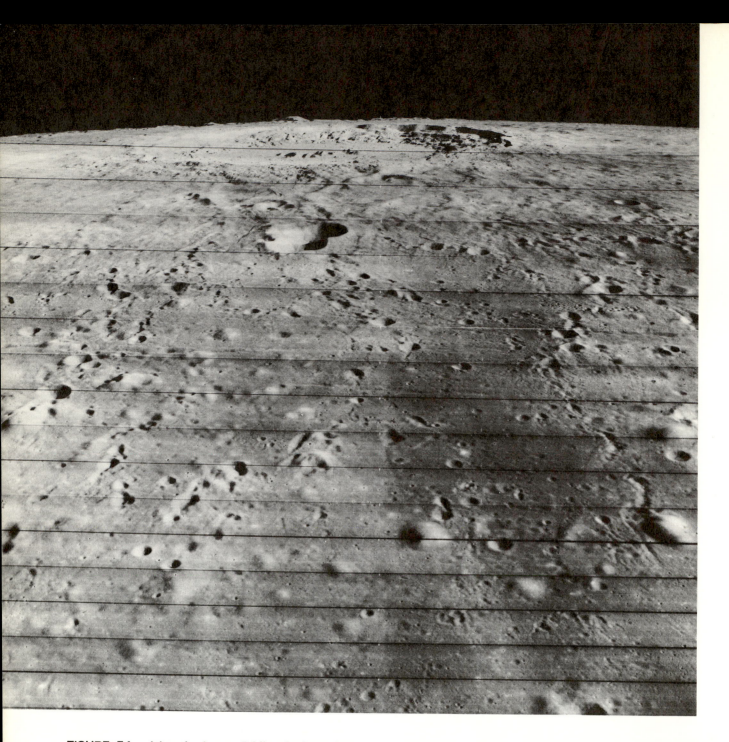

FIGURE 54. (above) Lunar Orbiter 2 views Copernicus (60-mile diameter; 12,600-foot depth) on November 23, 1966, from a point 150 miles south of the crater's center and 28 miles above the lunar surface. This wide-angle or "medium-resolution" photograph shows the great crater occupying the central half of the frame just below the curved horizon. Copernicus's complex central mountain appears as a series of black dots across the middle of the narrow-ellipse crater outline. The terraced inner walls stand out in shadowed contrast on the right, but those on the left are "burned out" in the glare of direct sunlight. The conspicuous double crater near frame center is Fauth. Notice the numerous domes across the foreground, shaded on their left side whereas the craters are shaded on their right. (IV Dot NW 4.8 and Fig. 4)

FIGURE 55. (right) Telephoto or "high-resolution" photograph of Copernicus taken simultaneously with the preceding wide-angle view. The frame includes only the central quarter of the crater. Details of many landslides along the north inner wall are shown as well as outcroppings and boulders on the mountains rising from the crater floor. The bottom four strips of film record a portion of the south wall. The Carpathian Mountains mark the horizon topped by the characteristic black daytime lunar sky. This Lunar Orbiter 2 photograph was taken 28 miles above the lunar surface and shows an area measuring 150 miles from base to horizon and 17 miles across. (IV Dot NW 4.8 and Fig. 4)

FIGURE 56.   Details of the north wall and floor of
Copernicus recorded by Lunar Orbiter 5 on August 16,
1967, as it passed 64 miles directly above the crater.
The inner slope stretches about 16 miles from crest to
floor. The broader, clumpy outer slope is heavily
marked with radial grooves cut by large blocks that
skidded or rolled outward during and after the huge
explosion produced by the kinetic energy of the im-
pacting meteorite that excavated the crater. (IV Dot
NW 4.8 and Fig. 4)

FIGURE 57.   Lunar Orbiter 3 looks across Oceanus
Procellarum from 33 miles above the lunar surface.
Featured is the 20-mile crater Kepler with 9-mile-
wide Kepler A near the right border. The scattered
rough highland material was hurled there from 600
miles beyond the horizon by the great Mare Imbrium
excavating impact and explosion. Relatively smooth
areas in the foreground and background are lava flows,
part of the vast Oceanus plain. (VI Copernicus W 2.8)

FIGURE 58.   On February 19, 1967, the wide-angle camera of Lunar Orbiter 3 recorded a large section of the southern far side near the terminator from an altitude of 900 miles. Conspicuous near center in this heavily bombarded region is the flooded crater Tsiolkovsky, 155 miles across. It was one of the first craters of the far side to be named because it stood out among the few features definitely registered on the pioneer but primitive photographs made by the Russian Lunik 3 sent to swing around the moon in October 1959. Konstantin Tsiolkovsky (1857–1935), a self-taught scientist and provincial school teacher in Borovsk, Russia, designed rockets, built prototypes, and wrote about space travel.

FIGURE 59. Here is the telephoto camera view of Tsiolkovsky taken simultaneously with the previous picture. Detail is vastly improved, but the area covered is greatly reduced—to about the size of a postage stamp on the wide-angle photograph.

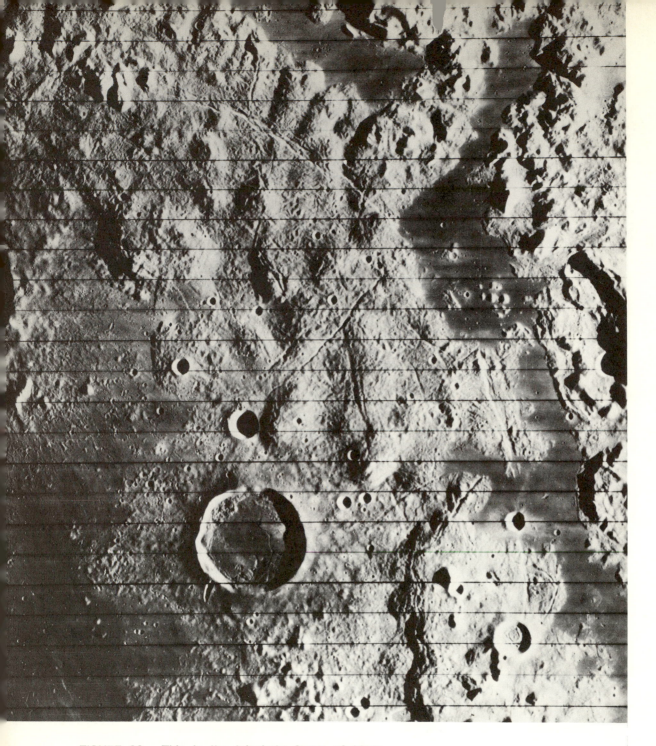

FIGURE 60. This is the telephoto frame of Mare Orientale taken by Lunar Orbiter 4 at the same time that wide-angle Figure 52, page 187, was made. Mare center is just beyond the lower left corner of the frame, and the mountains curving from right central to upper left corner are part of the Rook Range, upland areas bordered on the inside by a series of lava lakes. Only a slice of crater Maunder is seen along the left border, but smaller 25-mile crater Kopff shows well a little below center. Kopff appears to be an extinct volcano of the caldera type. Notice what is literally a tongue of lava hanging out over the south (lower) wall——lava which in the final stage of activity broke through the wall and slowly oozed outward about five miles before it congealed. (VI Gassendi W 2.7)

FIGURE 61.   Here Lunar Orbiter 3 directs its cameras northward over the Cavalerius Hills along the irregular western coast line of Oceanus Procellarum. The sharp crater on the mare plain is 10-mile-wide Galileo, and the slightly smaller one between it and the horizon is Galileo A. Notice the raked appearance of the left foreground——linear cuts and grooves running diagonally upward from the lower left margin——produced by enormous masses of rock blasted out in the creation of Mare Orientale 700 miles to the southwest. The spacecraft's altitude for this shot was about 38 miles. (IV Grimaldi N 2.0)

FIGURE 62.   The 97-mile-wide ancient limb crater
Riccioli, right center with diameter a little more than
half the frame width, recorded by Lunar Orbiter 4 on
May 24, 1967. A series of rills cross the floor from
lower right to upper left. The whole region has been
scarred heavily at right angles to the rills by ejecta
from Mare Orientale 500 miles to the southwest. Huge
blocks have grooved the surface, and large masses of
less coarse material have actually flowed across the
ground after landing. Later, black lava welled up to
cover the north portion of Riccioli's floor and obliterate
the damage. (IV Grimaldi NW 0.6)

FIGURE 63. In this Lunar Orbiter 4 photograph, another west-limb crater, 62-mile-wide Inghirami, is seen near the lower left corner in an upland region greatly disturbed by the Mare Orientale event. Part of the crater floor and the large valley above it resemble huge mudslide areas. There and elsewhere (upper center) masses of lunar material did slide or flow after the impact, but it probably was as dry as desert sand. Mare Orientale lies 670 miles northwest. (IV Tycho W 5.1 and Fig. 46)

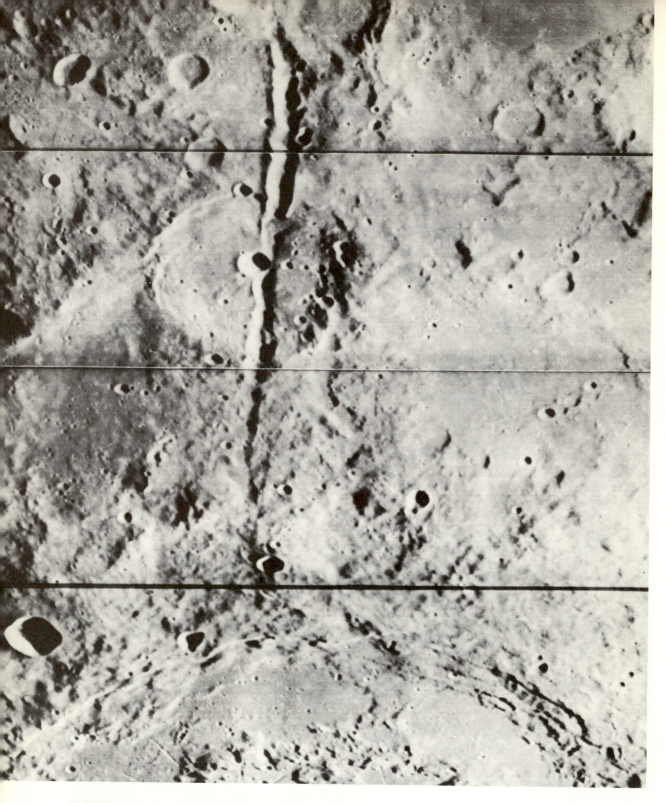

FIGURE 64. The Schrodinger Rill, 200 miles long
and 8 miles wide with a raised rim, photographed on
May 11, 1967, by Lunar Orbiter 4 as it crossed the
far side. Obviously, it is younger than the crater it has
cut through but older than several small craters which
have encroached upon its walls. The gigantic walled
plain partially shown at the bottom of the frame is
Schrodinger, 260 miles in diameter.

FIGURE 65. On August 15, 1967, Lunar Orbiter 5 took this photograph on the far side from an altitude of 768 miles. The large crater with flooded floor near the south border is Mare Moscoviense, 280 miles across. It lies nearly 1000 miles north of Tsiolkovsky, and it was named by you know who. The twin deep (almost totally shadowed) craters halfway from the Mare to frame top are Von Neumann and Ley. The large, heavily pocked, ancient crater with Ley astride its wall is Campbell, 180 miles in diameter.

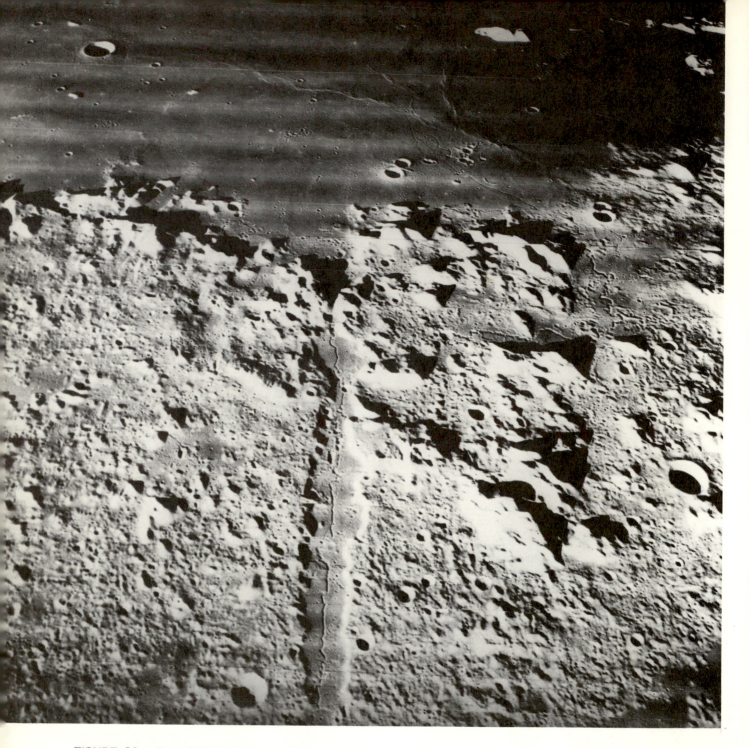

FIGURE 66.   Traveling southwest across the Alps,
Lunar Orbiter 5 photographed the Alpine Valley which
cuts through the range 110 miles from Mare Frigoris,
just below bottom of frame, to Mare Imbrium above.
Low sun emphasizes meandering rills and ridges on
the plain, and a large unnamed "ghost" crater (top
right) appears almost obliterated by the flow of lava.
Near the left upper edge of that crater rises bright
Pico, one of the few lunar mountains not part of a
range. The photo was taken on August 11, 1967 from
an altitude of 82 miles. (III Plato E 1.5, VI Plato S 0.8
and Fig. 9)

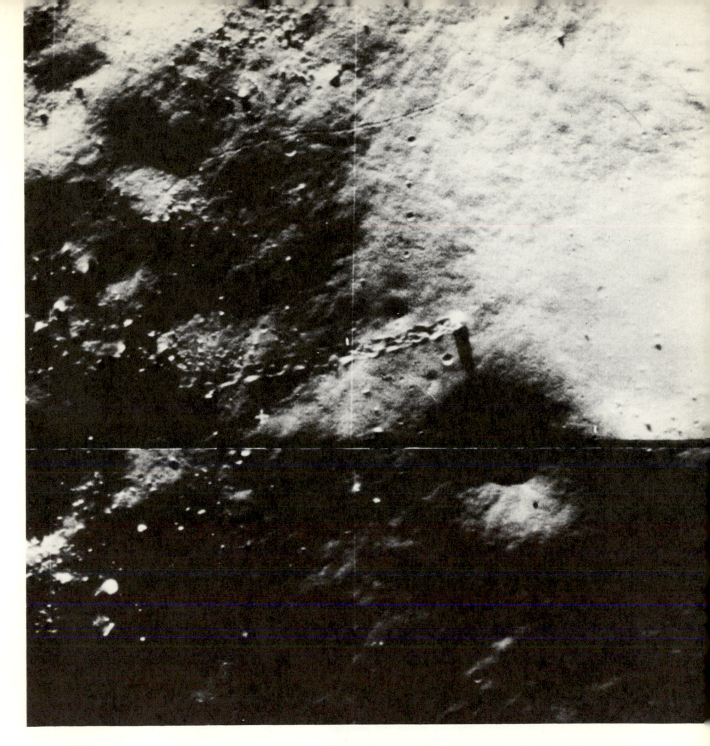

FIGURE 67. Changes do take place on the moon! Lunar Orbiter 5 recorded this evidence on the slope of the central peak of Vitello. In the upper half of this greatly enlarged tiny portion of the original frame, we see two boulders which have rolled down the mountain from left to right. The larger one, just above the small dark crater, is about 75 feet across and sufficiently irregular to have left a conspicuous tread-marked path some 900 feet long. It shines brightly and casts a long shadow into the crater. Near the upper border a 15-foot boulder with triangular shadow has left a weaving 1200-foot smooth trail. Numerous other boulders can be seen plus a few small craters showing opposite shading. (VII Gassendi S 2.4)

FIGURE 68.   The view an explorer might have looking northeast from a point halfway down the rugged
north outer slope of brilliant Tycho. The last of the
Surveyors (Number 7) made a soft landing there on
November 10, 1967, and one of its three footpads is
shown at the lower right corner nestled against a rock
big enough to have caused damage had the foot come
down on it. Surveyor's television camera stands at
about the height of a man's eyes above ground, and
this picture is a mosaic of some 100 frames transmitted back to NASA. The horizon just beyond the conspicuous gully is about 8 miles distant. The large long-
shadowed rock near the middle of the picture is 2 feet
across, and to its right is a rock-filled crater 5 feet in
diameter. Two-thirds of the way from that crater to
the horizon is a round boulder 50 feet across. (IV Dot
S 7.1 and Fig. 6)

FIGURE 69. "MEN WALK ON MOON" read the exceptionally large headline spread across the front page of The New York Times on July 21, 1969. Here Edwin Aldrin is photographed by Neil Armstrong as the two astronauts take a moonwalk. Notice the footprints resembling, at this point, those made by boots in a four- or five-inch snow cover. Aldrin's space gear and backpack protect him from the cold vacuum of the lunar environment and the searing solar radiation and enable him to converse with Mission Control at Houston as he works. The instrument packages he carries will make and report measurements of lunar conditions for many months. (II Theophilus N 2.5 and Fig. 10)